Four Dimensional Analysis
of Geological Maps

Four Dimensional Analysis of Geological Maps

Techniques of Interpretation

Clive A Boulter
Department of Geology
University of Southampton

Illustrated by

Josie Wilkinson
Department of Geology
University of Nottingham

John Wiley & Sons
Chichester · New York · Brisbane · Toronto · Singapore

649886
S

Library of Congress Cataloging-in-Publication Data:

Boulter, Clive A.
Four dimensional analysis of geological maps: techniques of
interpretation/Clive A. Boulter: Illustrated by Josie Wilkinson.
    p.   cm.
  Bibliography: p.
  Includes index.
  ISBN 0 471 92162 9
  1. Geology—Maps.   I. Title.
QE36.B68   1989
550′.22—dc19

GH18130856

88-33369
CIP

British Library Cataloguing in Publication Data

Boulter, Clive A.
  Four dimensional analysis of geological maps:
  techniques of interpretation.
  1. Geological features. Analysis. Use of
  geological maps.
  1. Title
  55.18

ISBN 0 471 92162 9

90 04

Phototypeset by Dobbie Typesetting Limited, Plymouth, Devon
Printed and bound in Great Britain by Courier International Ltd, Tiptree, Essex

To Geological Survey mappers the world over.

CONTENTS

5.3   Other photographic remote sensing   62
        5.3.1   Colour air photographs   63
        5.3.2   Infrared photography   63
        5.3.3   Space photography   64
  5.4   Multispectral scanner remote sensing   67
  5.5   Active remote sensing systems   72
  5.6   The future   73

CHAPTER 6   THE FOURTH DIMENSION — CHRONOLOGY   74

  6.1   Stratigraphy/historical geology   74
  6.2   Geochronology   76
  6.3   Stratigraphic procedures   78
  6.4   Stratigraphic correlation   79
  6.5   Lithologies that transgress time (diachronous)   86
  6.6   Stratigraphical information on maps   89
  6.7   Further reading and important sources of information   94

CHAPTER 7   DEFORMATION BASICS   95

  7.1   General comments   95
  7.2   Displacement categories   98
        7.2.1   Translation   99
        7.2.2   Rotation   100
        7.2.3   Strain   101
        7.2.4   Dilation   102
        7.2.5   General case   102

CHAPTER 8   CONTINUOUS DEFORMATION   103

  8.1   Fold morphology and description   103
  8.2   Minor structures associated with folding   120
  8.3   Outcrop patterns of folds   125

CHAPTER 9   DISCONTINUOUS DEFORMATION — FAULTS   138

  9.1   Absolute beginnings   138
  9.2   Description = separation   142
  9.3   Displacements across faults   144
  9.4   Slip/separation: what's the difference?   146
  9.5   Fault calculations   151
  9.6   Faults and outcrop patterns   154
  9.7   Fault planes and fault zones   156
  9.8   Fault patterns and associations   157
        9.8.1   Contractional faults   160
        9.8.2   Extensional faults   170
        9.8.3   Strike-slip faults   179
  9.9   Transtension/transpression   188
  9.10 Polyphase brittle tectonics   190

PREFACE

'There is no substitute for the geological map and section — absolutely none. There never was and there never will be. The basic geology still must come first — and if it is wrong, everything that follows will probably be wrong.' (Wallace, 1975, *Mining Engineering*, **27**, 34 – 6.)

By the end of this book students should be well grounded in the basic techniques of geological map analysis. This is primarily an exercise in 3-D geometry coupled with a need to appreciate the time element which is central to the study of geology. Geology is emphatically four-dimensional! The level of the text is introductory, targeted at first-year university students, though it may be of use to those in the last year or so of secondary education and should be a useful reference work throughout an undergraduate course. Because university courses vary considerably in style and emphasis, the attempt at comprehensive coverage here is probably too large for any one first year. However, the great diversity of courses will also ensure that for some topics in some courses the depth of coverage will be inadequate. The fundamentals given for each topic should satisfy all. Beyond this introductory stage each topic could fill several volumes. None of the coverage could be called advanced but in dealing with the basics a primary criterion has been one of correctness. A major failing of many texts in this field is that simplification has led to misleading or incorrect ideas being put forward. The basics cherished by most are dealt with but in the framework of modern approaches. Another justification for this book is that its many predecessors are now getting on in age and are often badly outmoded. Terminology and methods have changed considerably in the 1970s and 1980s particularly in dealing with stratigraphic procedures, and folded and faulted rocks. Also earlier texts are certainly weak in structural geology and this may partially account for the large number of practising geologists who show a very basic lack of understanding in this area. Stratigraphers would probably agree with these sentiments.

As in all branches of science the explosion of activity has generated a vast store of new knowledge. What may once have been dealt with at university, is now found in school courses. In geology these changes are probably more acute than in many other sciences following the effects of our recent Kuhnian scientific revolution. Plate tectonics forms the core of this upheaval but other influences are apparent. Many specialists in chemistry, physics and life sciences, are now rigorously applying their disciplines to geology, rapidly moving the subject further away from its qualitative roots. One of the tangible results of these pressures was the introduction of many new courses into undergraduate teaching programmes during the 1970s. More traditional subject areas were contracted with perhaps stratigraphy, macro-palaeontology and map analysis courses being the greatest sufferers. Some time back students would probably have spent several hours a week throughout their whole undergraduate course working on geological maps. This is now commonly confined to the first year with perhaps very little to follow in years two and three. Field mapping still figures fairly prominently in university degree courses, but the courses that provide the understanding of the patterns mapped have been eroded. Mapping techniques as such are not discussed in any detail and there is very little on how to recognize small-scale tectonic structures, sedimentary features, etc., in the field. However, where such information is important for the analysis of map scale relationships it is included. It must be realized that you have to be capable of understanding a geological map before you can go out and make one. It is no wonder that I have talked to students at the completion of their degrees who felt uncertain about analysing a geological map properly. All geology students must have a good grounding in the skills needed for geological map interpretation. With much diminished time in the curriculum there is all the more need for an up-to-date text that will carry students well on the way to the goal of being at ease with geological maps. The increase in activity and the change of style in geology has its feedback into maps and mapping. With heightened tectonic understanding a new generation of mapping is being undertaken to test hypotheses arising from the scientific revolution. It may seem odd to suggest that apparently factual maps are strongly influenced by ideas but we are all products of our times. Mapping is also influenced by new techniques as has been shown by satellite-based remote sensing and new geophysical methods. New concepts, such as inversion tectonics, have required a re-examination of old mapping and clearly much of this is now inadequate, demanding remapping. Despite all the advances, field relations, both spatial and temporal, remain fundamental.

The novel approach claimed for this book is the integration of remote sensing and map analysis. In this context the remote sensing is primarily a tool to provide some reality to what would otherwise be a lot of abstract representations on paper. This combination may also help to alleviate some aspects of geological future shock in that remote sensing should be taught to all geology students at some time in 'core courses' and this is one way of achieving this (or at least introducing the subject).

Because the text is primarily aimed at the first year I would be well satisfied if relevant second-year courses were introduced as follows, 'Boulter's book is a good (and correct) foundation and we shall take it from there. . . .' Raising the threshold level at which second-year courses take off is a worthwhile ambition. There are no co-authors to blame if things go wrong and the buck definitely stops here. All that remains to be done is to sit back and wait for the reviews (and keep an eye out for examples to put in the second edition). A concerted effort has been made to avoid words like opaque and indigestible appearing in any review though the subject-matter is difficult in places. My own biased review might start: 'This is a most welcome student text covering a vital and fundamental area of geology with a new style. A good grasp of the basic principles is conveyed forming an excellent springboard for advanced work. . . .'

ACKNOWLEDGEMENTS

Having become an academic author I can now sympathise with all the cries of anguish I have read in similar sections of other books. The thanks to tolerant families because of the intrusive demands of a manuscript can now be appreciated—gestation and birth are not an easy process and much goodwill is required to see the task through. Secretarial support via Jean Pearson, Dulcie Bryant and Jean Angell was ever present and in great supply. Deadlines were considerably eased by the assistance of this helpful trio. Nottingham Geology Department played its part by providing an office with a lousy view, thus minimising distractions through the long hard hours of writing and illustrating. Obtaining remote sensing images was an educational process. Some organisations overwhelmed me with generosity, particularly the National Space Science Data Center, Maryland, USA for Shuttle Imaging Radar. Brian Amos, Frank Habgood, and others at the British Geological Survey, Officers at the Ordnance Survey, and Frank Moseley at the University of Birmingham, all provided much information about sources of images. The Somali Government kindly allowed me to use any of the Royal Air Force coverage of their country, a source that could form the basis of a photogeology book on its own. Regrettably some organisations proved less than helpful and it appears that collections of magnificent air photos have now been lost.

I have taught in three (soon to be four) universities (Tasmania, Western Australia, Nottingham) and was taught in another (Sheffield). I am the product of all these experiences plus others. To those who see bits of their work in this text, I hope they will not be disturbed at how their material has been incorporated. Most likely to figure in this category are Max Banks and Professor Carey (Tasmania), Nick Archibald, David Groves and Rod Marsden (Western Australia), and Bill Cummins and Tony Dickson (Nottingham). Special thanks go to Emyr Williams (Geological Survey, Tasmania) and Chris Powell (Macquarie) for their guidance during my first steps into scientific writing. Mike Rickard (Australian National University) generously allowed access to his unpublished manual on map interpretation. As ever, any bias is my responsibility and readers will soon appreciate my structural background—stimulated by Jack Soper (Sheffield). My other consuming interests of Archaean geology and gold mineralisation are less relevant to this text and hopefully appropriate topics have been fairly treated.

The final hurdle was eased by the speedy response of the John Wiley team. In contrast to getting a piece of research into a journal, the book side of the business is pretty quick; though some painful stages are eliminated ('No passion in the world is equal to the passion to alter someone else's draft'—H. G. Wells) the size of a book creates its own burdens in the later stages of generation. Special thanks go to my father for help with proof reading over Christmas (!), following the one hiccup in the production schedule.

A final thought—If geological maps are supposedly factual why is the key to their analysis labelled 'Legend'?

CREDITS

The following material has been copied directly or modified from other works:

Figure 1.1   *Geologic Maps: Portraits of the Earth*. US Department of the Interior, Geological Survey.

Figure 2.1   *U.S. Geological Survey Professional Paper No. 1395*, J. P. Snyder, Map projections: A Working Manual, 1987. US Department of the Interior, Geological Survey.

Figure 4.2, 4.11   D. M. Ragan, *Structural Geology: An Introduction to Geometrical Techniques* (3rd Edn.). Copyright (1985), John Wiley and Sons, Inc.

Figure 4.12   B. E. Hobbs, W. D. Means and P. F. Williams, *An Outline of Structural Geology*. Copyright (1976), John Wiley and Sons, Inc.

Figure 4.13, 4.14   D. D. Haynes, 1964, *Geology of the Mammoth Cave Quadrangle (GQ351)*. US Department of the Interior, Geological Survey.

Figure 4.15   F. K. North, *Petroleum Geology*. Copyright (1985), Unwin Hyman.

Figure 4.16   D. M. Ragan, *Structural Geology: An Introduction to Geometrical Techniques* (3rd Edn.). Copyright (1985), John Wiley and Sons, Inc.

Figure 5.1   C. Mueller-Wille and R. M. Smith, *Images of the World*. Copyright (1981), Westermann Schulbuchverlag GmbH.

Figure 5.2   *Aerofilms Stereo Pairs Catalogue*. Copyright Hunting Aerofilms.

Figure 5.3   This air photograph (LANSDOWNE, SE52-5, RUN 6, CAF 4068, FRAME 6533) is Crown Copyright and has been reproduced by the permission of the General Manager, Australian Surveying and Land Information Group (AUSLIG), Department of Administrative Services, Canberra, Australia.

Figure 5.4   F. Moseley, *Methods in Field Geology*. Copyright (1981), W. H. Freeman & Company. Reprinted with permission.

Figure 5.6   *Aerofilms Stereo Pairs Catalogue*. Copyright Hunting Aerofilms.

Figure 5.9(a)   V. C. Miller and C. F. Miller, *Photogeology*. Copyright (1961), McGraw-Hill.

(c)   A. N. and A. H. Strahler, *Environmental Geoscience*. Copyright (1973), John Wiley and Sons, Inc.

Figure 5.10   These air photographs (V2 683A—RAF307/2 22 NOV 52 FRAMES 5164, 5165) are Crown Copyright (RAF Photographs) and have been reproduced by permission of the Controller of Her Majesty's Stationery Office.

Figure 5.11   These air photographs (SVY842 HENBURY RUN 7 20/APR/50 FRAMES 5156, 5157) are Crown Copyright and have been reproduced by the permission of the General Manager, Australian Surveying and Land Information Group (AUSLIG), Department of Administrative Services, Canberra, Australia.

Figure 5.13   C. Mueller-Wille and R. M. Smith, *Images of the World*. Copyright (1981), Westermann Schulbuchverlag GmbH.

Figure 5.14   Landsat image E 30169–01285-5, National Aeronautics and Space Administration, USA.

Figure 6.2   H. D. Hedberg, *A Guide to Stratigraphic Classification*. Copyright (1976), John Wiley and Sons, Inc.

Figure 6.3   These air photographs (SVY842 HENBURY RUNS 7 20/APR/50 FRAMES 5155, 5154) are Crown Copyright and have been reproduced by the permission of the General Manager, Australian Surveying and Land Information Group (AUSLIG), Department of Administrative Services, Canberra, Australia.

Figure 6.5   H. D. Hedberg, *A Guide to Stratigraphic Classification*. Copyright (1976), John Wiley and Sons, Inc.

Figure 6.6   E. J. Sawkins *et al.*, *The Evolving Earth*. Copyright (1974), Macmillan Publishing Co, Inc.

Figure 6.7   D. L. Eicher, *Geologic Time*. © 1976, p. 73. Reproduced by permission of Prentice-Hall, Inc, Englewood Cliffs, NJ.

Figure 6.8   F. K. North, *Petroleum Geology*. Copyright (1985), Unwin Hyman.

Figure 6.9(a)   J. H. Maxon, *Geologic Map of the Bright Angel Quadrangle, Arizona*. Grand Canyon Natural History Association.

(b)   A. E. Roberts, 1964, *Geology of the Chimney Rock Quadrangle. Montana (GQ-257)*. US Department of the Interior, Geological Survey.

(c),(d)   Reproduced from *British Geological Survey Sheet ST45, 1:25,000 Cheddar* by permission of the Director, British Geological Survey: NERC copyright reserved.

(e)   N. K. Flint, *Geologic Map of the Southern Half of Somerset County, Pennsylvania*. Reproduced by permission of The Pennsylvania Geological Survey.

Figure 6.10   *Geological Society of America Bulletin*, Vol. 98, 1985, J. H. Craft and J. S. Bridge, Shallow marine sedimentary processes in the Late Devonian Catskill Sea, New York State.

Figure 6.11   *1:100,000 Map, Seigal, Northern Territory*, (1981). Reproduced by permission of the Director, BMR.

Figure 7.2   J. G. Ramsay. The measurement of strain and displacement in orogenic belts. In: *Special Publication No. 3*, Geological Society of London, pp. 43–79. Copyright (1969).

Figure 7.3   A. C. Duncan, Curtin University, Western Australia. All rights reserved.

Figure 7.5   R. R. Compton, *Geology in the Field*. Copyright (1985), John Wiley and Sons, Inc.

Figure 8.1   J. G. Dennis, *Structural Geology*. Copyright (1972), The Ronald Press Co.

Figure 9.21   *Journal of Structural Geology*, vol.4, 1982 R. W. H. Butler, The terminology of structures in thrust belts. Copyright (1982), Pergamon Press plc.

Figure 9.22   *American Association of Petroleum Geologists Bulletin*, vol. 66, 1982, S. E. Boyer and D. Elliott, Thrust Systems. Reprinted by permission of American Association of Petroleum Geologists.

Figure 9.23   *Journal of Structural Geology*, vol. 6, 1984, R. W. H. Butler, Balanced cross-sections and their implications for the deep structure of the northwest Alps: reply. Copyright (1984), Pergamon Press plc.

Figure 9.24(b)   *Petroleum Geology of Taiwan*, J. Suppe and Y. L. Chang, Kink method applied to structural interpretation of seismic sections, western Taiwan. Copyright (1983), Chinese Petroleum Corporation.

Figure 9.25   *American Association of Petroleum Geologists Bulletin*, vol. 70, 1986, S. Mitra, Duplex structures and Imbricate thrust systems: Geometry, structural position, and hydrocarbon potential. Reprinted by permission of American Association of Petroleum Geologists.

Figure 9.27(a)   L. D. Harris. In: Fischer *et al.* (Eds) *Studies of Appalachian Geology: Central and Southern*. Copyright (1970), Wiley-Interscience.
(b)   J. Suppe, *Principles of Structural Geology*. © 1985, p. 345. Reprinted by permission of Prentice-Hall Inc, Englewood Cliffs, NJ.

Figure 9.29   *Journal of the Geological Society, London*, vol. 143, 1986, R. A. Chadwick, Extension tectonics in the Wessex Basin, southern England. Copyright (1986).

Figure 9.30(a),(c)   *Philosophical Transactions of the Royal Society of London, Series A*, vol. 317, 1986, M. A. Etheridge, On the reactivation of extensional fault systems.
(b)   *Geology Today*, Sept./Oct. 1986, P. L. Hancock, The latest on faulting. Copyright (1986), Blackwell Scientific Publications Limited.

Figure 9.31   *Journal of the Geological Society, London*, vol. 143, 1986, R. A. Chadwick, Extension tectonics in the Wessex Basin, southern England, Copyright (1986).

Figure 9.32(a),(c)   *Geology*, vol. 15, 1987, K. R. McClay and P. G. Ellis, Geometries of extensional fault systems developed in model experiments. Reproduced by permission of Dr K. R. McClay.
(b)   *Journal of the Geological Society, London*, vol. 141, A. D. Gibbs, Structural evolution of extensional basin margins. Copyright (1984).

Figure 9.34   *Geology Today*, July/Aug. 1987, L. Frostick and I. Reid. A new look at rifts. Copyright (1987), Blackwell Scientific Publications Limited.

Figure 9.35   *Proceedings of the Geologists' Association*, vol. 98, 1987, C. M. Powell, Inversion Tectonics in S.W. Dyfed. Copyright (1987), Geologists' Association.

Figure 9.36   *Tectonophysics*, vol. 141, 1987, C. J. Ebinger, B. R. Rosendahl and D. J. Reynolds, Tectonic model of the Malawi rift Africa. Copyright (1987), Elsevier Science Publishers B.V.

Figure 9.38   *Tectonophysics*, vol. 95, 1983, Z. Reches, Faulting of rocks in 3-D strain fields. II Theoretical analysis. Copyright (1983), Elsevier Science Publishers B.V.

Figure 9.39   *American Association of Petroleum Geologists Bulletin*, vol. 63, 1979, T. P. Harding and J. D. Lovell, Structural styles, their plate-tectonic habitats and hydrocarbon traps in petroleum provinces. Reprinted by permission of American Association of Petroleum Geologists.

Figure 9.40(a),(d),(e)   *Journal of Structural Geology*, vol. 8, 1986, N. H. Woodcock and M. Fischer, Strike-slip duplexes. Copyright (1986), Pergamon Press plc.

(b)   J. Suppe, *Principles of Structural Geology*. © 1985, p. 29. Reprinted by permission of Prentice-Hall, Inc., Englewood Cliffs, NJ.

(c)   T. P. Harding, R. C. Vierbuchen and N. Christie-Black, Structural styles, plate-tectonic settings, and hydrocarbon traps of divergent (transtensional) wrench faults. In: *Special Publication No. 37*, Society of Economic Paleontologists and Mineralogists. Copyright (1985) Society of Economic Paleontologists and Mineralogists.

(f)   LARGE FORMAT CAMERA ROLL 4 FRAMES 0877, 0878, US Geological Survey, EROS Data Center.

Figure 9.41   Copyright Dr P. Link, Oil and Gas Consultants International, Inc.

Figure 9.42(a)   *Geological Society of America Bulletin*, vol. 98, 1987, J. N. Carter, B. P. Luyendyk and R. R. Terres, Neogene clockwise tectonic rotation of the eastern Transverse Ranges, California, suggested by palaeomagnetic vectors.

(b)   *Tectonics*, vol. 5, 1986, C. Nicholson, L. Seeber, P. Williams and L. R. Sykes, Seismic evidence for conjugate slip and block rotation within the San Andreas fault system, southern California. Copyright (1986), American Geophysical Union.

(c)   *Journal of the Geological Society, London*, vol. 143, 1986, D. McKenzie and J. J. Jackson, A block model of distributed deformation by faulting. Copyright (1986).

(d)   S. D. Knott, 1988, *Structure, sedimentology, and petrology of an ophiolitic flysch terrain, Calabria, Southern Italy*, Unpublished D. Phil. Thesis, Oxford.

Figures 9.43, 9.45   *Bulletin des Centres de Recherches Exploration—Production Elf-Aquitaine*, vol. 7, J. Angelier and F. Bergerat, Systeme de constraintes et extension intracontinental. Copyright (1983), Elf-Aquitaine.

Figure 9.44   *Journal of Structural Geology*, vol. 6, 1984, D. J. Sanderson and W. R. D. Marchini, Transpression. Copyright (1984), Pergamon Press plc.

Figure 10.3   *1:250,000 Map, Lansdowne, SE 52–5*. Reproduced by permission of the Director, BMR.

Figure 10.4   These air photographs (SE 52–5 LANSDOWNE CAF 4069 RUN 3 FRAMES 6957, 6958) are Crown Copyright and have been reproduced by the permission of the General Manager, Australian Surveying and Land Information Group (AUSLIG), Department of Administrative Services, Canberra, Australia.

Figure 10.5   J. Suppe, *Principles of Structural Geology*, © 1985, p. 222. Reprinted by permission of Prentice-Hall, Inc, Englewood Cliffs, NJ.

Figure 10.6   E. B. Bailey, *et al.*, The Tertiary and post-Tertiary geology of Mull, Loch Aline and Oban, 1924. *Memoir Geological Survey Scotland*, British Geological Survey.

Figure 10.7   R. Thorpe and G. Brown, *The Field Description of Igneous Rocks*. Copyright (1985), Open University Press.

Figure 10.26   *Precambrian Research*, vol. 9, 1979, B. Chadwick and A. P. Nutman, Archaean structural evolution in the northwest of the Buksefjorden region, southern west Greenland. Copyright (1979), Elsevier Science Publishers B.V.

Figure 10.27   Vertical air photographs 239-C FRAMES 78, 79, 80, 1959. Geodaetisk Institut, Denmark.

Figure 10.28   *Geological Society of America Bulletin*, vol. 91, 1980, J. McLelland and Y. Isachsen, Structural synthesis of the southern and central Adirondacks.

Figure 10.29(a)   *Journal of Structural Geology*, vol. 2, 1980, J. G. Ramsay, Shear zone geometry: a review. Copyright (1980), Pergamon Press plc.
            (b)   *Journal of Structural Geology*, vol. 6, 1984, G. S. Lister and A. W. Snoke, S-C Mylonites. Copyright (1984), Pergamon Press plc.
            (c)   J. G. Ramsay and M. I. Huber, *The Techniques of Modern Structural Geology*, vol. 2. Copyright (1987), Academic Press.
            (d)   *Journal of Structural Geology*, vol. 5, 1983, P. Choukroune and D. Gapais, Strain pattern in the Aar Granite (Central Alps): orthogneiss developed by bulk inhomogeneous flattening. Copyright (1983), Pergamon Press plc.
         (e),(g)   *Precambrian Research*, vols. 40/41, 1988, J. Beeson, C. P. Delor and L. B. Harris, A structural and metamorphic traverse across the Albany Mobile Belt, Western Australia. Copyright (1988), Elsevier Science Publishers, B.V.
            (f)   *Tectonics*, vol. 5, 1986, B. Le Gall and J. R. Darboux, Variscan strain pattern in the Palaeozoic Series at the Lizard front, SW England. Copyright (1986), American Geophysical Union.

Figure 10.30   *Tectonophysics*, vol. 135, 1987, J. Carreras and J. M. Casas, On folding and shear zone development: a mesoscale structural study on the transition between two different tectonic styles. Copyright (1987), Elsevier Science Publishers B.V.

Figure 10.31   B. E. Hobbs, W. D. Means and P. F. Williams, *An Outline of Structural Geology*. Copyright (1976), John Wiley and Sons, Inc.

Figure 11.6   B. Simpson, *Geologic Maps*. Copyright (1968), Pergamon Press plc.

Figure 11.7   Reproduced from *British Geological Survey, Sheet ST 45 1:25,000 Cheddar* by permission of the Director, British Geological Survey: NERC copyright reserved.

Figure 11.8   Reproduced from *British Geological Survey, Special Sheet 1:63,360 Assynt* by permission of the Director, British Geological Survey: NERC copyright reserved.

Figure 11.10   These air photographs (SE52-5 LANSDOWNE CAF 4067 RUN7 FRAMES 6427, 6428) are Crown Copyright and have been reproduced by permission of the General Manager, Australian Surveying and Land Information Group (AUSLIG), Department of Administrative Services, Canberra, Australia.

Figure 11.11(a)   LANDSAT image E30169-01285-5, National Aeronautics and Space Administration, USA.
            (b)   SIR-A photography, Principal Investigator Dr Charles Elachi, World Data Center-A for Rockets and Satellites, Goddard Space Flight Center.

Figure 11.12   R. M. Mitchum, P. R. Vail and S. Thompson, Seismic stratigraphy and global changes of sea level Part 2. In: *American Association of Petroleum Geologists Memoir No. 26*, 1977, pp. 53–81. Reprinted by permission of American Association of Petroleum Geologists.

Figures 12.1, 12.2, 12.3 Reproduced from *British Geological Survey, Sheet 233 1:50,000 Monmouth* by permission of the Director,British Geological Survey: NERC copyright reserved.

Figure 12.4 P. R. Thomas, New evidence for a Central Highland Root Zone. In: *Special Publication No. 8*, Geological Society of London.

Figures A1.1, A1.4, A1.13 B. E. Hobbs, W. D. Means and P. F. Williams, *An Outline of Structural Geology*. Copyright (1976), John Wiley and Sons, Inc.

1   INTRODUCTORY COMMENTS

Geology is four-dimensional (4-D) yet the most common format for the presentation of data — the **geological map** — is two dimensional (2-D). It is quite a feat to gain from a geological map an image of the three dimensional (3-D) solid form of the geology together with a sense of the time factor — the **geological history**. Thoughts of geometry send some scuttling for cover but you cannot get going in geology without it, especially 3-D geometry. Field mapping, for instance, relies upon thinking in 3-D all the time. The training offered in this book provides all the basic techniques that will allow this feat to be performed on a regular basis (though not always without perspiration). The rationale behind this study is quite simple. Geological maps are the most fundamental units of geological data transfer and without the 3-D picture they convey and the history they present it would be impossible to sensibly explore the metals, hydrocarbons, coal, water, industrial minerals, etc., or to understand slope stability and related engineering problems.

Geological maps carry out a basic geographic function in that they portray the spatial distribution of different rock types. This is by no means the extent of their usefulness. They are a statement of our knowledge of the geology of an area giving information on the mutual geometric and stratigraphic relationships between contrasting rock masses which in turn allow us to analyse the geological history. Besides showing the distribution of distinct (mappable) rock units, a geological map on its own largely provides data on the geometrical interrelations of the rock masses. The map tells us whether boundaries between rock types are planar, curviplanar or irregular, but without a key (legend) we would not know if the contacts were intrusive (igneous), faulted (fracture) or stratigraphic (sedimentary) in type. Similarly, without a key we would have no information on the individual lithologies or their relative ages. A key is also required to explain the myriad of symbols used on geological maps especially as there is no such thing as a standard. Many symbols are used to convey the internal geometry of rock units, data which are important in understanding the form and evolution of the whole mass. The basic message is that despite the geometric information provided by a geological map, it is virtually useless without a key. Another problem with symbols is that they largely reflect the level of theoretical knowledge at the time of mapping. The only map available to you may have been mapped in the 1880s in a scientific environment very different to that of today. The symbols used may relate to old interpretations and before the modern reader can make sense of them you have to be aware how ideas have evolved. To ask students embarking on a course to appreciate the history and philosophy of geology is clearly unreasonable, but a feeling for changes in outlook has to be developed. Also not many maps will be as up to date as my readership so some reference to earlier (but now *passé*) methods will be necessary.

A major difficulty in geological map analysis is the problem caused by the 2-D representation of 3-D geological surfaces — the boundaries between different bodies of rock. An additional complication is that shape of the traces of the geological boundaries on the map are not solely controlled by the geology. The shape we see on the map is the result of the interaction between two 3-D surfaces (geological and topography) of varying degrees of irregularity. There is potential for complexity, but fortunately methodical analysis overcomes most of the problems. A systematic approach is recommended, particularly when dealing with published maps from geological surveys. Vast amounts of information are presented on such maps and they can be daunting documents. On each of many revisits to a map you may still find many new features that earlier had escaped attention. The approach we adopt here is to deal with individual and manageable subsets at the outset. Gradually the threads will be drawn together and earlier topics used as the foundation for more advanced aspects of map analysis. At the end of the course you should be able to tackle a complete geological map, analyse it in full and present an account of the whole map. An account involves a descriptive part giving the distribution of the rocks, their sequence of formation, lithological information, rock relationships and geometric form, and economic aspects. The account finishes with a geological history where inferences may be made about the processes which were responsible for what is seen on the map. Emphasis will vary considerably depending upon the purpose of the report, but most of the main headings will be present no matter what the circumstances. A company geologist interested in the concentration of copper at a certain place will produce a very different account of his mapping to a stratigrapher from an academic environment.

Much has been said of 3-D visualization, a topic that torments some students enormously. The aim in this text is to be as 3-D as possible which is hopefully reflected in the number of 3-D drawings used to illustrate each basic concept. To enhance this process I have enlisted the help of remote sensing. Overlapping coverage of aerial photographs may be viewed to give a stereoscopic 3-D image. Examples have been chosen where the landforms are strongly controlled by the geology which give us our best opportunity of converting a 2-D paper exercise into something approaching reality. The integration of map analysis and remote sensing is unusual but potentially very significant, particularly now that good resolution satellite imagery is available in stereopairs (SPOT, Metric Camera, and Large Format Camera, see Chapter 5). Lessons learnt from the stereoscopic models are valuable in themselves, but in the majority of examples they are specifically employed to help 3-D appreciation of geological maps. A feel for 3-D thinking may not come easily but do not give up as it is a vital attribute for all geologists even those who do not earn their living as geological mappers. In fact a very good grasp of 3-D relationships may be more critical in areas that cannot be conventionally mapped such as offshore hydrocarbon exploration where data points are very limited in number. It is essential that the most should be made of such expensive information in terms of lateral changes, structural patterns, etc.

Many maps and reports are illustrated by block diagrams, cross-sections, and rock-relationship diagrams. These are all aimed at increasing the reader's appreciation of the solid form of the geology which leads to a fuller (and quicker) understanding of the history. Much thought is now devoted to presentation of maps and reports and the best of modern practice will be used as examples.

A general geological map covers all aspects of geology — the only bias is that introduced by any inequality in the distribution of rock types, structures, etc. Some maps may abstract specific information and they will necessarily ignore many features. Examples are:

metallogenic — show distribution of ore bodies and highlight relationships between groups of metals;

engineering — give the location of possible landslip areas, depth of soil cover, physical properties of rocks, groundwater data, earthquake risk, etc.;

tectonic — large-scale crustal features.

neotectonic — highlight features generated since the establishment of the presently active stress field, particularly recent tectonic landforms such as fault scarps;

geothermal — heat flow, heat production, geothermal systems, isotherms in aquifers, Quaternary volcanic rocks.

We shall be mainly concerned with general/comprehensive geological maps that cover as many aspects as possible within the limitations set by the scale of presentation.

An incorrect impression may be gained from published maps in relation to the amount of exposure, i.e. rock observed at the surface (not covered by soil, unconsolidated deposits or vegetation). The majority of maps are highly interpretative: the reliability of the finished product can only be judged if the map indicates the extent of the actual exposure. In Figure 1.1c an interpretation has been made from the scattered factual data supplied by exposures mapped on Figure 1.1b. Two geologists given the same basic information may well produce two different completed versions: disparities multiply in relation to the complexity of the area and **the difference in age of the geologists**. The style of final published maps varies enormously. Most geological surveys do not accurately show the data base for map production, i.e. the exposure. The common practice is to use the distribution of symbols relating to measurements taken at individual exposures to give a qualitative impression of the amount of exposed rock. However, many regional maps are based on traverses where not every exposure is recorded, though here also the distribution of readings may give a good idea of ground coverage. Away from the traverses, the geology is normally inferred from remote sensing, perhaps satellite based or, more commonly, aerial photography. Maps at around 1 : 25 000 to 1 : 50 000 are typically based on mapping which involves visiting nearly all exposures. Scales less than 1 : 100 000 are mostly reconnaissance in style with only selected data being recorded. It would be instructive to look at a range of maps from 1 : 25 000 to 1 : 250 000 to see if you can detect the data base. The British Geological Survey gives some indication on its Drift maps which show the distribution of superficial deposits. The latter deposits in Britain are mostly glacial in origin, but the equivalent in a tropical area would show laterite, silcrete and similar materials. However, the 'solid' parts of the Solid and Drift maps still do not differentiate between rock seen at the surface and rock covered by a thin veneer of soil or vegetation. As your expertise develops you will start to notice inconsistencies in published maps. Some of these will be simple proofreading errors but some will be basic errors which indicate a lapse on the part of the mapper or perhaps a poor grasp of some aspect of geology.

Figure 1.1 also gives some idea of how a geological map is produced. A base map is chosen with a scale to suit the problem and will vary with area to be covered, time and people available, terrain and type of support including whether the ground has to be all walked or helicopters are available. Much of the world is still only mapped at 1 : 250 000 scale and the majority of geological mapping uses vertical air photographs as the base map. In some instances where maps are available only roads, rivers, bridges, position of towns, etc., will be shown on the map (**planimetric map**).

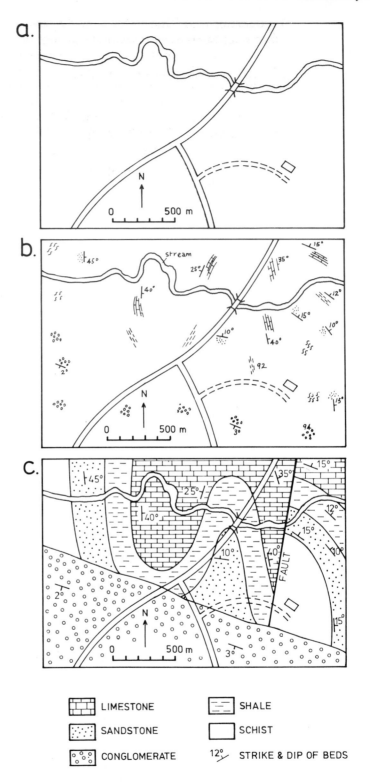

Fig. 1.1 The steps involved in making a geological map. (a) Choice of base map, in this case planimetric; (b) recording position of, and information, from exposures; (c) interpretation of the solid or bedrock geology from the incomplete data base. Version (b) is an exposure map; version (c) is an outcrop map

Topographic maps show relief in addition to planimetric detail and are most useful when small areas (tens of square kilometres rather than hundreds) are examined. In large areas or where the geological features are large, relief ceases to have much significance and ordinary planimetric maps are best. Some geological maps are overprinted in colour on a topographic base map, and though contours and cultural features are not easily seen they provide critical information on attitudes of geological boundaries. The way in which topography and geology interact can provide a great deal of information and we shall, therefore, provide topographic details whenever possible during the initial learning process. You will also be shown that careful reading of a topographic map on its own will give many clues as to the geology of an area.

2 BASE MAPS AND CARTOGRAPHY

Readers are assumed to have a certain familiarity with standard geographic maps and this chapter simply serves as a reminder of their most relevant and major features. An experienced geologist may often be able to interpret much of the geology of an area from topographic information alone because many landforms are strongly controlled by the geology. Also, without topographic information it may be difficult to assess how much of the geological form seen on the map is the result of the interaction of the geological surfaces with an irregular topography. Before attempting to interpret a geological map several aspects of base maps have to be understood.

2.1 PROJECTION

A fundamental problem for the map-maker relates to the representation of part of the spherical earth on a flat map. This is achieved by projecting from the sphere on to a flat sheet or another surface that can be opened out to lie in a plane, for example, a cone or a cylinder (Figure 2.1). In addition to these fairly direct geometric projections there are a large number of contrived examples derived mathematically. When dealing with large areas (which implies small-scale maps) it is very important to be aware of the properties of various projections and to use them appropriately. The Mollweide projection (mathematical) preserves areal accuracy and would be of use to a geologist measuring the relative areas occupied by different rock types in a particular tectonic zone. The cylindrical Mercator projection preserves directional accuracy and would be used for any comparison of directional properties of structures over a wide area. For maps of tens to hundreds of square kilometres the nature of the projection is not significant and areal or orientation distortion will be slight.

2.1.1 SCALE

This is the ratio of the length of an object measured on the map to its actual length. The simplest way of indicating the scale on the map is by writing it as a **representative fraction** (RF), e.g. $1/63\,360$. This has several advantages. It has the value of graphically conveying the difference between large-scale maps (1:10 000) and small-scale maps (1:250 000) because $1/10\,000 \geqslant 1/250\,000$. Because of the seemingly large number involved, calling a 1:250 000 scale map small

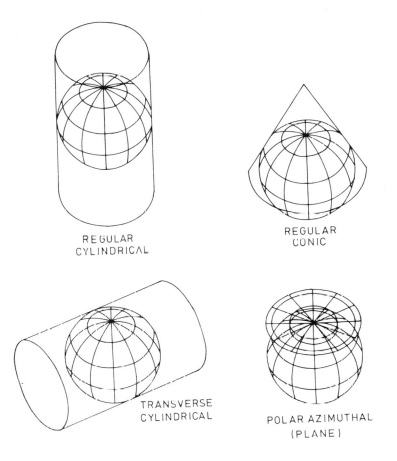

Fig. 2.1 Map projections. The earth's latitude – longitude grid can be projected on to a plane (polar azimuthal), or on to cones (regular conic) and cylinders (regular and transverse cylindrical) that can be cut and flattened out to planes. These are examples of geometric projections but many projections are mathematical to reduce specific distortional effects

scale tends to grate one's sensibilities but this has to be conquered. The method is also international: a person used to the metric system can appreciate the fact that 1/63 360 means 1 cm on the map represents 63 360 cm on the ground (though they may have trouble comprehending why it is used!). A graphic scale, in which a line representing convenient distances is drawn on the map, is useful because it remains true during reduction or enlargement. Descriptive scales, which make statements like 'four inches to one mile' or 'four miles to one inch', tend to be confusing but they are in common use. As international metrication proceeds map publication is being standardized at 1/25 000, 1/50 000, 1/100 000, 1/250 000, etc., scales.

The metric conversion process will cause problems for decades (?) to come because many old maps in imperial units will still be the only ones available for some areas. Whilst there are many conversion tables few give details of the imperial system itself and many modern students simply do not know 12 inches equals 1 foot, 3 feet equals 1 yard, and that 1 mile contains 1 760 yards. Such information is necessary when making constructions on old maps with metric graph paper and rulers or when converting old descriptive scales. Some British maps torment modern students; there are 1:25 000 maps with contours in feet! When centimetre graph paper is used, section drawing can be quite traumatic. In this mixed system, careful sums will lead to a happy ending, e.g. at a scale of 1:25 000 the following identities hold:

$$1 \text{ inch} = 25\,000 \text{ inches}$$
$$1.54 \text{ cm} = 2083 \text{ feet}$$
$$1 \text{ cm} = 820 \text{ feet}$$

2.2 MAP REFERENCES

In describing a map it is often necessary to specify the location of a point or a geological feature in an exact way. Likewise, if in a report you read that the gold mine you are interested in is found at a specific location it is convenient to have a quick method of finding the mine on the map. Both cases are satisfied by using a coordinate grid system consisting of a network of equally spaced straight lines superimposed on the map. One coordinate in a west to east direction (left to right) is known as the **northing**: the other is a south to north (bottom to top) coordinate called an **easting** (Figure 2.2). The origin of a grid system is conventionally taken as its most south-western point. A grid reference for a locality should give the number of metres east and north from the origin but in a large country or for a global system this would generate a very big number. Most grids are, therefore, subdivided into 100 km × 100 km squares which are identified on a regular pattern by letters or a combination of letters and numbers which reduces the size of the grid reference. To identify the relevant 100 km^2 grid square requires local knowledge which cannot be provided here. The United States is moving towards widespread application of the Universal Transverse Mercator Grid (Merrill, G. K. 1986, *Geological Society of America, Bulletin*, **97**, 404 – 9). The British National Grid, which follows a Transverse Mercator projection, is explained in outline on most Ordnance Survey maps. One point to note is that eastings on grids rarely parallel true north and angular differences may amount to several degrees.

Once the 100 km^2 square is identified the rest of the grid reference, for a point or small feature, is determined as follows. All values are first read from west to east (easting) and then from south to north (northing). The mnemonic 'read **right**, **up**' may be useful. Firstly locate the vertical grid line to the left of the point and read the number of the line. Then estimate tenths of the grid spacing to the point (four on Figure 2.2). For the northing locate the horizontal line below the point and read the number of the line. Again estimate tenths from the grid line to the point. This results in a six-figure grid reference which is appropriate for 1 : 50 000 and 1 : 100 000 scale maps giving a precision of ±100 m. A large-scale map (1 : 10 000) can give an eight-figure grid reference (precision ±10 m) where the size of the kilometre grid squares means that they may sensibly be divided into tenths and hundredths. If you mostly use maps of only two or three scales it is well worth the effort of constructing a grid reference reader — a romer — on tracing paper or drafting film. For each scale subdivide kilometric grid squares in tenths and hundredths if appropriate) to give a fast and accurate methods. Reference to 1 × 1 km grid squares is made by quoting the coordinates of its south-west corner (9722 in Figure 2.2). This is useful when locating larger features.

An alternative referencing system, **geographic coordinates**, uses latitude and longitude. On large-scale maps a point may be located to the nearest second giving a

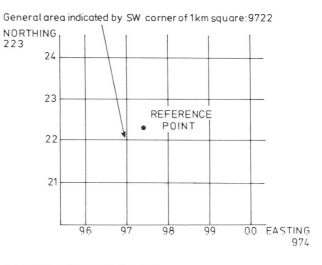

General area indicated by SW corner of 1km square: 9722

6 FIGURE REFERENCE 974 223

Fig. 2.2 The elements of a grid system. Northings run east – west and eastings run north – south. To give a six-figure grid reference for the point quote the easting immediately to the left (97) and estimate the tenths of the grid spacing to the point (4); the second half is the northing below the point (22) and again the tenths of the grid (3). The complete grid reference is 974223 which is always symmetrical so a full stop, oblique or whatever in the centre is unnecessary. A 1 × 1 km grid square is located by referring to the coordinates of its south-west corner — 9722 in this example

precision of about ± 30 m. By convention, latitude is quoted first then longitude, e.g. 20°46′20″S, 118°48′25″E. (Remember there are 60 seconds in a minute and 60 minutes in a degree.) The latest regional maps (1 : 250 000) of the British Geological Survey use for the offshore regions grid lines based on geographic coordinates and also offshore show the intersections of every 10 × 10 km grid of the Universal Transverse Mercator system. Onshore these maps show the National Grid of Great Britain which, though on a Transverse Mercator projection is differently numbered and located. Also onshore intersections of 10′ × 10′ grid lines of the geographic coordinates are marked. Indexing the variety of grids can be quite an exercise!

2.3 TOPOGRAPHIC MAPS

Relief may be represented in three ways; oblique illumination, hachuring, or contouring. The first two styles are schematic and give no quantitative information which is the reserve of contouring. Oblique illumination and hachuring have sometimes been used as bases for geological maps but rarely with success and hence are not considered further. A topographic contour may be defined as a line joining points of equal elevation on the surface of the ground (Figure 2.3). Elevations are measured above a selected datum plane, usually sea-level. A topographic contour at 100 m may also be thought of as a line of intersection between a horizontal plane at 100 m and the ground surface. Map representations of contours are in fact showing the vertical projections of contours on to a horizontal reference plane and not the actual contours themselves (Figure 2.3). The role of projection in mapping contours

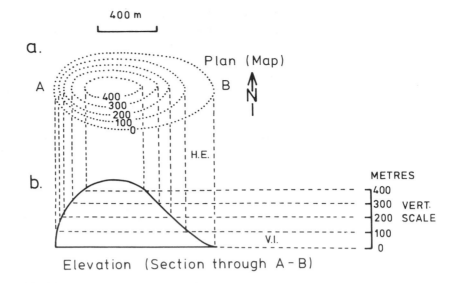

Fig. 2.3 (a) Topographic contours for an island. Readers are expected to be well experienced in interpreting landforms by this method and particularly to appreciate slope variations portrayed. Spacing of contours on the map is the horizontal equivalent (H.E.) which is inversely proportional to slope. (b) A vertical cross-section from A to B on the map. The vertical interval (V.I.) between contours is constant at 100 m. Note the convexity of the western side of the island and how it relates to the contour map pattern

must be kept firmly in mind as it is equally important in interpreting topographic and geological contours.

Contours may be drawn at any constant interval, the vertical distance between two successive contours being known as the **vertical interval** (VI). The latter is commonly multiples of 10, 25, or 100 but as a result of the traumas of conversion from imperial to metric you may come across intervals of 8 m and the like. Another variable is that map-makers may use one interval for lowlands and another for uplands on the same map. It will be assumed that the reader is familiar with interpreting contour patterns in terms of landforms. If not a diversion to an introductory text on map reading or geomorphology is a must. It is absolutely essential that you should be able to view a pattern of contours and quickly see the shape of the land being portrayed and the variation in slopes involved.

2.4   CARTOGRAPHY

In constructing a map or when examining a published map, a certain amount of information must be present for it to be useful as a scientific document. Some optional information may be incorporated depending upon circumstances. In the check list below essential data are marked by an asterisk:

1. **Title.*** The title gives the subject of the map, usually the name of the district and perhaps the type of map (e.g. solid geology, solid and drift, hydrological).
2. **Orientation.*** The direction of true north should be indicated by an arrow. As a matter of convention most maps are drawn so that north is at the top. Magnetic

north is a useful reference in many cases. Mine plans can be very disorientating as their grids commonly relate to the elongation of the deposit and grid north may be many tens of degrees away from true north. On virtually all maps grid north, true north and magnetic north, are in different orientations and the relationships should be studied carefully especially when you are involved in making a map. Also magnetic declination varies with time hence it is important to date the map.

3. **Scale.*** A vital piece of information best shown as a bar.

4. **Legend (key).*** This is an explanation of the symbols and colour scheme used on the map. Different rock types are represented on published maps by colours sometimes backed up by a letter and/or number code. Symbols on geological maps are the source of much anguish. The number of different systems is enormous and even within one country there is no such thing as a standard. The state of disarray is such that groups of related features do not even have the same style of symbols from system to system. The basic message is, treat every new map source warily and analyse their approach to symbology — never assume anything.

5. **Compilation**. It is standard practice to show on the map both the names of the person and organization responsible for compiling the map. In the case of multi-author maps some surveys show on a small index map particular regions covered by each geologist.

6. **Map projection**. Information of this nature is essential for small-scale maps of large areas but of little significance for maps greater than 1 : 50 000.

7. **Index maps**. This shows the location of the map sheet in relation to a larger more readily recognized region (see Figure 10.30) and is a feature that should be used more often. Many maps show the reference numbers of adjoining maps but this may be of little help if not put into the larger context.

8. **Coordinate system**.* Except for maps of very small areas, all maps should show at least two lines of latitude and longitude though these may be represented by ticks at the margin. A regular grid, taken from the appropriate national grid, is essential for ease of reference to localities (see for example Figure 11.8).

9. **Reliability**. Some idea of the style of mapping should be given. Some maps were completed by sparse ground traverses with much interpretation based on air photographs, whereas others will have involved mapping of every exposure and comprehensive walking of the ground. These differences are most commonly marked on reconnaissance maps by means of reliability diagrams. More detailed maps tend not to have such information and the reader is left to assume the extent of the data base. The rate of mapping is never given though this may vary from over 400 km^2 per week to less than 1 km^2. The reliability of such maps is vastly different.

3 TWO DIMENSIONAL PRESENTATION OF 3-D GEOLOGY

A highly artificial and simplified model will initially be used to demonstrate the fundamental principles of this chapter. A layer-cake sequence of rocks allows us to concentrate on the basics and branch out later to more realistic stratigraphies. In a layer-cake, each distinct layer (in this case of rock) has constant thickness though this may vary from layer to layer (Figure 3.1). We are not much concerned at the moment with the choice of particular boundaries, that is, lithostratigraphy (see Chapter 6). Two main factors influence this selection process. A mappable unit or **formation** is normally of sufficiently distinct characteristics (grain size, colour, composition, texture) to allow it to be readily distinguished from its neighbours. More difficult situations arise with gradational boundaries where arbitrary lines are drawn but then hopefully kept to. Another influence is scale of representation. A unit of conglomerate beds 30 m thick is easily represented on a 1:10000 map but at 1:50000 the same unit may only occupy a width of 0.6 mm in its least favourable attitude. A formation at one scale is not necessarily a formation at a smaller scale.

3.1 ATTITUDES OF PLANAR SURFACES

Mappable units may be sedimentary rocks, sequences of lava flows or perhaps complexes of metamorphic or intrusive material. Our first steps will consider concepts applicable to sedimentary rocks as these are the simplest to deal with. Many sedimentary layers are deposited horizontally with approximately planar bounding surfaces, particularly when viewed at mapping scales less than 1:10000. If subsequent deformation tilts the layer-cake then we need a system to accurately describe its new attitude. **Structural contours** fulfil this need by using an approach that is very similar to that of topographic contours. The main difference is that,

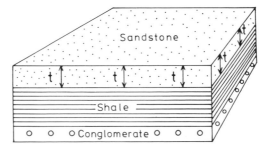

Fig. 3.1 The layer-cake, a simple starting-point. Layers are of constant thickness (*t*) measured perpendicular to the bounding surfaces — also known as stratigraphic thickness

instead of the land surface, we contour geological surfaces of any kind (sedimentological, intrusive igneous, or fracture). A structure contour is a line of equal elevation drawn on a geological surface (Figure 3.2a). A graphic illustration of a structure contour is the intersection of a bedding surface and a body of water (Figure 3.3). Though we are here applying the method to sedimentary formations, it is equally useful in describing the shape of an intrusive body or a fracture surface. For this reason the old term stratum contour is best replaced by the more general term structural contour.

Figure 3.2a shows the top surface of an inclined bed of sandstone where several structure contours have been drawn in. Again, as with topography, it is wise to consider a regular vertical interval to be represented by the contours; here it is 100 m and each 100 m rise or fall of the surface has a structure contour. Because our layer-cake has planar bounding surfaces to the formations, it follows that the structure contours will be straight lines. An important step in understanding the nature of structure contours comes from appreciating that their appearance on the map is a function of vertical projection (Figure 3.2) as is the case with topographic contours.It is the process of projection that allows 2-D representation of 3-D shape. For the tilted layer-cake the structure contours on one surface will be parallel and evenly spaced when projected on to the map. To cut down on verbiage the projected nature of the structure contours will not usually be mentioned though it always has to be borne in mind. Structure contours, which are horizontal lines, may trend along any bearing of the compass and hence their orientation should be specified. In Figure 3.2b the bearing is N75° W, E15° S, 285° or 105°, depending on what system you use to specify compass bearings. By definition the bearing of a structure contour is called the **strike** (Figures 3.2 and 3.3). Note that each end of any straight horizontal line such as a structure contour has a bearing and that these are 180° apart; methods will be given later that will allow us to uniquely specify just one end. The sandstone surface we are examining (Figure 3.2) has a maximum inclination from the horizontal, this angle being known as the **true dip** (or dip for short). The spacing of the structure contours on the MAP (the horizontal equivalent) is directly related to the angle of the dip of the surface in the same way that the spacing of topographic contours relates to the slope of the land surface. Widely spaced contours mean a gentle dip and a closer spacing means a steeper dip.

The dip angle may be calculated by using simple trigonometry; of the SOCATOA mnemonic (sine is opposite over hypotenuse, etc.) you only have to remember tangent is opposite side divided by adjacent. The true dip is found in a direction at right angles to the strike (AC in Figure 3.2b and XZ in Figure 3.2b and the inclination measured in any other direction (e.g. AB in Figure 3.2b and XY in Figure 3.2c) is a lower value known as an apparent dip (Figure 3.4). In the direction of true dip (towards 195°) the sandstone surface (Figure 3.2b) is seen to fall 200 m from C to A in a ground (horizontal) distance of 500 m and therefore each 100 m structure contour has a ground spacing (horizontal equivalent) of 250 m. Tangent of the dip angle is 200 ÷ 500, making the dip 22°. To claim a precision better than to the nearest degree is a pointless exercise and geological realities often continue to give several degrees of uncertainty. Because we are dealing with a planar surface the 250 m ground spacing of the structure contours on Figure 3.2b is constant.

Several shorthand notations have been established to describe the attitude of planar surfaces. The most direct is to quote the dip amount and the direction that the surface dips towards, which for Figure 3.2a/b gives 22 → 195. By giving the direction of the fall of slope a unique specification of attitude is produced. Other systems quote dip and strike but, because strike is double ended, confusion may arise. One group of methods records the general quadrant of the dip direction and the strike by various

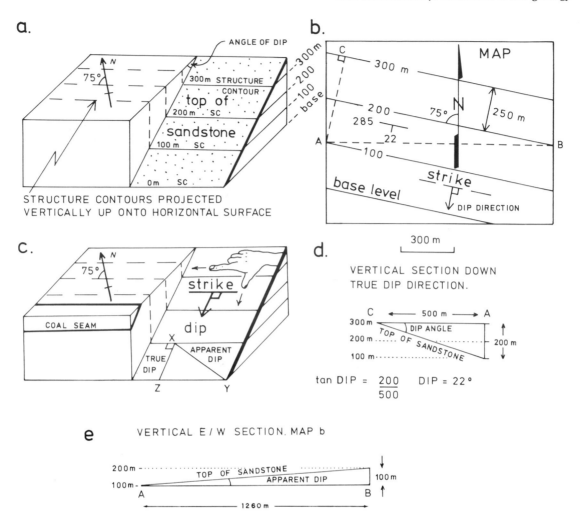

Fig. 3.2 (a) Structure contours drawn on the exposed surface of a sandstone layer dipping towards the south-south-west. When the structure contours are vertically projected on to a horizontal surface, a map view is created. (b) Map view of (a). Because the sandstone has planar bounding surfaces the structure contours have a constant horizontal equivalent of 250 m. The slope in the direction AC is the maximum on the surface — the **true dip**. Any other direction is a lower slope known as an apparent dip, e.g. AB. The orientation of the structure contours is the **strike**. (c) Horizontal structure contours have two bearings one at either end. The right-hand rule uniquely specifies one of these for a dip and strike reading. With the palm of the right hand on the surface and the thumb down the dip, the strike to choose is pointed to by the index finger. The front face of the block is a strike section and the coal seam parallel to the sandstone has a horizontal trace in this view, i.e. zero apparent dip. XY is an apparent dip direction between the strike section and the true dip direction (XZ). (d) A vertical section along AC (b) to show the true dip of the sandstone. Tangent of the dip is the 200 m fall from C to A of the surface divided by the horizontal distance (500 m). (e) A vertical east — west section along AB (b) to show the apparent dip of the sandstone in this direction. Tangent^{-1} (200/500) ⩾tangent^{-1} (100/1260). Apparent dips are always less than the true dip

means, e.g. 22°SW, N75°W or 22° SW, 105; the part before the comma is the dip and the quadrant of the dip direction, and that after the comma is the strike. A more reliable approach establishes a convention which consistently gives the bearing of one end of the structure contour. Both the right-hand rule or the clockwise convention achieve the same result. Following the right-hand rule, imagine that you have placed your right hand palm down on the surface with your thumb pointing down the dip, you then record the strike bearing your index finger is pointing to

Fig. 3.3 Sandstone bedding surfaces dipping steeply into the sea (on a calm day a horizontal planar surface). The intersection of these two planes defines the 0 m structure contour and the bearing of this intersection is the strike

(Figure 3.2c and 3.5). The clockwise convention records the strike bearing that is clockwise from the dip direction. With either style, Figure 3.2a/b would be recorded as 22/285. If two digits are used for the dip and three for the strike there can be no confusion between strikes less than 090° and dips. A surface dipping 5° to the north-west and striking N30° E would be recorded as 05/030. Mistakes can arise when using these conventions and a dip direction 180° in error may be recorded either by carelessness or unfamiliarity. A not so well oriented person may also make mistakes with dip and dip direction though this is impossible with compasses specifically adapted for this system. When structures become more variable I believe that dip and strike symbols (Figures 1.1b, c, 3.2b) are more graphic than dip and dip direction in highlighting changing trends, and for this reason prefer to record dip and strike.

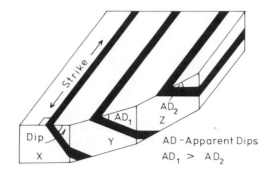

Fig. 3.4 Three beds of constant dip and strike and thickness seen in differently orientated vertical sections (X, Y and Z). X is at right angles to the strike and down the true dip (commonly abbreviated to dip). The orientations of Y and Z progressively move away from the dip direction towards that of strike sections and hence the apparent dip on Z (AD$_2$) is less than that on Y (AD$_1$)

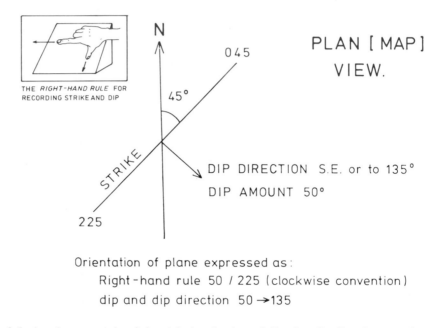

Fig. 3.5 Another example of the right-hand rule and dip plus dip direction specification of attitude for a plane. The plane dips at 50° towards 135°. Its strike is either 045 or 225. According to the right-hand rule (clockwise convention) 225° is quoted and full specification is 50/225

3.2 ATTITUDES OF LINEAR STRUCTURES

Without necessarily realizing it we have already considered, at least in part, the attitude of lines. Structure contours are linear features and many other linear structures, both real and constructed, will be met as we progress through the book. A line inclined to the horizontal is said to **plunge** (Figure 3.6). It is helpful to have a different nomenclature to that of planes such that the important difference between

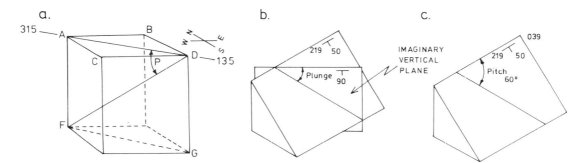

Fig. 3.6 (a) The line DF is plunging at *P*° towards 315°. Its bearing is angle BDA. (b) The bearing of a linear structure is the strike of an imaginary vertical plane that contains the linear. The plunge is also measured in this plane as the angle between the linear and the horizontal. (c) A linear structure resting on a plane could have its orientation specified either by plunge and bearing (b) or by the angle between the linear and the strike of the plane measured on the plane, i.e. the pitch. Here the pitch is 60° from 039° on 50/219.

lines and planes is emphasized. To specify a line's plunge and bearing involves the somewhat awkward step of imagining a vertical plane containing the line (Figure 3.6b). The strike of this plane is the bearing of the line and of the two ends of the strike we quote the direction the line is plunging towards. It is also within this vertical plane that the maximum inclination of the line — the plunge — may be measured. For Figure 3.6a the plunge and bearing is 45→315 (two digits→three digits). The arrow between the plunge and the bearing is to make the distinction (in field notebooks, reports, etc.) between lines, and planes which have an oblique between dip and strike. The non-plunging structure contours of Figure 3.2 would be written as 00→285 or 00→105 because these are identical. If the linear feature is resting on a plane (e.g. ripples on a sandstone surface) or striations on a fault surface) its attitude may be specified using the **pitch**. This quotes the angle between the line itself and the horizontal direction within the plane (the strike) and in the field would typically be measured by placing a protractor (or Silva-type compass) on the planar surface. The smaller of the two pitch angles is conventionally quoted (60° not 120° in Figure 3.6c) but in some circumstances the larger angle may be useful. In Figure 3.6c the 60° pitch may have been from the south-west end of the strike (219°) instead of the north-east end (039°) of the strike and the true orientation must be clearly stated. One means of achieving this is to quote the strike bearing that encloses the measured pitch angle, 039° in the case being discussed. A line's attitude is only properly specified by pitch if the plane's dip and strike is also quoted. The full statement for Figure 3.6c is a pitch of 60° from 039 on 50/219, though 120° from 219 on 50/219 would be equally clear and valid. The linear features in Figure 3.6b and c have the same plunge and bearing (42→087) but are represented differently to show the basis of attitude specification by both plunge and bearing, and pitch.

A common but rather unsatisfactory means of quoting pitch is to say, for example, a line pitches 80° to the south-east on a particular plane. In the case of Figure 3.6c pitches of 80° from 039 and 80° from 219 both pitch to the south-east and quoting the general quadrant of the pitch does not differentiate lines with 30° between their bearings. Rake has been used as a synonym for pitch but this practice is no longer favoured. At this point I should mention another example of the transatlantic separation of English-speaking peoples by a common language. In the United States rake is often used instead of pitch for the angle between a line and the strike of the containing plane. Part of the problem is that they sometimes use pitch in the place of plunge.

3.3   APPARENT DIPS

If Figure 3.2b were to be sectioned down the dip of the beds then the maximum inclination of the surface would be seen as in Figure 3.2d. A vertical slice in any other direction would give an apparent dip which is a linear feature with plunge and bearing but where the plunge is always less than the true dip. On an exposed portion of the sandstone bed, a walk from west to east (A to B of Figure 3.2b) would be up a 5° slope significantly less than the 22° true dip. As a linear feature the attitude of this apparent dip is quoted as 05→270 which can be readily calculated knowing the height and position of two points on the line. A vertical slice parallel to the strike (**a strike section**) shows no inclination of the surface for any value of true dip (except 90°). Such an apparent zero dip is shown by the coal seam on the front face of Figure 3.2c. Simple trigonometry shows the relationship between apparent and true dip. On Figure 3.2b, a fall of 100 m down the true dip of the sandstone is achieved in a ground distance of 250 m. The same fall of 100 m in an east – west section occurs along a ground distance of 1260 m giving a much lower dip (compare Figures 3.2d and e). Following the same procedure it is a simple matter to calculate the apparent dip in any direction with any surface dip and strike. Rather than use a construction in every case the following relationship links true and apparent dips:

$$\tan A = \tan B \sin C$$

where angle A is the apparent dip, angle B is the true dip and angle C is the difference in bearing between the strike and the apparent dip direction. For the east – west section of Figure 3.2b, the angle C would be 15°. In regions where strike varies, a single vertical section cannot always show the true dip of the beds and apparent dips have to be calculated or constructed. It is also advisable to have a feeling for how apparent dips change as the section direction moves away from the true dip. For example with a true dip of 80°, a vertical section on a bearing 45° from the dip direction would show an apparent dip of 76°; 80° from the dip direction the apparent dip has reduced to 45°.

Several situations arise where two apparent dips on a plane are known and we need to calculate the attitude (dip and strike) of the plane. The method depends upon drawing the 'horizontal equivalent' for a 100 m drop of the surface for each apparent dip. (Note that stereographic projection — see Appendix 1 — is a quicker solution.) **Method** (Figure 3.7): Apparent dips 20→330 and 24→053 both rest on a plane. What is the dip and strike of the plane?

1.   From a single point (O) draw two lines representing the bearings of the two apparent dips.
2.   Along the 330° bearing measure off (using any reasonable scale) the distance OA = 100 ÷ tan 20° and along the 053° bearing mark off OB = 100 ÷ tan 24°. From O to A and O to B the surface has fallen 100 m (compare with parts d and e of Figure 3.2).
3.   A line joining A and B is the strike (measure orientation on map) and is the projection of a structure contour 100 m lower than that running through O. (Note only relative heights are important.) Draw a line parallel to AB through O to define the horizontal equivalent of the structure contours. Use a protractor to measure the strike (110°/290°).

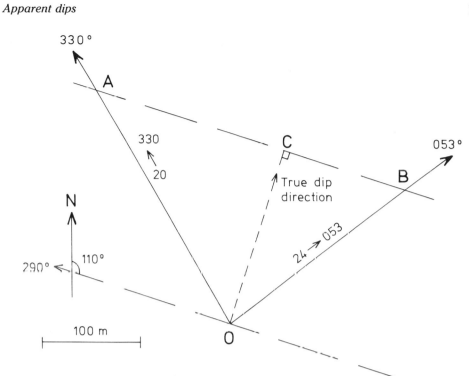

Fig. 3.7 Calculation of true dip and strike from two apparent dips (20→330 and 24→053). Select any convenient scale and choose a point to represent O. Draw OB and OA to represent 100 m falls of the surface along the apparent dip directions. The line AB is the strike and OC is the horizontal equivalent which allows calculation of the true dip

4. The perpendicular distance between the two structure contours gives the true dip $= \tan^{-1} (100 \div OC)$ which in this case is 30°.

∴ Surface dip and strike is 30/110 (right-hand rule).

All of the examples described in this chapter relate to planar surfaces. More complex shapes are also interpreted by constructing structure contours as will be discussed in later chapters

4 GEOLOGY AT THE EARTH'S SURFACE

We shall now grasp the nettle and study the interaction between geological surfaces and the ground surface. Both may be complex shapes and the resultant interplay may be hard to predict. Rather than running before walking, in the initial stages some of the variables will be controlled and for the moment we shall be faithful to the layer-cake unless clear statements to the contrary are made. At the outset, simplifications will be along the lines of planar horizontal topography interacting with dipping beds or horizontal beds with real topography. Coming to terms with the next step of the dipping layer-cake intersecting a varied topography forms the basis for much map interpretation. Many of the more complex structures can be broken down into simple elements of this style. Training and practice will show how to recognize the elements and to integrate them to determine the form of the whole structure.

4.1 OUTCROP AND EXPOSURE

Common usage would suggest that the words outcrop and exposure have similar meanings. However, in the scientific context they are very different. A formation's **outcrop** is the pattern produced by the intersection of the formation boundaries with the earth's surface. The formation is **exposed** where it is not covered by superficial deposits, vegetation or water. This definition has an element of the qualitative because superficial deposits are treated differently by different geologists. It is unfortunate that many geologists are somewhat lax in their use of these terms, using them interchangeably, which can be very confusing. The distinction between outcrop and exposure is critical if we are to describe and discuss maps and mapping sensibly. Most areas have very little exposure and, to make maps intelligible, rock units are shown where they are only thinly covered such that their subsurface presence (outcrop) may be readily interpreted. Very few published Geological Survey maps make the distinction between exposure and outcrop, and estimating the amount of the former requires educated guesswork. Symbols relating to measurements made at exposures are normally the only hint of the data base that allowed the map to be constructed. Even in semi-arid areas the amount of exposure can be very limited (see Figures 5.3, 5.10, 5.11 and 5.14).

We have already considered outcrop inadvertently whilst dealing with apparent dip. The top surface of Figure 3.4 gives the map view of the dipping succession, that is, the outcrop pattern. Under natural conditions most of this plateau-like surface would be at least covered by soil. Mapping the outcrop depends upon exposures for identification of different rock types but the formation boundaries may be mainly located by changes in soil and/or vegetation. Bare rock exposure is extensive only in some recently glaciated and desert terrains.

4.2 WIDTH OF OUTCROP

Outcrop width is measured on the earth's surface, and as with structure contours the map view involves a projection (all points on a map are portrayed as if viewed from vertically above). In order to standardize the process of measurement, the outcrop width is measured in a direction perpendicular to the strike. For dipping beds on a horizontal topography, the outcrop width and its projection are the same (Figures 3.4, 4.1a). There is a simple relationship on such topography between dip of the beds (D), thickness of the beds (T), and outcrop width (W); $T = W \sin D$ (Figure 4.1a). Unit thickness is the perpendicular distance between the bounding surfaces of the formation and for sediments is equivalent to the stratigraphic thickness. A vertical drill-hole through a dipping unit would see a different thickness to the true value (t, Figure 4.1a). On a plateau surface, outcrop width is clearly a function of unit thickness and dip; Figure 4.1b shows beds of constant thickness at different dips. Vertical beds have outcrops as wide as the unit thickness (T) but as dips decrease outcrop widens significantly. A horizontal bed at the height of the plateau would outcrop over the entire plateau.

The simple model of the Triangular Hill (Figure 4.1c) illustrates how topographic slopes influence outcrop width (W) and also the plan (projection) view (W'). To obtain a standard view, the area has been sectioned down the dip direction of the beds. Though the bed is of constant thickness and attitude, the outcrop varies considerably in width. For equal slopes and constant unit thickness, outcrop width is always greater when the beds dip with the slope of the ground (Figure 4.1c). During map analysis all the influences on outcrop width have to be borne in mind to understand any variations that may be observed. Most commonly in the field outcrop width is used to measure unit thickness where systematic stratigraphic changes may have significance for sedimentation studies or tectonic processes. Figures 4.2a, b, c show the relationship between dip (D), slope (S), thickness (T), and outcrop width (W). The general formula is $T = W \sin |D \pm S|$, the absolute value of $D \pm S$ overcomes the difficulty of negative thickness when $D < S$. If dip and slope are opposite in sense then the sum of D and S is used, if both dip and slope are in the same direction then the difference is appropriate. In practice in reading maps unit thickness is calculated by directly measuring the projection of the outcrop width (W') and if the true outcrop width is of interest it would have to be calculated once the slope was determined. Figures 4.2d, e, f show this map-based approach using W', the dip and v which is the vertical distance between the top and bottom contacts measured down dip. On large- and medium-scale maps v is easily measured if contour information is provided. The general formula is $T = W' \sin D \pm v \cos D$, where the sum applies to dip and slope in opposite directions and the difference if they are in the same direction.

Detailed study of variations in stratigraphic thickness is very time consuming and the number of calculations performed will depend upon the nature of the exercise. Given a published map to write up in two or three practical sessions will focus your attention on significant variations. If there are considerable thickness changes, then the cause may be of importance for your understanding of the evolution of the area. On large-scale maps ($>1:10\,000$) outcrop width is strongly controlled by topography

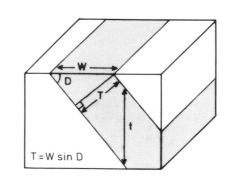

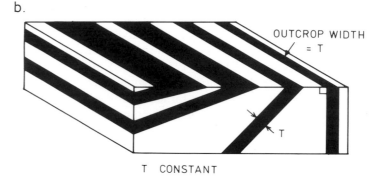

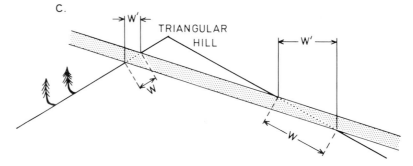

Fig. 4.1 (a) The relationship between thickness (*T*) and dip (*D*) of a layer and its outcrop width (*W*) for a smooth horizontal topographic surface (e.g. a plateau). (b) Variation in outcrop width for a layer of constant thickness but differing dip. Note that for the vertical layer *T* = *W*. (c) The section through Triangular Hill is parallel to the dip direction of the layering (i.e perpendicular to the strike). The outcrop width (*W*) is a function of how this single thickness layer interacts with the topographic slopes. *W'* is the projected outcrop width as would be seen on a map

and it may take many calculations to relate this to thickness variations. As map scales decrease, the influence of relief diminishes to the extent that for 1:250000 maps *T* = *W'* sin *D* will give a good approximation of unit thickness except for very low dips (<10°). One proviso is that in extremely mountainous areas, even small-scale maps may show pronounced changes of outcrop width related to topographic controls. Most measurements of unit thickness are probably taken from vertical cross-sections, though with vertical units we meet a problem on sections (Figure 4.3) similar to outcrop width on slopes. When a section is at a low angle to the trend of a vertical unit the cut-effect greatly exaggerates the apparent width — see the south-south-east-trending dyke on Figure 4.3. Fortunately, when the unit dips at 90° the

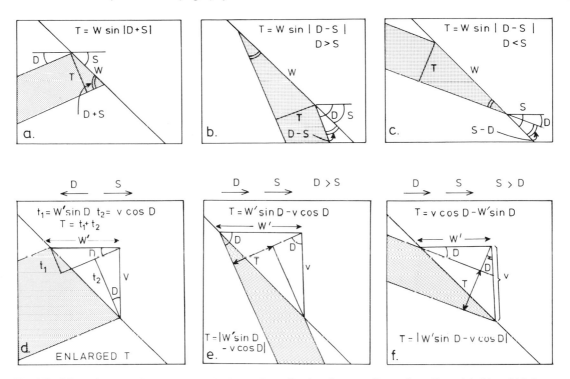

Fig. 4.2 All sections are drawn perpendicular to the strike; i.e. they are dip sections. Parts (a), (b) and (c) show how layer thickness (*T*) is calculated from knowing the layer dip (*D*), the topographic slope (*S*) and the outcrop width (*W*). This approach would be used in the field. Parts (d), (e) and (f) show how layer thickness is calculated from a map where the projected outcrop width (*W′*), layer dip (*D*) and the height difference between the top and bottom outcrops (*v*), are more readily determined. Note that the use of absolute values simplifies some of the equations

plan view of the outcrop width is equal to the unit thickness (Figure 4.1b). As ever, lax usage of terms may cause confusion. Very few geologists make the distinction between outcrop width (*W*) and the projected view (*W′*) as seen on the map. The general assumption is that there is not much difference, but on large-scale maps, and some small-scale maps, it does become important.

4.3 OUTCROP AND REAL TOPOGRAPHY

Until now we have largely dealt with rather idealized ground surface forms such as plateaux and planar slopes. The methods we shall now introduce will cope with any relief and our brief look at outcrop on slopes will have demonstrated the important controls exerted by topography. This is further emphasized in Figure 4.4 where the geology on the three maps is kept constant and only the relief is changed. For each map the sandstone layer is in the same position, has the same thickness, and does not change dip or strike. With a varied topography (Figure 4.4c), a fairly complex outcrop pattern is generated from an evenly dipping sandstone layer. A simplistic equation can express the general relationship:

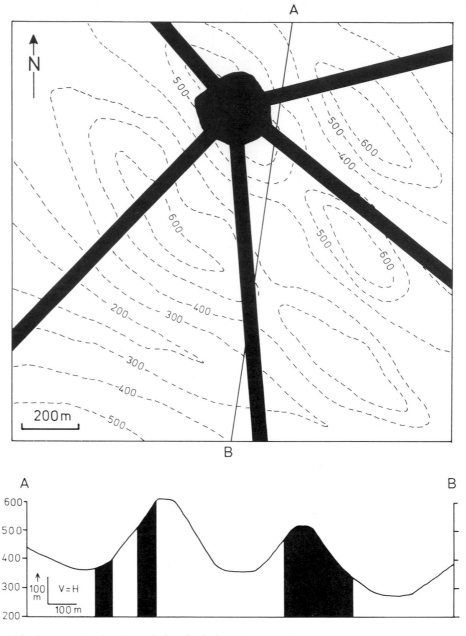

Fig. 4.3 A vertical volcanic neck (a cylinder) and a suite of radiating vertical dykes (intrusive tabular-shaped bodies) are shown in map view. The cross-section shows how the cut orientation with respect to the dykes controls the apparent width of the dykes even though their true widths are the same. Also clearly displayed is the way vertical units cut across topography which in this case does not influence the outcrop pattern

Geology shape + Geology orientation + Topography = Outcrop pattern

(as seen in plan view)

Any change either of the variables on the left-hand side, must bring about a change in the outcrop. The T-shaped outcrop of the sandstone and the isolated oval areas of shale outcrop (Figure 4.4c) have been created by topographic variation. In the case of Figure 4.4c fairly careful analysis is required because similar patterns

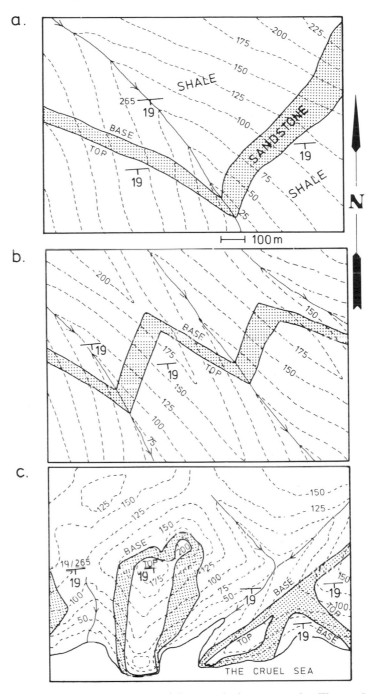

Fig. 4.4 The only variable in this series of diagrams is the topography. The sandstone layer in (a), (b) and (c) is of constant thickness and attitude. As the topography becomes more varied, the outcrop pattern becomes more complex

could be generated by complex variations in the layering attitude (see Figures 8.32 and 10.25/10.28). With a more regular pattern of valleys and ridges simple outcrops are generated of continuous swathes of each formation (Figure 4.4b).

The most straightforward interaction between geology and real topography occurs when formation boundaries are horizontal (Figure 4.5). Under these circumstances any one boundary will be found at a constant height coincident with a particular topographic contour. The outcrop pattern will be in perfect harmony with the contours

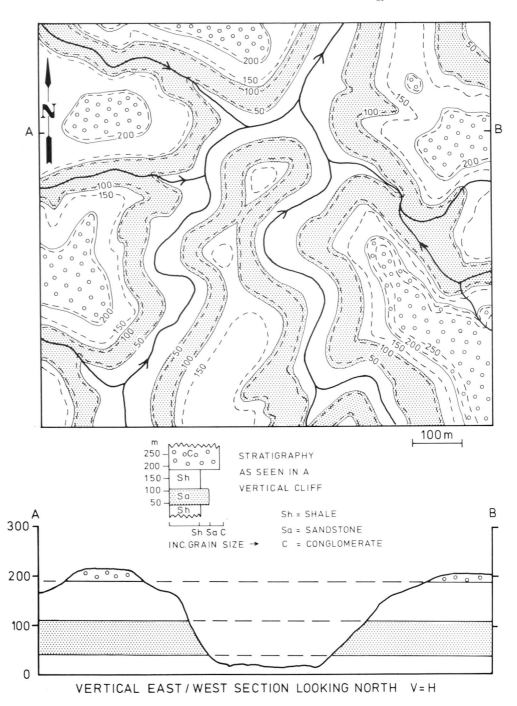

Fig. 4.5 The outcrop pattern of horizontal layers. Here the outcrop pattern follows the
topographic contours and is totally controlled by topography

and some visually striking maps result from this combination (see Figure 5 . 10). Every
variation in the shape of the ground is reflected in the outcrop and complex patterns are
possible despite the very simple nature of the geology. In real life it is rare for perfectly
horizontal beds to occur over a large area and dips of a few degrees commonly affect
the outcrop pattern (a good example is the Grand Canyon map published by the Grand
Canyon Natural History Association, see also Figures 4.13 and 4.14).

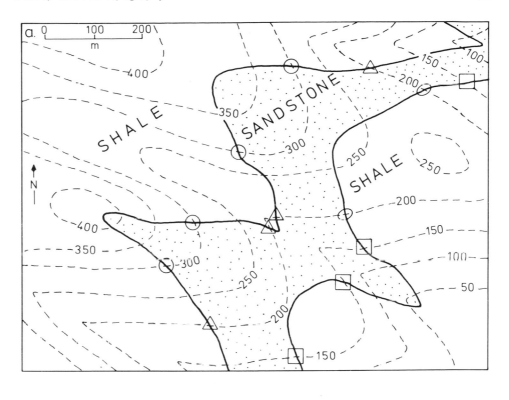

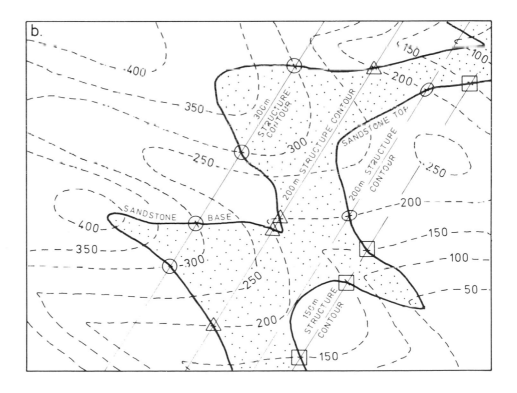

Fig 4.6 a,b (Caption on following page.)

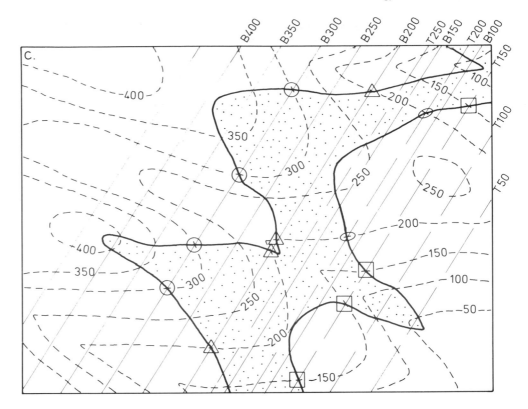

Fig. 4.6 The interaction between geology and topography allows the attitude of the layers to be determined. (a) The circled points are all localities where the western sandstone/shale contact is at 300 m above datum level. These points are defined by the outcrop of the western sandstone/shale contact crossing the 300 m topographic contour. The triangular marked points are the same surface at 200 m. For the eastern sandstone/shale contact the points in the squares are at 150 m and in the ellipses at 200 m. (b) Joining the circled points defines the 300 m structure contour (SC) for the western sandstone/shale contact, and the triangles define the 200 m SC. The points in the squares lie on the 150 m SC for the eastern contact. The structure contour pattern shows that the layering dips south-east, and the spacing allows the dip to be calculated (34°). The full specification of attitude is 34/210 (right-hand rule). Also it can now be seen that the western sandstone/shale contact is the base of the sandstone and that the eastern shale is the uppermost. (c) By extending the procedure demonstrated in (a), more SCs can be constructed. Note that two surfaces (top and base of the sandstone) are being plotted hence careful labelling is necessary

We have already noted the strong contrast between outcrop in topography with a regular pattern (Figures 4.4a, b) and outcrop in an area of irregular relief (Figure 4.4c). In the former, fairly simple zigzag outcrop results from interaction with valleys and ridges. We can understand how this pattern is generated if we study the interaction of structural and topographic contours. In Figure 4.6 a sandstone layer is sandwiched between two thicker units of shale. The outcrop shape has the elements of the simple zigzag of Figure 4.4b but a somewhat more irregular relief is clearly reflected on the map. We shall initially work on the westernmost contact of the sandstone (Figure 4.6). To fix the position of a structure contour we need at least two points on the western surface at the same height and these are found where the trace of the surface cuts topographic contours. The circled points on Figure 4.6a are all at a height of 300 m and they lie on a single straight line which is the map projection of the 300 m structure contour (Figure 4.6b). For the same sandstone/shale surface the triangular points define the projection of the 200 m structure contour. To demonstrate that this surface is planar we need at least another structure contour because on planes all structure contours are parallel with the same horizontal equivalent. The next contour could be defined by the boundary at 250 m or 350 m or any other convenient heights. Once these additional structure contours are drawn in, a regular spacing and parallel arrangement is seen which proves the interface is planar (Figure 4.6c). Using the scale on the map and simple trigonometry, the dip may be calculated (see Figure 3.2d). The strike of the surface is the orientation of the structure contours and the attitude (right-hand rule) is 34/210. Having defined the attitude of the sandstone it should be fairly easy to picture the 3-D situation (a hand may be useful for visualization) and from this you should see that the western boundary is the sandstone base (Figure 4.6b).

Structure contours for the eastern sandstone/shale contact (sandstone top) are constructed in the same way. The points marked by squares (Figure 4.6b) define the 150 m contour and the elliptical points the 200 m contour (Figure 4.6b). Again it is possible to locate extra contours at 50 m vertical intervals (Figure 4.6c). The orientation and spacing of structure contours for the eastern contact is the same as that for the western contact; this proves we are dealing with the layer-cake, i.e. the sandstone is of constant thickness. If the sandstone body was wedge shaped, then the contour spacing of the two surfaces would be different (and the strikes of the top and bottom surfaces might be different). An irregular shape to the sandstone body would generate curved structure contours with variable spacing.

4.4   RULE OF Vs

The zigzag or V-shaped outcrop patterns of Figures 4.4a and b represent a very common style. In the majority of areas, the geological surfaces dip at angles greater than the slope of valley bottoms (the streams) and most topographic gradients. Under these circumstances the attitude of the geology exerts a greater influence on outcrop shape than does relief. If we follow, in Figure 4.6, the relationship between structure contours, topographic contours and outcrop, you should see why the outcrop of dipping beds runs obliquely up valley and ridge sides. The combined effect of this oblique relationship on two sides of a valley is to generate a V shape. In the case of Figures 4.4a and b, the Vs always point downstream which is the dip direction of the beds. This simple observations will be used over and over again in map analysis, map-making, and remote sensing. In the vast majority of situations the beds dip in the direction of V in the valleys. Note that ridges give the opposite sense of V.

This, however, is not a complete expression of the rule as is shown by Figure 4.7d where a sharp V points upstream but the beds dip downstream. Such a reversal of fortunes is generated by a stream gradient greater than the dip of the layering, a situation most commonly met where the geology is gently dipping (<5°) but watch out for steep mountain streams. Fortunately, the relative slopes of the stream and the layering can be judged from the outcrop pattern without having to calculate the slope of the valley floor. A complete statement of the rule of Vs is as follows:

'If a geological boundary descends the valley sides to form a V-shaped outcrop, the geological surface dips in the same direction as the closure of the V.'

In both Figures 4.7a and b, walking along the outcrop trace from ridge to the valley V takes you downhill, i.e. descending to a lower elevation. In contrast the

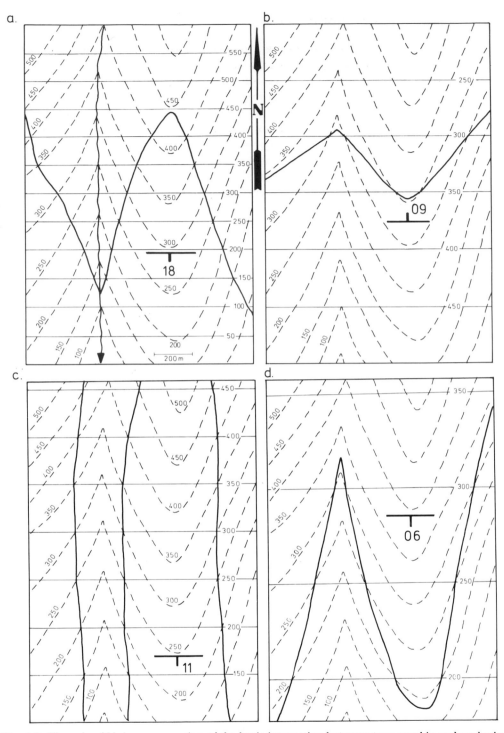

Fig. 4.7 The rule of Vs is an expression of the basic interaction between topographic and geological surfaces. In the majority of cases the V of the outcrop in a valley gives the dip direction but some situations require a fuller statement (see text). In (a), (b) , (c) and (d) the geometry is simplified with regular topography and the layering striking at right angles to the valleys. (a) The V outcrop of a thin bed in the valley is reached by going downhill (descending), therefore the V points in the dip direction. Check with the structure contours and dip + strike symbol. (b) Again the V in the valley is reached by descending hence the V gives dip direction. (c) The layer is dipping downstream at about the same angle as the valley bottom. This creates subparallel outcrops on either side of the valley and no V. (d) This is the exception to the simple statement of the rule of Vs. Here the V points upstream but the dip is downstream. To reach the V in the valley you have to walk uphill (ascend) from the ridge outcrop position

same walk on Figure 4.7d, to the closure of the V, is uphill, i.e. ascending. Hence if you have to ascend a valley (gain height) to reach a V pointing upstream, then the bed dips in the direction opposite to the V (downstream). Figure 4.7c illustrates the outcrop pattern of a layer dipping downstream at about the same angle as the stream. Subparallel outcrop traces result. All of the examples used in Figure 4.7 strike at right angles to the valley and ridges. A more oblique relationship would give rise to skew outcrop patterns (see Figures 4.4a and b where the strike is not perpendicular to the valley trend).

The rule of Vs deals with the intermediate amount of interaction between topography and geology. The two extremes are horizontal beds where outcrop patterns are totally sensitive to topography, and vertical beds that have outcrop patterns totally independent of topography (cf. Figures 4.3 and 4.5). Structure contours drawn on vertical units can easily explain the latter outcrop style. On Figure 4.3 take any bounding surface of the dyke trending to the south-west and draw the 600, 500, 400, 300, and 200 m structure contours. You will find that these are all stacked one on top of the other, an expression of a vertical surface. This arrangement of structure contours also means that there will be no development of V patterns in valleys or over ridges. Remember that for a dipping surface the horizontal equivalent of the structure contours is inversely proportional to the dip. When the gap between the structure contours disappears the surface is vertical.

4.5 THREE-POINT PROBLEM

If you know that a surface is planar, then height information from three points is sufficient to determine its dip and strike and predict its complete outcrop. Figure 4.8a gives a fairly typical situation. You may be told the thin layer exposed at A, B and C, is planar or that this should be assumed. In a field situation you could measure dip and strike at these three points and perhaps at isolated bedded exposures in the surrounding rocks. If these measurements were in reasonable agreement, a plane will be a best fit to the observations. Even if the geology has a constant dip and strike, measurements at small exposures may vary for a variety of reasons; measurement error, soil creep rotating 'apparent' exposures, irregular sedimentary structures, local magnetic effects, etc.

The three-point method for a planar surface relies upon the fact that the gradient in any direction (apparent dip) on the surface will be constant for that direction; the actual value of apparent dip, however, will vary with orientation. Hence, if we know the map position of a point on the surface at 200 m and another at 400 m, then half-way along the join between these two is the projection of a point at 300 m. Any join between two known points may be subdivided proportionately. The procedure is outlined in Figure 4.8. Of the three points, at least one must be at a different height to the other two. Two at the same height simply define the structure contour for that height and the strike.

Following and adapting the basics of the three-point method, an outcrop pattern may be completed given one exposure of a surface or thin bed together with the dip and strike. With this information a structure contour is drawn through the one exposure in the direction of the strike. Structure contour spacing on the map is

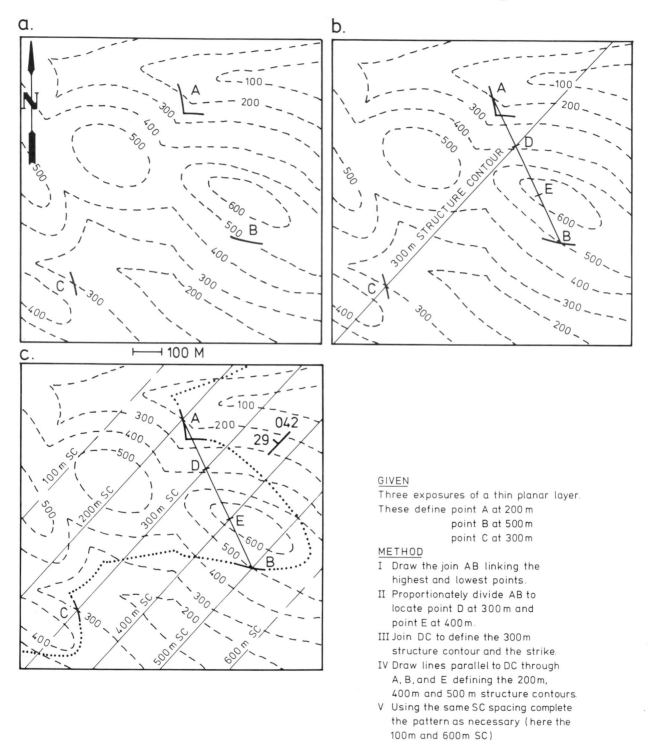

GIVEN

Three exposures of a thin planar layer.
These define point A at 200 m
                   point B at 500 m
                   point C at 300 m

METHOD

I   Draw the join AB linking the
     highest and lowest points.

II Proportionately divide AB to
     locate point D at 300 m and
     point E at 400 m.

III Join DC to define the 300 m
     structure contour and the strike.

IV Draw lines parallel to DC through
     A, B, and E defining the 200 m,
     400 m and 500 m structure contours.

V Using the same SC spacing complete
     the pattern as necessary (here the
     100 m and 600 m SC)

VI Complete the outcrop based on
     intersections of SC and topographic
     contours.

Fig. 4.8 The three-point problem. From three exposures of a single surface, known to be of constant attitude, the complete outcrop may be predicted. At least one of the points must be of different altitude to the other two. Method is given on the diagram

governed by the dip. Extrapolation, based on as little information as this, is obviously adventurous and you should be prepared for signs that the original assumption has broken down, e.g. that the dip, strike or both have changed.

4.6   CROSS-SECTION CONSTRUCTION

The method discussed in this section is relevant to large-scale maps where structure contours can be accurately derived from outcrop and topographic contour data. It should be noted here that the exercise maps (e.g. Figures 4.4, 4.6 and 4.9) are artificially regular; even the most regular natural example will show some sinuosity in the structural contours. A different approach is needed to handle planimetric or small-scale relief maps. Figure 4.6 will be used as the basis for the section construction exercise and the section line chosen is down the dip of the beds (Figure 4.9). Traditionally, a section is a vertical slice through the region, though there are a fair number of situations that are better suited to sections inclined at less than 90° (see fold profiles in Chapter 8 on Continuous Deformation). The construction method is as follows:

1.   Place the straight edge of a piece of paper along the line of section and mark off accurately the ends of the section. These should be labelled (with grid references if provided). Where the paper crosses topographic contours place a tick on the paper and note the contour height. Valleys and summits should be noted, e.g. the high point at 390 m near A (Figure 4.9).
2.   Draw, on a piece of graph paper, a horizontal line equal to the length of the section (labelling each end) and at each end draw vertical lines. On the latter mark off altitudes using the same scale as the map. This provides the x, y positional framework for the section.
3.   Place the paper with the contour data horizontally along the graph paper section below the lowest point of the section. Put a point on the cross-section, for every contour mark, vertically above the mark using the appropriate altitude.
4.   Join the points to complete the topographic profile.
5.   Return the paper edge to the line of section and mark off the position where structure contours are crossed. If several surfaces are being plotted, careful labelling is required.
6.   Reposition the paper on the cross-section and, vertically above each structure contour mark, put a point at the correct altitude.
7.   The three points for the base of the sandstone fall on a straight line confirming the planar nature of the surface. The two points for the top of the sandstone form a parallel line showing the layer-cake nature of the succession.
8.   Make sure that the section is properly labelled. It is sometimes useful to quote the viewing direction which in this case is towards the north-east.

Because the above method introduced both topographic and geologic section drawing, the simplest was taken first, i.e. the topographic part. As a general approach when working with large-scale maps (1 : 10 000 or better) the geology should be plotted before the topography. Basically, this is justified by the irregularity of the typical topography and the lack of detailed control information on its shape. With

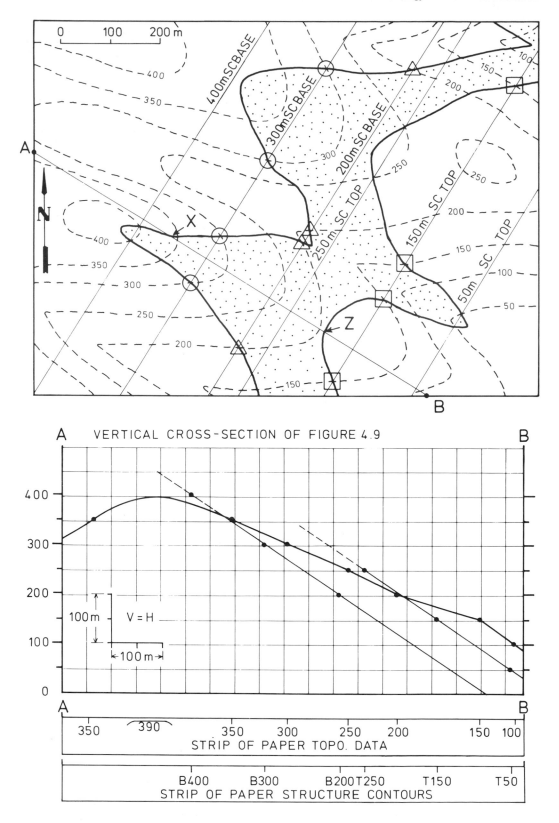

Fig. 4.9 Detailed construction of a vertical cross-section (topographic and geological) from large-scale maps where structure contours have been drawn. See text for procedure

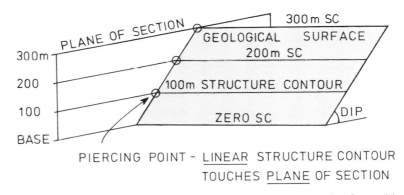

Fig. 4.10 Structure contours are transferred from a map to a cross-section by positioning the intersection points of the linear structure contours (on a geological surface) with the planar cross-section

some section lines, a class of 30 students can produce quite a range of topographies even if all the data points (where the section cuts contours) are correctly positioned. Between the fixed points there can be scope for very variable interpretation of slopes and highest plus lowest heights reached. At large scales it is better to draw on the geology and then the topography which is further constrained by the outcrop pattern, i.e. make the extrapolation of topography fit the position where the section cuts geological boundaries. Some authors recommend an alternative technique which is in fact quite unsuitable for large-scale maps. This less desirable method would calculate the dip and strike of the layering, draw the topography, and use the points X and Z on Figure 4.9 to fix the position of the base and top of the sandstone. From X and Z as located on the section the sandstone-bounding surfaces would be drawn in using the dip data. In this approach the positioning of the geological surfaces on the section depends upon the interpretation of the topography which may not be well constrained and thus different representations of the geology may be found from person to person. If the section is not down the dip of the layering then the apparent dip has also to be calculated (Section 3.3). This alternative technique is to be used on published Geological Survey maps at scales of 1:20 000 or less where dip and strike information presented is good enough to indicate the overall attitude of a layer. It should, however, be remembered that, at scales between 1:20 000 and approximately 1:60 000, structure contours could still be constructed and a section drawn from them.

The structure contour method is the most accurate one available. It is also very adaptable. By following the same procedure for a north – south section on Figure 4.9 the apparent dip is automatically taken into account without need for calculation. The basis for the method is shown in Figure 4.10. Each structure contour is a line which intersects the plane of section at a point. The method simply locates these points of intersection (effectively recording the *x*, *y* coordinates of these piercing points) and joining two or more points defines the trace of the geological surface on the section.

Many cross-sections use a vertical scale larger than the horizontal. This practice of vertical exaggeration has largely fallen into disrepute because of the distortions that are introduced. In the early days of plate tectonics virtually every tectonic cross-section had a considerable vertical stretch to the extent that few people had any real idea of true cross-sectional shapes of island arcs, subduction zones, continental margins, etc. Even a very low vertical exaggeration of ×2 brings about considerable shape changes in relatively simple geological structures (Figure 4.11). The main problem is that all dips are increased, but if dips were initially variable then the

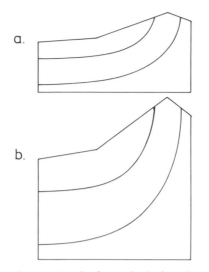

Fig. 4.11 Vertical exaggeration or stretch of a geological section is a common practice, but it involves distortions not readily appreciated by many geologists. The simple fold in (a) is strongly modified by only a two times stretch (b). Only use vertical exaggeration when it cannot be avoided, e.g. metres-thick sediments in sections hundreds to thousands of metres long. Workers dealing with thin Quaternary and Recent deposits have good reason to stretch their vertical scales

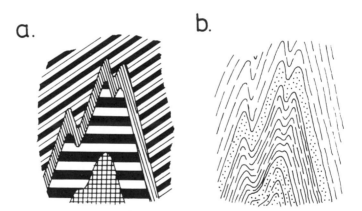

Fig. 4.12 (a) This is an illustration (it could be a map, a cross-section, exposure sketch) where the ornament has been deliberately designed to obscure the geology. This unfortunately is like many student early efforts at graphic work. (b) An example of ornament in harmony with the geology. This is a simple style of graphics which can be much more elaborate

change in dip is not linear. Features like layer thickness are also variably altered with differing dips and I doubt that many people could mentally reconstruct Figure 4.11a from 4.11b given the exaggeration factor. The layer in Figure 4.11a is of even thickness (a very significant factor in analysing fold mechanics) but after distortion thickness varies considerably, suggestive of a very different mode of fold formation. Note that where the layering is vertical, its thickness is not modified by a vertical stretch.

When completing a cross-section it is worth while paying some attention to graphic artistry. Good use of ornament can enhance the final product whereas poor finishing may lead to disaster (cf. Figures 4.12a, b). Generally, ornament should emphasize layering, not be in conflict with it.

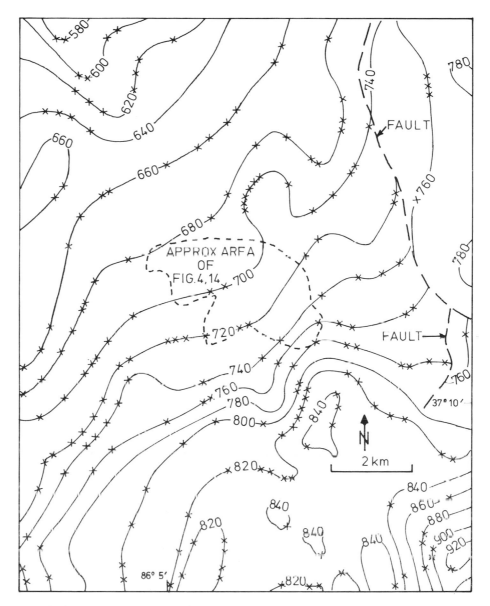

Fig. 4.13 Structure contour map for the base of one formation (the Big Clifty Sandstone Member) Mammoth Cave Quadrangle map of the United States Geological Survey. The outcrop pattern shows the layering to be subhorizontal but the regional structure contour map clearly shows a regional dip (0.4°) towards the north – west. Crosses mark data points where the formation base cuts a topographic contour

4.7   REAL STRUCTURE CONTOURS

Dealing with a very regular layer-cake stratigraphy has greatly simplified the geometry we have had to handle. In nature, structure contours are rarely parallel, and evenly spaced, straight lines, though some examples approach this pattern. Figure 4.13 shows a typical structure contour pattern in a region with minimal tectonic overprint. The base of one formation (the Big Clifty Sandstone Member) has

38

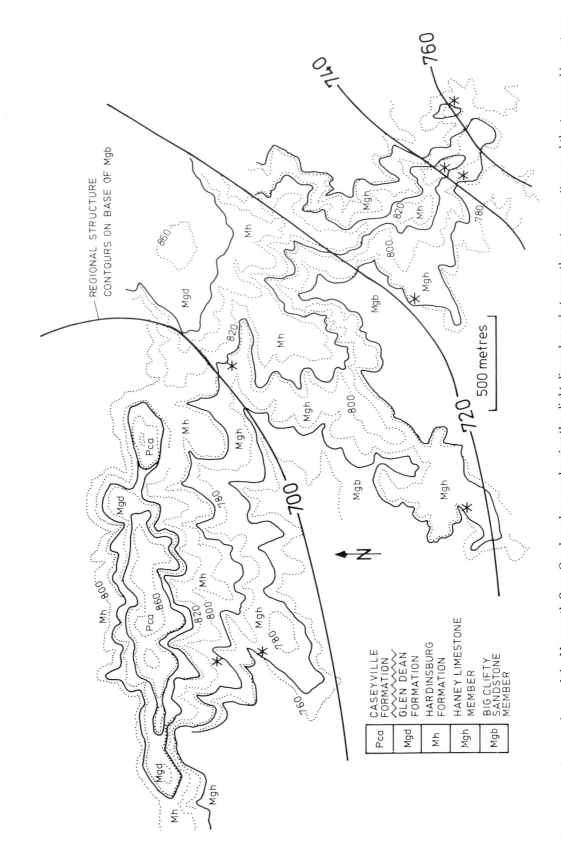

Fig.4.14 An abstract of part of the Mammoth Cave Quadrangle map showing the slight discordance between the outcrop pattern and the topographic contours. The systematic cutting of the contours by the formation boundaries (marked by asterisks) shows that the gentle dip is regional in extent

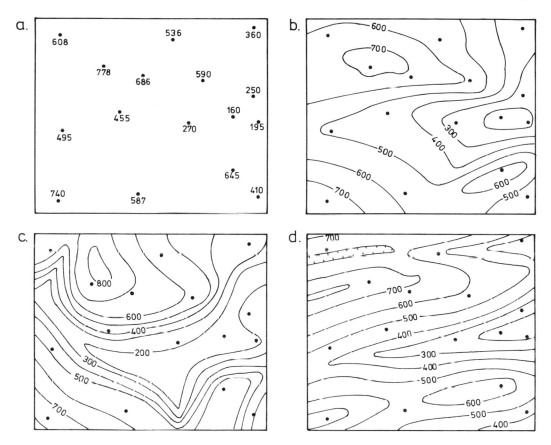

Fig. 4.15 (a) A limited number of topographic spot heights; (b) contours based on even gradients between spot heights; (c) contouring of the same data based on the assumption of a youthful topography in a humid climate; (d) contouring of the same data assuming that the topography is controlled by strike ridges

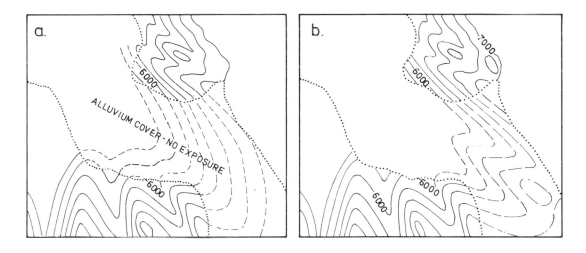

Fig. 4.16 Structure contours in two exposed regions separated by a zone of no information. Part (a) is a fairly direct extrapolation of the contours, whereas (b) follows the style seen in both of the areas of data and is to be preferred

been contoured and the crosses mark the data points where the base of the unit cuts the topographic contours (equivalent to the points marked by circles, squares, triangles, and ellipses, on Figure 4.6). The outcrop pattern on the map (Figure 4.14) is very strongly influenced by the topography and is very similar in style to that of Figures 4.5 and 5.10; subhorizontal layering is indicated. A detailed examination of the relationship between outcrop pattern and topographic contours shows the two to be inclined at very low angles (asterisks on Figure 4.14 mark cross-overs). A regional dip of about 0.4° to the north-west is shown by the overall structure contour pattern with a maximum dip of 1° over less than 2 km. Interrupting the regional trend are several irregularities. In the south-east corner, the structure contours define closed-loop patterns which should be interpreted in the same way as topographic contours. Most obvious are loops equivalent to topographic highs in the form of very gentle domal structures. Clearly the shape of the Big Clifty Sandstone is intricate. The fault along the north-east edge of the map marks a discontinuity in the structure contours because the blocks on either side of the fracture have undergone relative movement.

In this example there are a large number of data points (crosses on Figure 4.13) and several workers, given this same information independently, would probably draw similar structure contour maps. Given a sparser data set much greater variations would be expected (Figure 4.15). The most straightforward approach to contouring is to establish linear gradients between all points and draw in contours at a convenient interval (Figure 4.15b). However, if something is known of the regional pattern or if additional data are available from another source (e.g. geophysical) then a very different style may be appropriate (compare Figures 4.15b, c, and d). Figure 4.16 shows structure contour data for two exposed regions and different attempts at extrapolation into the covered ground to make the link. Part (a) is a fairly direct extrapolation, whereas part (b) looks more sensible as it matches the style of the structure contour pattern in both exposed areas. Differences in approach to contouring probably varies more with experience than with any other factor.

Geological interpretation of the Mammoth Cave region (Figure 4.13) would involve a regional tilt of the layering and explain the domal structures in terms of slight compression (folding). Some of the irregularities in the contour pattern might, however, relate to initial variations in layer thickness and geometry at the time of deposition. Such finer-scale resolution would require the construction of an **isopach** map which shows the stratigraphic thickness (as contours) of lithostratigraphic units. Isopach maps are major tools in regional stratigraphic studies because variations in unit thickness can say a lot about depositional conditions and palaeogeography. A somewhat related map is an **isochore** map which shows the variation in vertical thickness of units. For regions with low dips (<5°), isopach and isochore maps will be virtually identical.

5 REMOTE SENSING

5.1 OVERVIEW

Like all too many terms in earth sciences, remote sensing means many things to many people and the definition is commonly subdiscipline dependent. In our context we are dealing with systems that sample various parts of the **electromagnetic spectrum** (EMS) reflected or emitted by objects. Hence some of their characteristics are determined without physical contact; the essence of **remote sensing**. Human beings, using their eyes, act as quite sophisticated remote sensors but are limited to a narrow portion of the EMS, effectively defined by our capabilities as visible light (Figure 5.1). Many other techniques, mainly geophysical, involve investigation without touching the subject (sonar, magnetic, electrical, gravity) but these are generally excluded from definitions of remote sensing. Such restrictions are largely a matter of convenience and some texts on remote sensing include sonar, particularly as images of the sea-floor from this method are very useful in geological interpretations. Geophysicists dealing with magnetic and gravitational force fields are increasingly referring to their activities as remote sensing (perhaps an attempt to cash in on politicians' willingness to fund satellite remote sensing ahead of many other geological activities!). Before the late 1960s remote sensing was a relatively simple subject, but since then the introduction of satellite-based systems has brought about an explosion of activity and led to diversification into many totally new aspects of remote sensing. The whole growth process continues to mushroom, leaving those on the sidelines further and further behind in the wake of technology (also emphasizing that geology is now a big science in terms of expenditure required). Despite all the recent activity, traditional photography from aircraft is still the most used technique mainly because so many field geologists use air photographs every day in the field as base maps, and also as a means to interpret the geology before going into the field and whilst they are in the field. Satellite imagery is not yet at the correct scale and resolution to supplant air photos from this role, though developments along the lines of the French SPOT system may change the position. It is interesting to note that the US military establishment has opposed the development of high-resolution sensors but have now been forced to step aside by the competition and we may soon have 5 m resolution images; SPOT's best is 10 m and the current best from the US is 30 m. Currently available satellite images are not for day-to-day mapping but their large field of view emphasizes the big structures and major components of a region which help to put the geologist on the ground into the bigger picture. The computerized nature of satellite data collection gives enhancement and manipulation capabilities far beyond mere photographic methods.

Having hinted at the wonders of modern technology the sensible place to start is with seemingly dull black and white vertical air photographs. There are several

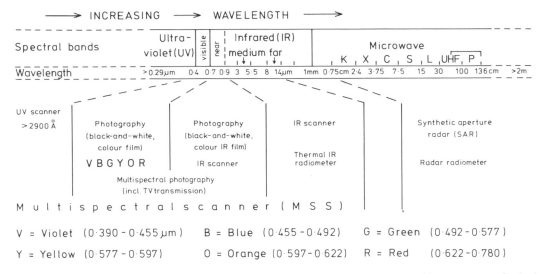

V = Violet (0·390 – 0·455 μm)   B = Blue (0·455 – 0·492)   G = Green (0·492 – 0·577)

Y = Yellow (0·577 – 0·597)   O = Orange (0·597 – 0·622)   R = Red   (0·622 – 0·780)

Fig. 5.1 In wavelength terms, this is the part of the electromagnetic spectrum of interest to geological remote sensing. The approximate ranges of the major remote sensing systems (photography, infrared scanner, radar, multispectral scanner, etc.) are indicated

compelling reasons for the order adopted. The technology is so familiar to us all that little in the way of theoretical introduction is required even though many of us do not appreciate the fundamentals of the photographic process. Even with no training and an unusual viewing direction (straight down), many features may be readily identified including roads, population centres, watercourses, vegetation patterns and a variety of landforms. With only a little insight, the clearer geological features may be easily understood, all facilitated by our frequent use of the photographic medium in day-to-day activities. The reality conveyed by vertical air photographs greatly enhances our analysis of geological geometry and geography; I would claim the integration of mapwork and air photo interpretation is the most distinctive and valuable contribution of this text.

For logistical reasons Chapters 4 and 5 are presented separately, but hopefully the modular treatment within this chapter will allow readers to pick out the short section on photogeology and to use this whilst working on Chapter 4. The greatest advantage of vertical air photos is their ability, with a special viewer (stereoscope), to generate a **stereoscopic** or 3-D model (Figure 5.2) of the earth's surface from overlapping photographs. This takes us rapidly from the realm of flat images and gives us the third of our four dimensions and is another reason for the popularity of traditional air photographs over satellite images where, until SPOT, the stereoscopic capabilities were very, very limited. The stereoscope allows each eye to look at just one image (photograph) and then requires us (the observer) to persuade the visual system that our eyes are being used in the normal way; this is easily done with very little practice and the mind fuses the images taken from different positions to create a 3-D view. Using one photograph as a base, we can trace out on a transparent overlay the pattern generated by formation boundaries, faults, intrusive contacts, that is, make a geological map. At the same time, because the tracing is done on a transparent overlay, we can see the shape of the land from the **stereomodel** and see what aspects of the pattern result from geological surfaces interacting with topography and also appreciate how this interaction works. The stereoscopic view is particularly good at providing information about geological geometry — layering attitude and the morphology of lithological boundaries including intrusive (igneous) and fault (fracture) contacts. I maintain this is a considerable improvement over purely map-based analyses because 3-D thinking is dramatically brought home. For many regions, particularly the arid and semi-arid areas, a very good geological map may be made from air photographs without even visiting the region. In almost every type of terrain when used before and in conjunction with ground mapping, photogeology greatly speeds up the mapping process thus saving money. Much first-phase reconnaisance mapping around the world is based on photogeology with only limited vehicle or helicopter traverses. Also, later more detailed mapping may be following up old surveys of variable quality (for cost or logistical reasons) and modern air photographs may considerably assist the remapping. Some early surveys were biased towards particular economic considerations and through time these needs may have changed, thus necessitating remapping (perhaps suggesting that comprehensive surveys are the best).

5.2 PHOTOGEOLOGY

Aircraft (and space vehicle) photography is either vertical or oblique, describing the attitude of the camera axis. Oblique conditions are equivalent to a view from a high lookout where we see the terrain in perspective. Even though this is a normal way of viewing the land it is of little value for detailed measurement or geological analysis. Overlapping ground photographs of vertical cliffs, quarry faces or escarpments do provide a useful record of such exposures and may be the only way, other than rock-climbing, of analysing these difficult to get at bits of information. The overlap gives the opportunity of stereoscopic viewing and the 3-D effect helps the interpretation of folds and intersecting planes (joints/joints; faults/beds; dykes/beds; etc.) Stereoscopic oblique photography of exposures also has been effectively employed by C. McA. Powell in a teaching manual, *Structures in Stereo* (Macquarie University, Sydney, 1973), to illustrate such forms as folds, boudins and joint patterns. This made use of classical examples and was a partial substitute for field-work as few courses can include all the necessary field-work.

5.2.1 FUNDAMENTALS OF AIR PHOTOGRAPHS

All further comments will relate to photographs taken with the camera pointing vertically downwards. The aircraft flies as straight a line as possible (a **run**) and exposes film at a predetermined rate to give a 60 per cent or better overlap between successive photographs (Figure 5.2). Modern photographs normally have about 30 per cent sidelap between adjacent runs to ensure complete coverage of the ground. Some old photography (1940s, 1950s) has irregular-shaped flight paths and variable to non-existent sidelap so be prepared for different quality and style; these flights were sometimes used to train navigators! At one time radar beacons guided the planes and the photos are parts of concentric strips around the beacon. Negatives are typically about 25×25 cm and contact prints are the standard product though enlargements may be employed depending on the type of survey. Modern prints contain quite a bit of data (Figure 5.3) including various reference numbers. CAF 4068 on Figure 5.3 is a film number as the primary reference point, though this may be a sortie number in some systems. E52-5 is the international code for the 1 : 250 000

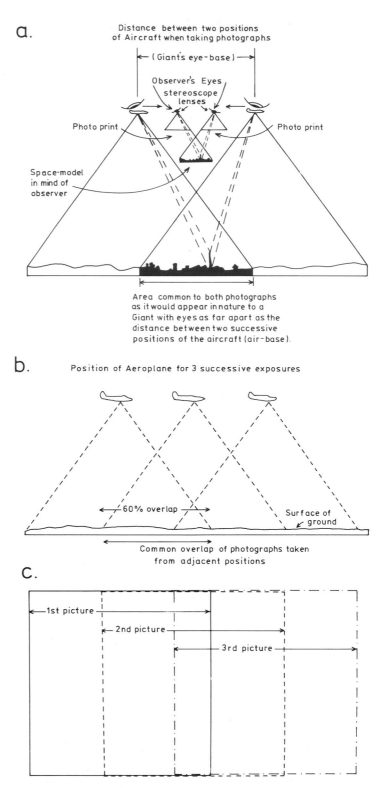

a.

Distance between two positions
of Aircraft when taking photographs

(Giant's eye-base)

Observer's Eyes
stereoscope
lenses

Photo print   Photo print

Space-model
in mind of
observer

Area common to both photographs
as it would appear in nature to a
Giant with eyes as far apart as the
distance between two successive
positions of the aircraft (air-base).

b.   Position of Aeroplane for 3 successive exposures

60% overlap   Surface of
                                               ground

Common overlap of photographs taken
from adjacent positions

c.

1st picture

2nd picture

3rd picture

Plan of area covered by 3 successive pictures

map sheet which in this case is Lansdowne in the Kimberley region of Western Australia. The photograph, frame 6533, is from run 6, it was taken at 10.20 a.m. on 29 July 1968. Most modern vertical air photography is taken ±2 hours of high noon to avoid losing too much information in long shadows. Crucial information is the flying height of the aircraft above sea-level (H=25 000 feet) and the focal length of the camera (f=88.45) quoted in this case in millimetres though the units are not given on the titling strip! Lines drawn between opposing fiducial marks (Figure 5.3) cross at the **principal point** which was vertically below the camera when the exposure was made.

Some basic appreciation of the physics of the photographic method is necessary even for straightforward photogeology. Several distortions are found on air photographs and perhaps the most surprising is that the scale varies throughout a single photograph unless it is of a smooth level surface. Scale on the negative is a function of the distance between the camera lens and the object, hence topographic variation changes the scale (Figure 5.4). From simple geometry, angles ALB and CLD are the same and therefore triangles ALB and CLD are similar giving the following relationship.

$$\frac{AB}{CD} = \frac{f}{H-h}$$

where f=camera focal length, h=average height of terrain above sea-level. Because the ratio of the size of a distance or object on a photograph to the actual distance or size (on the ground) is the scale, the above equation can be rewritten:

$$\text{Average scale (representative fraction)} = \frac{\text{Focal length of camera}}{\text{Flight height above the terrain}}$$

A more direct derivation of the representative fraction (expressed in the more familiar ratio style) is given by the formula:

$$1 : \frac{\text{Flight height above the terrain}}{\text{Focal length}}$$

For the original contact print of Figure 5.3 the plane flew at 25 000 feet (H) over an average terrain height (h) of 1000 feet and the camera focal length is 88.45 mm. The scale, as a representative fraction and converting feet to millimetres, is 1 : (25 000-1000×304.8) ÷ 88.45, which on simplification equals 1 : 82 704. It must be noted that for publication the original 23×23 cm contact print has been reduced to about 14×14 cm so the scale as presented is about 1 : 136 000 — the formula works for contact prints only. To emphasize the variability of scale on air photos consider the effect of a step up of about 1000 feet from one level to a higher-level plateau (Figure 5.4). If the plane was flying at 8000 feet above sea-level, the average scale on the

---

Fig. 5.2 (on page 44) Vertical air photographs under ideal conditions are taken in straight lines (**runs**) at a predetermined interval to give a fixed amount of overlap between successive exposures (at least 60 per cent but may be 80 per cent). In the zone of overlap, objects are photographed from two different positions and when observed on the two prints using a special viewer (stereoscope) a 3-D image is created (inset in a). The flight path can be determined by overlapping common features on all adjacent prints (c). Cross-winds not adjusted for by the plane heading and/or camera orientation gives runs that step consistently one way or prints skew to the flight path

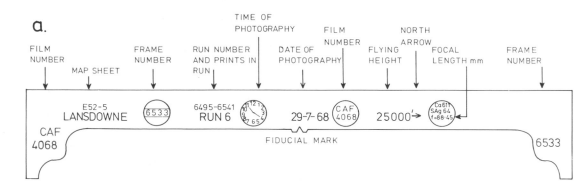

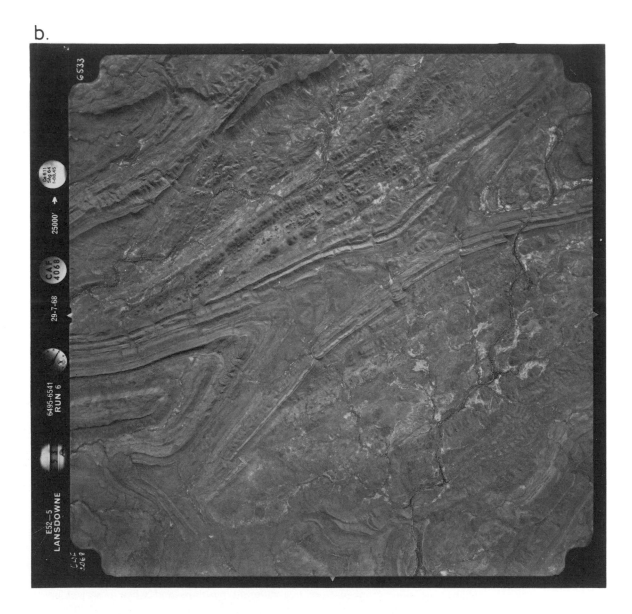

Fig 5.3 a, b

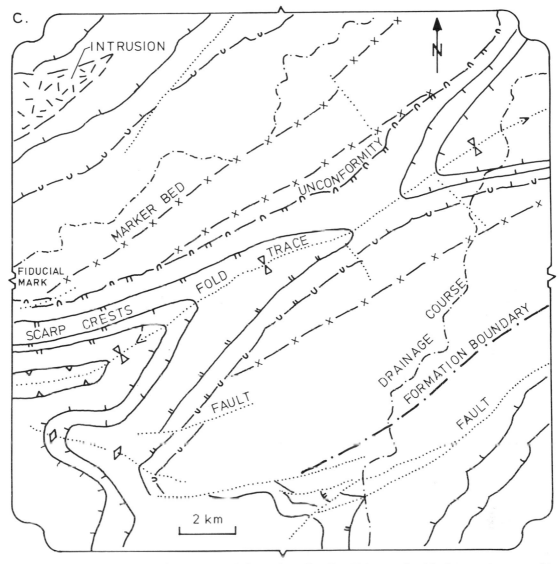

Fig. 5.3 Modern photographs are very informative. On the **tilting strip** (a) there are geographic references, flight details, time, date and camera focal length data. The interpretation (c) of the geology of this example is based on stereoscopic analysis using the photographs on either side of frame 6533 (b). Reference to this interpretation is made in later chapters after features such as folding, faulting and unconformities, have been studied. The symbols on this interpretation are in part non-standard to compensate for the lack of colour schemes that are normally employed (see Figure 5.12)

plains around CD for a focal length of 152.4 mm (6 inches) is (in feet) $1:(8000-1000) \div 0.5$ which is $1:14000$. In contrast the average scale on the plateau is $1:(8000-2000) \div 0.5$ which gives $1:12000$. Scale increases on higher ground and air photos are geometrically different to maps, though if terrain variation is not great the differences will not disturb our photo interpretations of the geology.

Maps show all points as though they were viewed from vertically above, but for each air photo only the principal point fulfils this condition. The distortion on air photos relative to maps is shown by the point X (Figure 5.4) which on a map would be the scale equivalent of the distance D+Xl from D but on the air photo would be displaced to D+Yl. This style of distortion, known as **radial displacement**, is dramatically shown by clusters of high-rise buildings, chimneys or power station

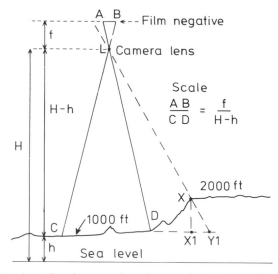

Fig. 5.4 On contact prints, the photograph scale is easily determined because triangles CLD and ALB are similar, hence scale= $f \div H\text{-}h$, where f is the camera focal length, H is the flying height above sea-level and h is the terrain height. Scale varies with h and a step up from one plateau to another increases the scale. In an irregular terrain, scale is varying from place to place when represented on a photograph. This figure also shows the non-planimetric nature of air photographs. On a map, point X would be shown as viewed from directly above but radial distortion displaces its position on the print outwards relative to its location on a map. (From *Methods in Field Geology* by F. Moseley. Copyright©1981 W. H. Freeman and Company. Reprinted with permission.)

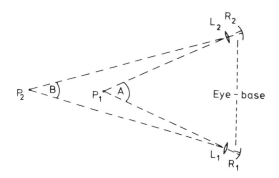

Fig 5.5 Binocular vision allows us to see a 3-D world for the several hundred metres immediately around us. The image of P_1 and P_2 fall on different parts of the retina ($R_1 R_2$) giving an estimate of relative distance based on different angles (A>B). A larger eye-base would extend the range of stereoscopic vision

cooling towers. Likewise, a vertical photograph of a forest of tall pine trees would show all the trees leaning away from the principal point in a radial fashion. The degree of apparent lean increases with distance from the centre of the photograph. Knowing the optics involved it is possible to correct the distortions introduced by photography and many maps, including contouring, are now made from air photos.

5.2.2 ACHIEVING THE STEREOMODEL

Our ability to see the world in 3-D is largely dependent on having two separate viewpoints (eyes) about 40—60 mm apart, a distance known as the **eye-base**. If you observe two objects at different distances away from you, the two images are found

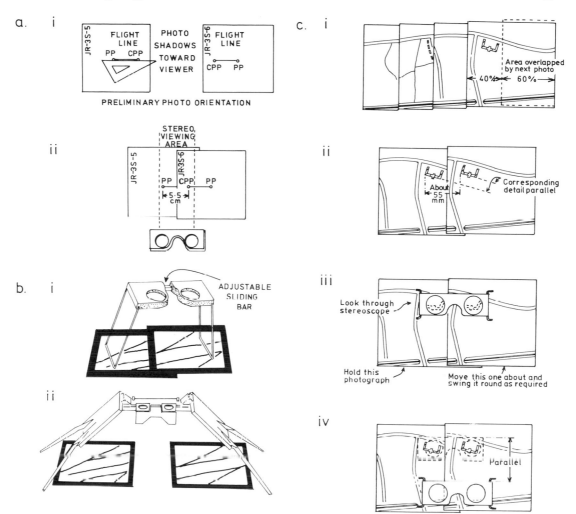

Fig. 5.6 (a) Precise alignment of air photographs for stereoscopic viewing requires the flight line to be established. On adjacent prints (a stereopair) locate the principal point (PP) at the cross-over of the fiducial mark joins. For each print locate the feature that is at the principal point on the neighbouring print — the conjugate principal point (CPP). The traces joining PPs to CPPs are the flight lines and the traces have to be parallel for stereo viewing. (b)(i) Pocket stereoscopes (mainly for field use) allow a narrow strip (approximately equivalent to your eye-base) to be examined stereoscopically. (ii) Mirror stereoscopes are much larger instruments that allow the entire overlap to be examined as a stereomodel. (c) With practice, setting up for stereo viewing with a pocket stereoscope can be rapidly achieved. Lay out the photo runs such that corresponding features are superimposed on top of each other (i). Take a stereopair and under the stereoscope draw the two photographs apart until identical objects on the two prints are separated by about 55 mm (ii). Small adjustments should then produce a stereo image (iii). All of the overlap area can now be viewed by moving the stereoscope (iv) and more can be studied by bending back the edge of the upper print

at different places on the retina which allows the brain to judge relative position (Figure 5.5). Note the relationships on this figure are considerably exaggerated for clarity, and that for much of our viewing we are dealing with much smaller angles mainly because our eye-base is small compared to object distances. The smaller angle B (Figure 5.5) relates to greater distance from the observer, but there is a limit to the resolution of angles by our eyes at 0.008° which for narrow objects occurs at 400 m. Beyond this, binocular vision does not give stereoscopic information and impressions of 3-D rely on shadows and perspective. For this reason, a person in a high-flying aircraft does not get information on topographic relief from the ground by using two eyes and the use of one eye would be just as effective. Of great practical interest for stereo viewing of photographs is the result of increasing the eye-base in Figure 5.5 but keeping P_1 and P_2 fixed in position. This would lead to a greater separation of the two images on the retina and the two points would appear to be further apart, that is, depth perception is increased.

At this stage it is useful to determine your own eye-base using a pocket stereoscope (which is illustrated on Figure 5.6bi):

1. The pocket stereoscope will either have a sliding bar or swivel to allow the distance between the lenses to be varied; adjust the spacing to comfortably fit your eyes;
2. Take a sheet of paper and mark and a dot on it;
3. Place the pocket stereoscope so that the dot appears immediately below the centre of the left eyepiece;
4. With both eyes look vertically down into the stereoscope;
5. Place a pencil to the right-hand side of the stereoscope;
6. Move the pencil point into the field of view until it appears to coincide with the dot and mark this position with another dot;
7. Remove the stereoscope and measure the distance between the two points to the nearest millimetre;
8. Three separate determinations should be made and averaged to give an accurate measurement of your eye-base.

From a photo run, if we choose two neighbouring photos, the 60 per cent overlap means that they have much ground in common. The shared area occurs on one print photographed from one point in space and on the other print recorded from a different position (Figure 5.2). A giant with an eye-base equivalent to these two exposure positions would be able to observe the terrain stereoscopically, but the giant's depth perception would be spectacular because of the huge eye-base. In using contact prints we have a scaled-down version of the giant's vantage-point, and to duplicate the viewing direct of the giant's eyes we use the stereoscope to require one eye to look at one print and the other eye only to look at the second print. When the photos and the stereoscope are correctly adjusted and aligned, the brain should fuse the two images to give a stereomodel. The mechanics of the setting up for a pocket stereoscope are (Figure 5.6):

1. Select two consecutive photos from a run (Figure 5.6c) and overlap them such that equivalent topographic or cultural features are close to being on top of each other.
2. Arrange the overlapping pair such that the shadows fall towards you and have the photo on your right on top.
3. On both the left- and right-hand prints locate the **principal points** (PP) and note what topographic or cultural features are at these points. The feature under the PP of the right-hand print will appear on the left-hand print but here it defines a **conjugate principal point** (CPP). Likewise, the PP feature of the left-hand print appears as a CPP on the right-hand print. On each photograph, a line joining the PP and CPPs gives the flight path (Figure 5.6ai). The traces of the flight lines on the two photographs must be parallel to achieve the stereomodel.
4. Separate the two photographs in a direction parallel to the flight path such that the PP on the left-hand print is about 5.5 cm from the CPP on the right-hand

print. For comfortable stereo viewing, identical features on adjacent photographs should be separated by a distance equal to your eye-base.

5. Adjust the stereoscope eye-base so that it comfortably fits your own eye separation. Remember most eye-bases are between 40 and 60 mm and there is not much difference in measurement between a frank honest look (wide eye-base) and a narrow opposite.

6. Position the stereoscope over the prints and place the midpoint between the lenses above the centre-line with the lenses over corresponding points on the two photos (Figure 5.6bi). In this position you should be close to getting a 3-D image but some nudging (millimetre scale movements) and slight rotation of the prints may be necessary to see a sharp image. When it is there it hits you in the eye as quite a dramatic image even in areas of low relief. If 3-D is not achieved fairly readily, concentrate on obvious features (e.g. a bend in a river, a distinctive-shaped hill) and place the lenses of the viewer over the two images of the same object. Mirror stereoscopes, by means of mirrors and prisms, increase your eye-base, hence photographs have to be much further separated relative to pocket stereoscope examination (Fig. 5.6bii).

Steps 1 – 6 above are a fairly careful approach to setting up the prints and with practice a more direct method may be developed. After step 2, separate the photos by about 5.5 cm, and place a finger of the left hand on a feature close to the visible edge of the left-hand photo. Place a finger of the right hand on the equivalent point on the right-hand photo, set up the stereoscope over these points, and keeping the fingers on the points move the photos until the two fingers appear to merge when viewed through the stereoscope. At this point only very small adjustments are needed to obtain a sharp image. After a little practice, setting up becomes very rapid and rigid adherence to the recipe is not necessary (e.g. starting with the left-hand print on top is equally as rapid a method of aligning the air photographs). A variation on the theme of rapid alignment of air photos for stereo viewing is shown in Figure 5.6c. With experience it is a simple matter to superimpose equivalent features on adjacent prints and then to draw out the photographs approximately along the flight path until a separation equal to your eye-base is reached; fine tuning from there will be needed to get a sharp image. Rigorous establishment of flight lines is only necessary where successive photographs are not in a reasonably straight line (usually caused by the plane's heading compensating for a cross-wind but the camera orientation not being adjusted in sympathy).

Many of the stereopairs in this book have been presented with a fixed separation of 55 – 60 mm. Align a pocket stereoscope perpendicular to the heavy line separating the two photographs with its centre over this line. Vary the distances between the stereoscope lenses (sliding bar or swivel) or skewing the stereoscope to be oblique to the centre-line until a 3-D image is obtained. If the stereomodel proves difficult to achieve, move the stereoscope until it is over the most prominent topographic features. Some relaxation of the eyes may be necessary before the 3-D effect clicks into place. If the stereomodel still proves elusive try closing alternate eyes while concentrating on a prominent feature, and hopefully image coincidence will come.

Once you see the stereomodel there is no mistaking it because differences in elevation are greatly expanding. This **vertical exaggeration** considerably steepens slopes, basically as a function of artificially increasing our eye-base to the distance between successive principal points in a photo run (the **air base**). This vastly increased depth perception makes even subtle topographic features stand out and can be a considerable advantage. For most stereoscopes the vertical exaggeration is between two and five times; Figure 5.7 shows the effect of a 3× stretch. If a bedding

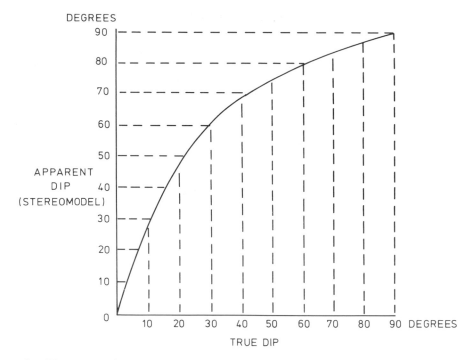

Fig. 5.7 The optics of stereoscopes induce a vertical stretch which we know from cross-section work can be hard to handle. Subtle topographic features are enhanced but slopes can be very unnatural. Where topography has etched out bedding planes (dip slopes) the apparent dip in the 3-D view has to be estimated and the graph will indicate the approximate true dip. The graph shows a ×3 vertical exaggeration but this will vary from stereoscope to stereoscope. Because of the difficulties of estimating dip and strike, symbols are of a ball-park nature (<10°, 10—30°, 30—60° >60°, see Figure 5.12)

plane dipping at 15° is exposed as a topographic surface, it will appear to be dipping at 40° if the vertical exaggeration is ×3. Clearly from Figure 5.7 the apparent dip in the stereomodel cannot simply be divided by 3; the formula describing these relationships is X tan (true dip)=tan (apparent dip), where X is the vertical stretch. Estimating true values of steep dips may be somewhat problematic, but these disadvantages are far outweighed by the enhancement of subdued variations in topography which may relate to important geological controls.

The amount of vertical exaggeration is a function of the air base for a given flight height, increasing with a larger base. Compare the exaggeration seen with 60 per cent overlap (adjacent prints) with that under 30 per cent overlap (miss out the adjacent print and take the next in the run. The extreme exaggeration of the latter is very unnatural and hard to handle. Also, for a given scale of photography, the amount of vertical exaggeration will also increase as the focal length of the camera decreases.

5.2.3   PROBLEMS YOU MAY ENCOUNTER

1. If the photos are set up in reverse order relative to the order of photography, viewing through the stereoscope reverses the topography—a pseudoscopic effect results where rivers apparently run along the top of sinuous ridges and

ridges are depressions. You should attempt this style of observation deliberately to be aware of the product so as to be forearmed for the day you make your first mistake. Strong shadows falling away from you may also bring about this reversal effect even with correctly aligned photos. If this happens try blinking or adjusting the lighting or take a brief tea-break to relax the eyes. Note that some photos appear very different when rotated to other orientations; this is typically a function of strong shadows.

2. Perhaps 5 per cent of people do not have stereoscopic vision for a variety of reasons and a similar proportion of the population have great difficulty in achieving the stereomodel with the pocket stereoscope, perhaps because of eyes of unequal strength. In most years, in a class of 30, I have one or two students who cannot see in stereo, but with some special coaching they are able to derive significant benefit from air-photo interpretation. This is partly because of the direct and familiar nature of air photos but is also dependent on extra guidance. Again one or two students have trouble with pocket stereoscopes but see the stereomodel fairly easily with large mirror stereoscopes.

3. If you wear spectacles try viewing with and without these aids to see which suits you best.

4. In very mountainous regions, large-scale differences between adjacent regions may make it difficult to obtain a 3-D image.

5. Note that objects changing position between exposures (cars, trains, boats) cannot be viewed stereoscopically.

6. The setting-up method for pairs of contact prints only gives direct access to a strip not much more than 5 cm wide for viewing with a pocket stereoscope, yet the total overlap between adjacent prints is commonly 12 cm or more. Manual dexterity and practice can overcome this limitation. By sliding the pocket stereoscope along the line of the flight path towards the area of overlap and keeping the prints in position, gentle bending up of the edge of the top print will progressively reveal the hidden overlap. As long as the curled photograph edge comes up between the lenses of the stereoscope it will not interfere with vision and helps to emphasize the optics of the stereomodel. The skill in this exercise rests in moving the stereoscope, curling one print and continuously fine adjusting the two prints to keep a sharp stereomodel — all with one pair of hands! When performing this feat, do not crease the prints as these will be permanent defects and the cracked emulsion will break off, thus losing information.

5.2.4 PHOTOGEOLOGICAL INTERPRETATION

We now come to the point of the exercise, that is, deriving geological information from vertical air photographs. In addition to geology, these images provide data on vegetation, landforms, hydrology, pedology, human activity and such miscellaneous features as bush and forest fires, grazing intensity, etc. The above characteristics define the nature of the terrain, some aspects of which obscure the geology, others reflect it, though none of these relationships are necessarily simple. It is only in very arid regions that you find extensive areas of bare rock, and even in semi-arid areas much geology is blanketed by a rudimentary soil. These complications together with the remote style of data gathering means that most photo-interpretation is inferential and with no ground-derived data ('ground-truth') we should be wary of presenting our story as the whole truth or of making too much of it. When reading my commentaries on images I hope by my style that you will be able to distinguish between knowledge derived from remote sensing and that given by ground surveys.

Tone is the basic element of a photograph and is a measure of the relative amount of light reflected by an object. The human eye is sensitive to 100—200 tones which on a photograph are registered as shades of grey ranging from black to white. Without tonal contrasts, photogeology would be limited to an analysis of landforms and their geological controls. Tonal variations are generated by changes in:

(a) the colour and colour intensity of the ground surface and vegetation;
(b) the smoothness of the ground surface; and
(c) the angle of the sun which varies seasonally and diurnally.

Contrasts in tone are more important than absolute value for interpretation. The correlation between tone and lithology is more consistent in thinly vegetated dry regions, whereas for humid areas vegetation may be more closely related to microclimate, soil acidity and drainage, than geology. Tonal characters in cleared, pastoral or cultivated areas, might be impossible to link to the underlying geology though soil colour variations may be helpful.

Texture refers to the general appearance of part of a photograph. As such it is difficult to define rigorously and, in using the term, the scale of the feature being described is important. It is possible to refer to the texture of a single bed and also of a large terrain element that contains this particular bed. Texture is the aggregate tone, shape, size, and pattern characteristics which may include drainage and vegetation influences. Examples of texture styles include mottled, speckled or smooth/even. Textures might be defined by abundant closely spaced fractures (joints) in one or several orientations, and differences in orientation and spacing will probably relate to a change of rock type. Many authors refer to texture when describing a drainage pattern using coarse drainage texture for widely separated watercourses and fine texture for close-spaced examples.

Patterns are the geometrical arrangement of any or all of the tonal, textural and landform elements in the region. The study of patterns is central to photography and largely depends upon the arrangement of topographic features, but vegetation and stream patterns may be equally important. The most obvious patterns are created by the distribution and orientation of linear features which may be joints, faults, dykes, sills, beds or unconformities (or even anthropogenic—roads, fences, crop boundaries, etc.). Drainage-system patterns often are a very good reflection of the larger-scale geological controls (Figure 5.8). Circular or oval patterns may be generated by domes or basins (folds) (see Figures 8.32 and 10.25), calderas (see Figure 10.18), batholithic intrusions (see Figure 10.7), impact structures, or eroded volcanoes (see Figure 10.22).

Shapes of individual landform components of the terrain commonly give very direct evidence on the type of lithology and more particularly on the structural stage (Figure 5.9). This is not the case for constructional landforms such as sand dunes, river terraces, raised beaches, moraines, eskers, drumlins, etc., though lithology may be inferred from the characteristics of a recent lava flow. The shapes of destructional or erosional landforms are very informative about the bed-rock geology at most stages of the erosional cycle. However, very mature peneplained landscapes may blanket the geology with superficial deposits.

Sedimentary, metamorphic, plutonic and extrusive rocks may all be layered, but the best definition of layering is in sediments and some lava flows. It is very unusual for the layers to be equally resistant to erosion and soft rocks will form subdued or negative relief features. Figure 5.10 shows horizontal beds of alternating hard and soft layers. The hard beds (dark tone) form small cliffs or zones of steep topographic slope and typically narrow outcrop width. The softer (light tone) beds are

DENDRITIC : the common-
est pattern. Indicates
uniform materials.

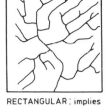

RECTANGULAR : implies
strong bedrock jointing
and thin soil cover. The
stronger the pattern,
the thinner the soil.

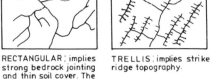

TRELLIS : implies strike
ridge topography.

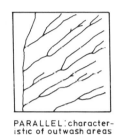

PARALLEL : character-
istic of outwash areas
of low topography, where
main stream may indicate
a fault.

ANASTOMOSING OR
BRAIDING : in alluvial
areas where sediment
load exceeds carrying
capacity of stream.

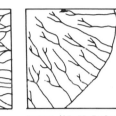

RADIAL (CENTRIFUGAL):
in isolated circular hill
masses.

INTERNAL : indicates
highly porous level
materials or karst
conditions.

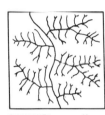

DERANGED : with many
ponds, bogs, or lakes.
Indicates flattish
landscape often glaciated.

PINNATE : generally
indicates high silt content
as in loess or on flood
plains.

ANNULAR: indicates
igneous or sedimentary
domes with concentric
fractures or escarpments.

DISLOCATED : due to
interruptions of
drainage lines by faults
or extrusions.

CENTRIPETAL: a
variation of the radial
pattern with drainage
towards a central point,
usually a sink or the centre
of an eroded antiform
or synform

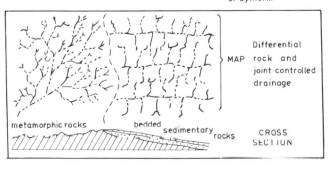

Differential
rock and
joint controlled
drainage

CROSS
SECTION

Fig. 5.8 Drainage patterns on air photographs (and topographic maps) often provide fairly direct evidence as to the nature of the underlying geology. Experienced interpreters can glean much information from such patterns alone, both about lithology and structure

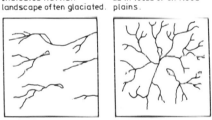

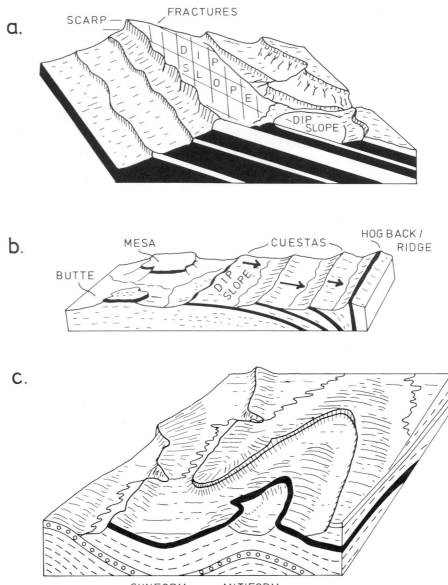

Fig. 5.9 Destructive landforms in many cases have a one-to-one relationship with the geology. Differential erosion can etch out even minor lithological differences between layers. Tilted layering may commonly have weak layers stripped off to expose bedding planes (dip slopes) (a). Even in humid terrains with good soil profiles, dip slopes can be very clearly defined. Because many joints in sediments are perpendicular to bedding, the up-slope termination of a dip surface is a cliff line (scarp slope) at a 90° angle to the dip amount. Where valleys cut across the strike, the trace of the scarp slope makes a V into the valley. Flat-topped hills typically form in areas of horizontal layering; dip slopes are best seen in regions with gentle to moderate layer dip which on steepening gives either limited dip slopes or eventually hogback ridges (b). Complex geology may be dramatically defined by destructional landforms (c)

Fig. 5.10 (on page 57) This fixed stereopair can easily be examined by a pocket stereoscope placed with its centre over the photograph dividing line. Separate the eye-pieces until the stereomodel is seen, perhaps skewing the viewer to sharpen focus. Horizontal layered sediments have been deeply incised by drainage courses. The dark layers are hard and resist erosion, forming cliffs in proportion to the thickness of the layering. The pale layers are soft and form benches. The lower part of the sequence is dominated by dark layers which form steep valley sides, whereas the upper part is an alternation of hard and soft producing a stepped landscape with flat-topped hills. Some dark layers show dramatic variations in the plan view of the outcrop width which, because layer thickness does not change, is purely a function of topographic slope differences. Note that there is some very minor faulting

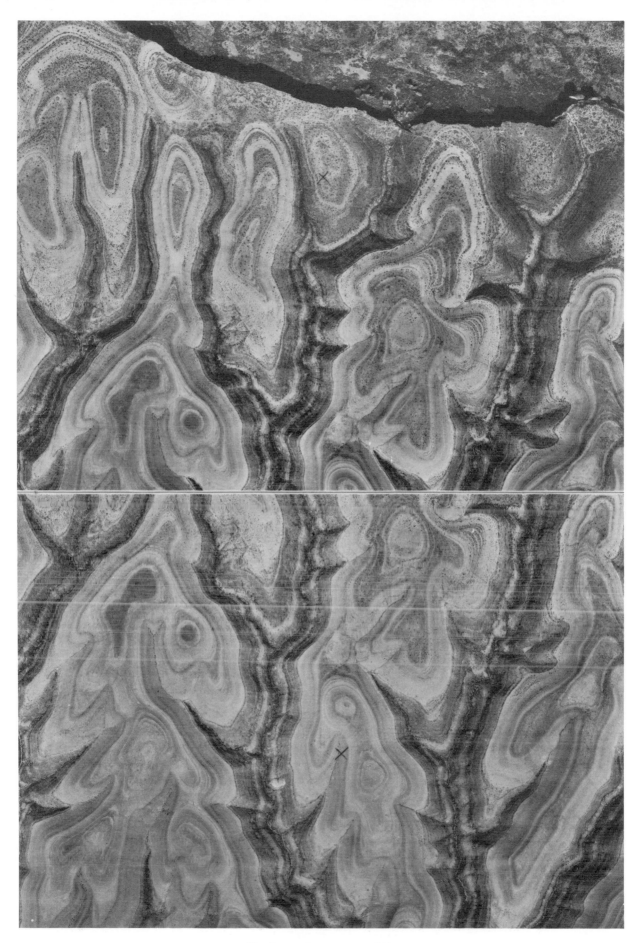

Fig. 5.10 (Caption on p.56.)

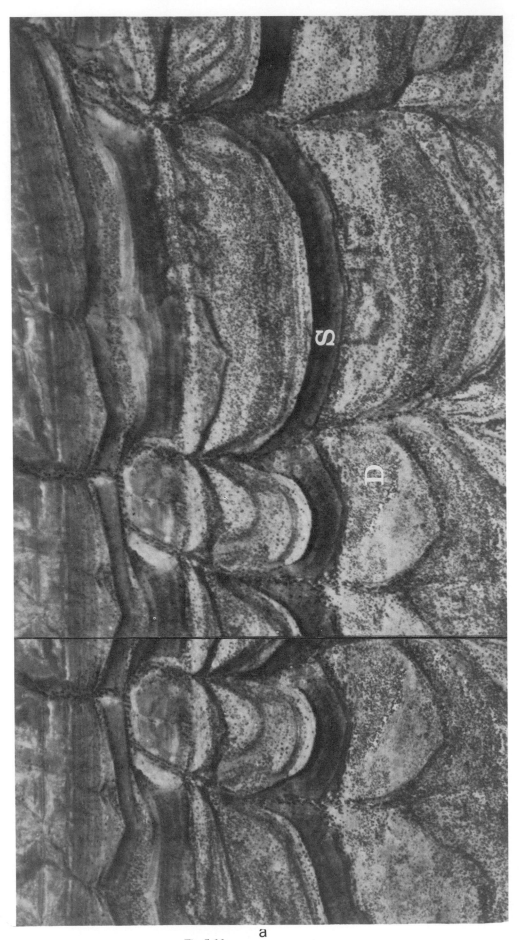

Fig 5.11a

characterized by gentle slopes and much wider outcrop. The complicated outcrop pattern reflects the subparallel relation between outcrop and topographic contours.

Inclined layered rocks generate characteristic landforms where resistant layers form linear or sinuous ridges as a function of constant or varying strike. In the majority of climates any significant differences between layers will be exploited to expose bedding planes — these in geomorphology are dip slopes (Figure 5.9a) and the relationship to the geology means that the dip amount may be estimated through the stereomodel (Figure 5.11). Most sedimentary layers have one or more sets of fractures (joints) sub-perpendicular to the bedding planes such that, when they are

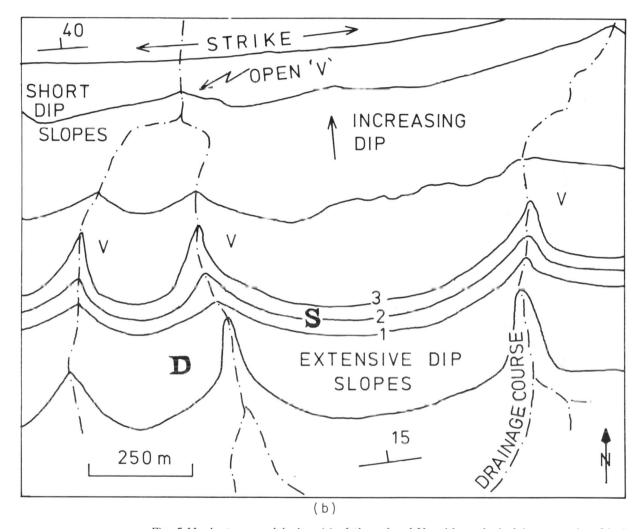

(b)

Fig. 5.11 A stereomodel view (a) of the rule of Vs with geological interpretation (b). A prominent scarp (S), created by three closely spaced sandstone layers, runs the length of the images. The uppermost of these three layers produces a very sharp mini-scarp which is in total shadow. Below the scarp extensive smooth surfaces closely follow one bedding surface as a dip slope (D) at about 15°. Dip slopes above the prominent scarp are more limited because of the close stacking of layers producing many scarp/dip steps. The dip increases to about 35 – 40° at the top of the image. At this edge, shales and siltstones appear as linear traces on a smooth near-flat topography to define the strike of the beds. Following any of the prominent sandstone beds at the centre of the image would generate a zigzag trace (V pattern) whose geometry depends upon the spacing and depth of cross-strike valleys, dip amount and resistance of beds to erosion. The prominent sandstones V in the direction of dip of the beds in the valleys and have opposite Vs on the ridges

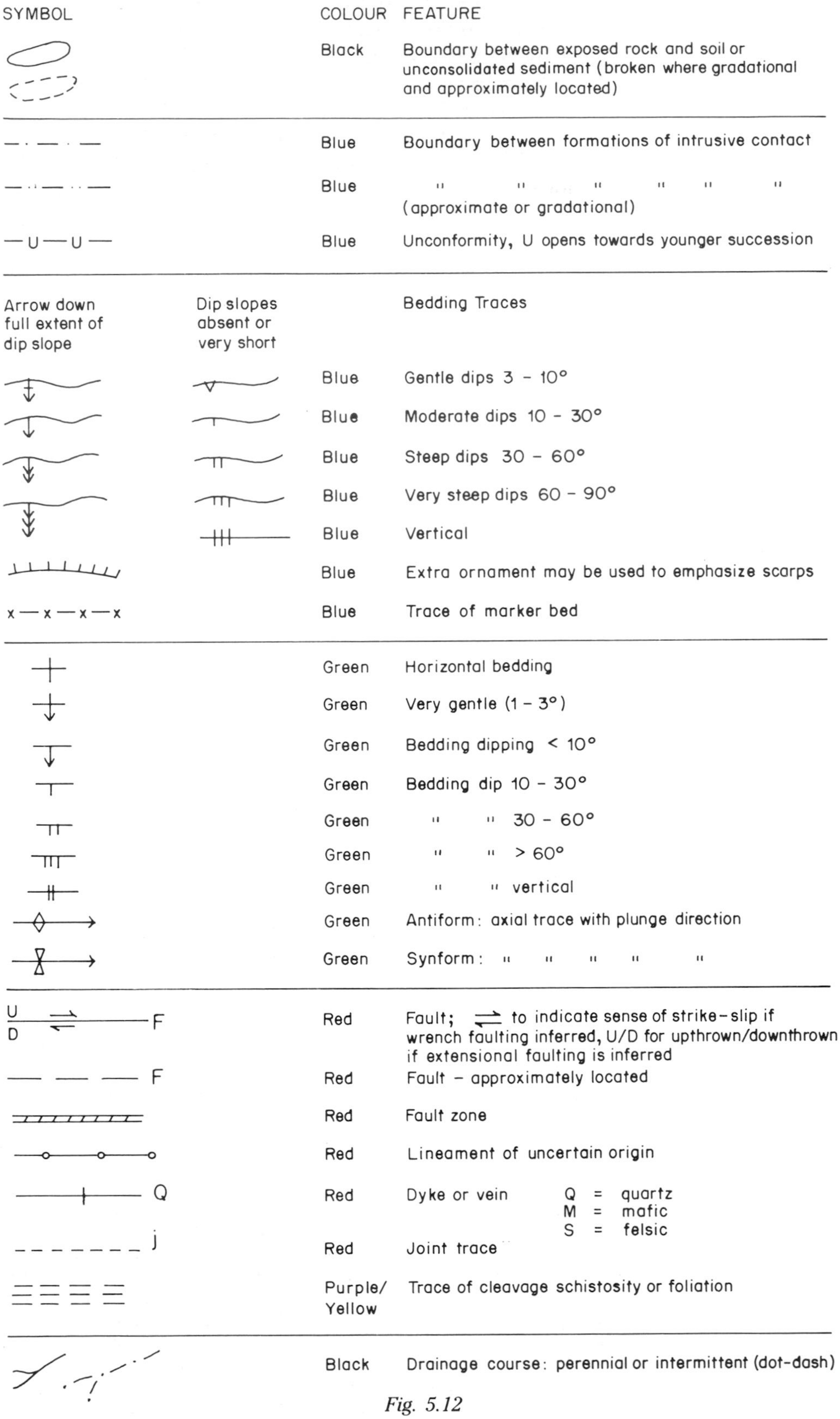

SYMBOL		COLOUR	FEATURE
(oval symbols, solid and broken)		Black	Boundary between exposed rock and soil or unconsolidated sediment (broken where gradational and approximately located)
— . — . —		Blue	Boundary between formations of intrusive contact
— .. — .. —		Blue	" " " " " " (approximate or gradational)
— U — U —		Blue	Unconformity, U opens towards younger succession
Arrow down full extent of dip slope	Dip slopes absent or very short		Bedding Traces
(arrow symbol)	(short tick symbol)	Blue	Gentle dips 3 – 10°
(arrow symbol)	(tick symbol)	Blue	Moderate dips 10 – 30°
(arrow symbol)	(double tick)	Blue	Steep dips 30 – 60°
(arrow symbol)	(triple tick)	Blue	Very steep dips 60 – 90°
(arrow symbol)	—+++—	Blue	Vertical
(scarp ornament)		Blue	Extra ornament may be used to emphasize scarps
x — x — x — x		Blue	Trace of marker bed
(cross symbol)		Green	Horizontal bedding
(cross with arrow)		Green	Very gentle (1 – 3°)
(T with arrow)		Green	Bedding dipping < 10°
(T symbol)		Green	Bedding dip 10 – 30°
(double T)		Green	" " 30 – 60°
(triple T)		Green	" " > 60°
(crossed T)		Green	" " vertical
(antiform symbol)		Green	Antiform: axial trace with plunge direction
(synform symbol)		Green	Synform: " " " " "
U/D (fault with arrows) —F		Red	Fault; ⇌ to indicate sense of strike–slip if wrench faulting inferred, U/D for upthrown/downthrown if extensional faulting is inferred
— — — F		Red	Fault – approximately located
(hatched line)		Red	Fault zone
—o—o—o		Red	Lineament of uncertain origin
—+— Q		Red	Dyke or vein Q = quartz M = mafic S = felsic
— — — — — j		Red	Joint trace
(double line ornament)		Purple/ Yellow	Trace of cleavage schistosity or foliation
(drainage symbols)		Black	Drainage course: perennial or intermittent (dot-dash)

Fig. 5.12

tilted and eroded, the opposite slope to the dip slope makes an angle of 90° to the dip. This is the scarp slope (Figure 5.9a). Material falling from the scarp slope forms a talus apron which may eventually obscure the full extent of the scarp particularly in areas of humid climate and thick vegetation. Steeply dipping (>60°) sediments will normally have very limited dip slopes and at very high angles of bedding dip (80—90°) hogback features are generated (Figure 5.9b). With very gentle dips (<5°) the dip slopes may be difficult to recognize on standard air photographs and scarps may not form depending upon the weathering conditions.

For readers who had some trouble envisaging the 3-D relationships in the rule of Vs (Chapter 4.4, Figure 4.7), Figure 5.11 should remove any lingering problems. Here we have reasonably constant dip and strike in a sedimentary succession where hard/soft alternations have led to the exposure of bedding planes. The strike ridge has been breached by valleys running subperpendicular to the length of the ridge probably controlled by fractures (joints). This extra dimension to the erosion pattern has interrupted the smooth dip slopes and created somewhat rounded-off triangular facets known as flatirons after the style of old-fashioned heavyweight irons. Between the cross-cutting valleys, the flatiron triangles (which are exposed bedding planes) have their apexes pointing upwards and in plan or map view they V up the ridge. In complementary style, the bed creating the flatirons will V in the opposite direction in the valley. Within the zone of overlap of Figure 5.11 two valleys are seen stereoscopically and the prominent sandstone bed marked on the geological interpretation creates two pronounced Vs pointing to the top of the image in these valleys. The more rounded V on the intervening ridge is a function of river spacing and the scale of the topography as well as the fracture (joint) control. It may be useful to sketch imaginary topographic and structure contours for a couple of cross valleys and an intervening ridge. This should help to make the link between 2-D representation (contours and outcrop patterns) and the 3-D real world.

Photogeological mapping requires integration of tone, texture, pattern and shape to determine as much as possible about the distribution of different lithologies, the nature of their boundaries, and the overall structural configuration. Experienced interpreters can be very successful at making geological maps in this way though success varies considerably depending upon how much the geology is obscured by soil, vegetation and man's activities. Experience is very much the key word and beginners may find it difficult to believe that the method is so useful. However, limitations must be remembered and it is rare for an interpreter, without additional ground-derived information, to be able to say what rock types he is looking at. Distinguishing sedimentary rocks from intrusive or high-grade gneisses may be easy, but to give a composition to most of the different layers will generally be impossible. Limestones, sandstones, and quartzite will be light in tone whereas mudstones, shales, and schists/slates will tend to be dark. Mafic to ultramafic lavas and intrusions are typically dark in tone but, in areas of strong hydrothermal alteration, basalts may be very pale. Igneous intrusions are rather homogeneous in that, within the body, one zone of several hundred square metres in extent will be fairly similar to an

Fig. 5.12 (on page 60) Symbols and colour code to be used on air-photo interpretation. Because of the difficulty in estimating vertically exaggerated slopes it is customary to use broad ball-park bands. Formation boundaries have a reserved symbol: within one formation there may be many resistant layers producing scarp and dip slopes. Where dip slopes are extensive, the top of the scarp should be marked as a continuous blue line and arrows drawn down the full length of the dip slope. Ornament on the arrow can show the dip amount. For small-scale photographs or areas of little topographic relief, the dip slopes are too short to be marked so an alternative style is adopted. Lithological boundaries (formation, intrusive, unconformable) could also have attitude data indicated in the same way

Table 5.1 Aircraft-based photography

Pluses	Minuses
High resolution (better than 1 m)	Restricted to a limited part of the EMS
Low cost	Restricted by lighting conditions, weather, atmospheric effects
Ease of operation (hire aircraft and camera)	Field of view limited, not suited to synoptic looks at large areas
Film provides much information	Generates many prints for a region, handling becomes difficult
	Computer manipulation of data limited

equivalent-sized zone elsewhere. They also have well-defined joint systems, but several other rock types produce strong joints so again it is a matter of using several attributes in combination for sensible discrimination. In producing a 'fact' map the first step is to define areas of exposure (where geological information may be gained) and to clearly distinguish these from areas of soil or unconsolidated sediments (alluvium, wind-blown sand, etc.). This process is normally very quickly achieved on air photographs because exposure and cover have very different tonal, textural, and shape characteristics — compare the valley bottoms of Figures 5.10 and 5.11 with the ridges and hillside.

Many photogeological mapping exercises will have some ground-truth to work with (earlier small-scale maps, company reports, reconnaissance studies, etc.) and in survey or commercial work all pre-existing data is incorporated (with caution). Other data bases may also be useful. Aeromagnetic maps may indicate the position and extent of magnetic bodies such as ultramafic intrusions, mafic dykes, kimberlite pipes, and banded iron formations. Alternatively, relatively non-magnetic bodies such as granitoids and sedimentary basins may be delineated by their quiet signature. Gravity anomaly maps may indicate gross structure such as sedimentary basins as well as individual granite plutons. Very dense rocks such as granulite facies, metamorphic rocks, and mafic/ultramafic intrusions may be discriminated if there is sufficient contrast with the country rocks.

Symbology for air-photo interpretations is not standardized and each organization has its own 'in-house' style. Having said that the variations are not huge and there is a fair amount of common ground. Figure 5.12 shows one scheme. The most usual variation on the theme is to combine the symbols for bedding traces or bedding scarps with the dip attitude information. In this style the bedding or scarp traces would then have a single arrow or one, two or three ticks depending upon the dip. Without the aid of measuring instruments, dips are estimated to a 'ball-park' range (<10°, 10—30°, 30—60°, >60°) because of the difficulties introduced by vertical exaggeration though photogrammetry can give accurate dip readings.

5.3   OTHER PHOTOGRAPHIC REMOTE SENSING

Only one of these methods will be employed in this text but readers should be made aware of what is available, the applicability of the different systems, and their relative merits. In a reasonably intensive look at traditional air photography we hardly discussed the sensitivity of film to the electromagnetic spectrum, i.e. the absolute basis of the whole process! Standard black and white film records EMS information in a single broad band from 0.3 to $0.9\,\mu m$ giving a spread either side of visible wavelengths into ultraviolet and infrared (Figure 5.1). Electromagnetic (EM) radiation is generated by the vibration of particles which gives temperature a dominant control on the type of radiation produced. Peak emitted radiation energy from the sun at 6000 K (degrees kelvin) occurs at $0.4 - 0.7\,\mu m$, the visible part of the spectral region, and shows a steady fall-off to wavelengths of about $100\,\mu m$. The much lower surface temperature of the earth generates peak energy with wavelengths around $9 - 10\,\mu m$, but at intensities 6 orders of magnitude less than the sun. Radiation from all objects on the earth's surface is a combination of that emitted in relation to its temperature, and that reflected from other sources. Standard black and white panchromatic film utilizes reflected radiation derived from the sun being unresponsive to the earth's emitted radiation. Between object and sensor (camera), the atmosphere may cause considerable scattering and, because this effect is greatest at the blue end of the spectrum, filters are employed to remove this part of the spectral region. Atmospheric interactions with EM radiation is an area of considerable research effort and is important in all aspects of remote sensing.

5.3.1 COLOUR AIR PHOTOGRAPHS

Colour film is composed of three emulsion layers sensitive to blue, green and red, the primary colours. Processing exposed colour film triggers a complex reaction to create this familiar product. The greatest advantage that colour photography has over black and white is the eye's ability to distinguish 200 000 colour hues. This allows maximum discrimination of different rock types and different soils. Some authors claim that such a large amount of information is too confusing and could lead to the wood being lost amongst the trees. Having used both styles of photography I would much rather use colour and did not feel I was losing the synoptic power of air photography but felt I was gaining useful information. Many mining companies exploring in arid and semi-arid regions seem to be of the same opinion, routinely flying colour runs over prospects and their immediate environs. Also aerial photography is used with many different purposes in mind and, for rapid reconnaissance mapping, the detail on colour prints might slow down progress when 50 or $100\,km^2$ may have to be mapped day in and day out. My experience in relation to colour photography is considerably influenced by a long stint in semi-arid regions where the clear air produced sharply defined prints every time; in contrast, a humid atmosphere causes much scattering and a blue haze significantly reduces the effectiveness of colour prints.

5.3.2 INFRARED PHOTOGRAPHY

Standard black and white film coupled with a red filter will only record reflected radiation in the range $0.7 - 0.9\,\mu m$, red and near infrared, because wavelengths shorter than $0.7\,\mu m$ have been cut out. One advantage of this method is the improved haze penetration given by the longer wavelengths resulting in sharper images. Near-infrared radiation is strongly absorbed by water which appears jet black on positive prints. This property provides excellent definition of shorelines and highlights details of drainage systems in densely vegetated areas and is particularly useful in the wet tropics. Healthy vegetation is a strong reflector of near infrared and differences between vegetation types are emphasized and best shown by infrared-sensitive colour film. The emulsion layers of this type of film are sensitive to longer wavelengths relative to normal colour film giving rise to the name **false colour photography**. Objects green to our eyes appear blue, red comes out green and infrared wavelengths are seen as red in this medium. This transformation may be a puzzle at first, but it is useful to appreciate the changes because the same colour system is used on the common Landsat satellite false colour composite images.

Forestry and agriculture applications dominate the use of false colour air photography because it gives pronounced discrimination of vegetation types. Deciduous trees are strong near-infrared reflectors and appear bright red; conifers are less reflective and are rendered purplish red to blue; dead foliage is bright green as no infrared is reflected. Geologists make use of the strengths of this process by studying geobotanical connections. Vegetation under stress because of growth on toxic soils of high metal content (Cu, Pb) near an ore deposit will contrast with surrounding regions as will plants specifically adapted to cope with such unusual conditions. For any one terrain this type of work would normally require an orientation survey over a known anomaly to determine the style of response to be sought. False colour photography may also emphasize thriving vegetation on permeable contacts and hence define faults and other fracture traces. Soil and rock will appear blue if saturated with water or show yellows or browns if dry.

Going further into the infrared region looks at emitted radiation which from the earth's surface is much weaker than solar radiation so these surveys are generally carried out at night to avoid being swamped. The detector, in fact, is a linescan system, not a photographic process, so not properly part of this section. Indirectly, the sun provides most of the energy for emitted radiation and the method effectively registers different temperatures. The best daytime absorbers are not necessarily the most obviously warm features on thermal infrared images because surveys are normally carried out just before dawn, and heat retention, a function of density, becomes an important factor. Pale limestones appear cooler than light-coloured sandstones because of density differences. Geologically, the main use is in the study of volcanic areas, particularly in prospecting for geothermal anomalies. Man-made effects, such as power stations, complicate the picture and routine geological mapping is difficult with **thermal infrared** because of the complex interaction of the variables. The advantages and disadvantages of aircraft-based photography are given in Table 5.1

5.3.3 SPACE PHOTOGRAPHY

To date we only have considered aircraft-based remote sensing systems and the step into space mainly concerns the same principles but with the limitations generated by the need to return exposed film to earth. Very early in manned space exploration, hand-held cameras recorded spectacular images of the earth and moon, helping to stimulate popular interest in all aspects of space research. Much of this photography was of little use for geological or cartographic mapping or other environmental studies until the adoption of vertical aerial survey techniques. The many stunning Shuttle photographs, taken on 70 mm film with Hasselblad cameras, are mostly low obliques, that is non-vertical but not showing the horizon. In 1973 Skylab carried an aerial survey type system but the camera specifications were not sufficiently rigorous

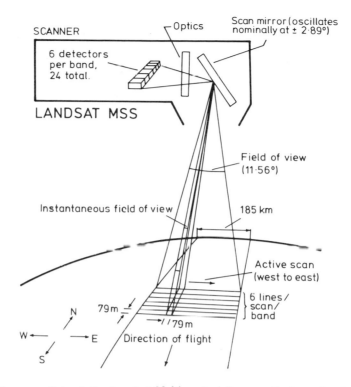

Fig. 5.13 The essentials of the Landsat Multispectral Scanner System. A mirror oscillates across the flight path and samples reflected radiation from 79×79 m pixels. Six lines are analysed at once. The optics split the radiation into four channels and focus the radiation on the radiometers. Intensity of radiation, from each pixel and in each channel, is recorded as a DN or BV

for mapping applications. After a ten-year gap, Spacelab 1 (Shuttle flight No. 9) performed the Metric Camera experiment which used a slightly modified aerial survey camera. The experiment was largely organized by West Germany through the European Space Agency and flown in conjunction with the National Aeronautics and Space Administration (NASA). About 1000 photographs were taken in a series of widely scattered runs with either black-and-white film or false-colour near infrared film. Negatives 23×23 cm cover 190×190 km and at flight height of 250 km, a 305 mm focal length gives a mean scale of 1:820 000. A 60 per cent overlap was standard for stereoscopic viewing though 80 per cent was needed over mountainous areas to give better height precision in the photogrammetric production of maps. Resolution on these prints is about 20 m, less than expected because of poor lighting conditions during the flight which was delayed by several months. The main interest in this programme is from the map-makers, though the stereoscopic capability is generating some attention from the geological community.

After one reflight of the Metric Camera, NASA developed a similar but upgraded system, the Large Format Camera (LFC), first flown in 1984. One extra film type was tested (colour) and, by compensating for the forward movement of the Shuttle during film exposure, a 10 m resolution was achieved. This allows 1:50 000 contour maps to be constructed and could greatly speed up coverage of underdeveloped areas. To show the experimental nature of these programmes, three flight heights (370, 272 and 239 km) were tried out to see which gave the best results. From the first runs, which again are very limited in extent, several exploration companies are studying the images to assess their usefulness in

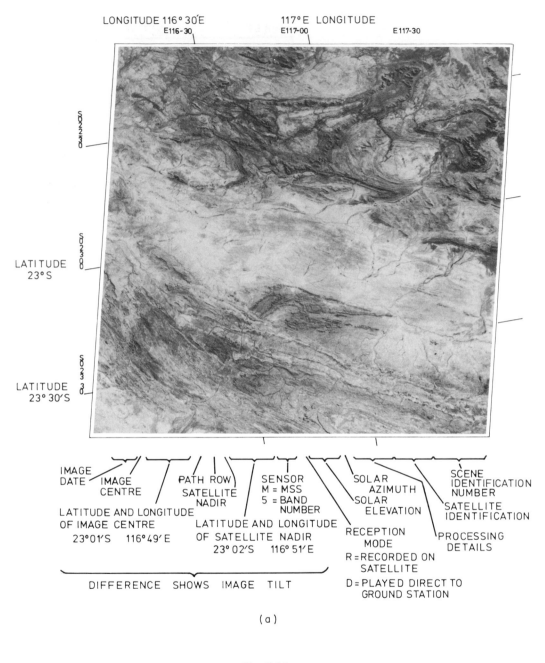

Fig. 5.14 a

photogeological exploration. The LFC products have good resolution, but not too much to get lost in the detail, and they have very large fields of view to give information on regional structures (see Figure 9.40f). They are potentially very useful geologically. It is interesting to note that virtually all Russian Sojus—Saljut missions carry cameras taking 10 m resolution photographs. The main limitations of the systems at the moment is the irregularity of the flights (even before the ill-fated Challenger disaster), their limited latitudinal range and the competition from other experiments for time and space on the Shuttle.

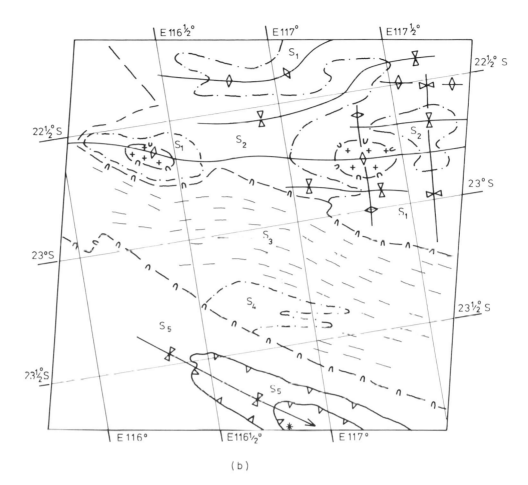

(b)

Fig. 5.14 (a) An example of an MSS standard product together with an explanation of the title strip. The DN data for Band 5 is shown in shades of grey. The image is a parallelogram because during the 25 seconds it takes to record a scene, the earth rotates beneath the satellite. Such a systematic distortion can be corrected by computer methods. (b) The geological interpretation (using symbols as Figure 5.12) deals with the regional features and subdivides the scene into major elements. There are three major divisions. To the north is a complexly folded sequence of well-layered rocks and varied stratigraphy ($S_1 + S_2$ + the cores to the domes — cross pattern). The central region is fairly uniform (S_3) with a strong structural grain and one zone of lithological variety (S_4). Along the southern margin is a gently folded sequence of layered rocks (S_5) with dips typically around 10° — very long dip slopes kilometres in length define the gentle fold structure. Note that even in this region it is difficult to gain an impression of the 3-D shape of the landforms and any analysis of geometry must be treated with caution. The area imaged lies just on the south edge of the tropics, but being a midwinter scene the sun is well to the north and shadows help to define the structure. At the point marked by an asterisk, the scarps face south (in shadow) and hence dip slopes face north. Regions like this would normally be studied ×4 to ×6 enlargement in the context of exploration or mapping work

5.4 MULTISPECTRAL SCANNER REMOTE SENSING

Perhaps the first message to convey here is that we have now moved away from photographic systems even though the final copy examined is commonly a photographic print. Also, this style of data gathering is very much dominated by

Table 5.2 History of the Landsat and SPOT satellite series including sensor specifications

	Sensor	Band numbers	Band width (μm)	Pixel size (m)	Ground swathe (km)	Radiometric resolution
Landsat	RBV	1	0.5 − 0.6	80×	183	
1, 2, 3		2	0.6 − 0.7	80		
1= 1972 (as ERTS) to		3	0.7 − 0.8			
1978						
2= 1975 to 1983	MSS	4	0.5 − 0.6	79×	185	64
3= 1978 to 1983		5	0.6 − 0.7	56		
920 km		6	0.7 − 0.8			
18D		7	0.8 − 1.1			
Landsat	MSS	1	0.5 − 0.6	82×	185	64
4+5		2	0.6 − 0.7	56		
(now EOSAT)		3	0.7 − 0.8			
4=1982 (TM to 1983)		4	0.8 − 1.1			
5=1984	TM	1	0.45 − 0.52	30×	185	256
705 km		2	0.52 − 0.60	30		
16D		3	0.63 − 0.69			
		4	0.76 − 0.90			
		5	1.55 − 1.75			
		6	10.4 − 12.5			
		7	2.08 − 2.35			
SPOT-1	HRV (×2)	1	0.50 − 0.59	20×	60	256
1986	(as mss)	2	0.61 − 0.68	20	(in tandem	
832 km		3	0.79 − 0.89		117)	
26D	HRV panchromatic	—	0.51 − 0.73	10× 10		

Note

Landsat 1+2—RBV virtually did not operate hence MSS was the main data source.
Landsat 3—the RBV operated in one broad band 0.51 − 0.75 μm and MSS had an extra channel, Band 8 (10.4 − 12.6).
Landsat 6—1988. Enhanced TM.
Landsat 7—more upgrades.
SPOT-2—1989.
Distances under satellite name=flight height; 16D, etc.=orbital cycle.

Table 5.3 Thematic mapper—waveband applications

Band number	Band name	Band width (μm)	Band application
1	Blue/green	0.45 − 0.52	Designed for water body penetration — coastal water mapping. Strong vegetation absorbance differentiates soil and rock from vegetation
2	Green	0.52 − 0.60	Designed to measure green reflectance of vegetation. Also turbidity of water
3	Red	0.63 − 0.69	Absorbs reflections from vegetation and detects ferric ions, useful in lateritic terrains in determining the underlying geology
4	Near infrared	0.76 − 0.90	Gives high land/water contrasts and very strong vegetation reflectance
5	Near − middle infrared	1.55 − 1.75	Differentiates clouds from snow and is very moisture sensitive
6	Thermal infrared	10.4 − 12.5	Of use in stress analysis, soil moisture discrimination, and surface temperature studies
7	Middle infrared	2.08 − 2.35	Discriminates rock types and detects hydrothermally altered zones around some mineral deposits

unmanned satellites, and although multispectral scanners have been used in aircraft this has mainly been to simulate the potential products of the spacecraft in trial programmes. Multispectral scanners, which are radiometers, simultaneously sample small elements of the earth's surface for their radiance characteristics in several well-defined wavelength bands (channels) though some can be switched to single broad-band coverage. The new French system (SPOT) looks at each 20×20 m area on the ground (a pixel) and records the amounts of radiant energy reflected in three narrowly defined segments of the EMS, $0.50 - 0.59\,\mu$m (green), $0.61 - 0.68$ (red), and $0.79 - 0.89\,\mu$m (near infrared). The same sensors can also operate in a single broader band $(0.51 - 0.73\,\mu$m) which is effectively a panchromatic mode similar to the sensitivity of black and white film. Acting in this mode, a 10 m resolution is achieved. When SPOT is used as a multispectral scanner, for each wavelength band the amount (intensity) of radiation from each pixel is recorded as a value in the range $0 - 255$ (low to high radiance). The more levels that can be handled the greater the radiometric resolution. This information may be displayed as a series of grey scales on a black-and-white print or as colour hues on a colour print. Also display on a computer visual display unit (VDU) is a common means of examination.

There are several styles of multispectral scanner and it is their common dependence on optical – electronic processes that sets them apart from photographic systems. The first routinely operated multispectral scanner, the Landsat series, scans across its flight path by means of an oscillating mirror (Figure 5.13). The acronym MSS is reserved for the initial type of multispectral scanners on the Landsat programme and new-style multispectral scanners have been given different names (e.g. Thematic Mapper, TM and Haute Resolution Visible, HRV – SPOT).

With the MSS sensor, six lines are scanned at once to cope with the rapid movement of the satellite, each line being 79 m wide and 185 km long. Four bands are sampled, the optics noted on Figure 5.13 splitting the incoming radiation into the wavelength bands in a similar fashion to a prism creating a spectrum of colours from white light. Some confusion is possible over the numbering of the MSS bands. The first three Landsat satellites each had three television-like cameras (return beam vidicon, RBV) on board to separately record the green, red and near-infrared bands. These were labelled Bands 1, 2 and 3 so numbering of the MSS bands started at 4 through to 7 (Table 5.2). However, on the Series D satellites, Landsats 4 and 5, the RBV cameras had been abandoned so the MSS band numbers were adjusted accordingly. This new Landsat series, in addition to the MSS sensor, carried a considerably improved multispectral sensor, the TM with seven bands of its own (Tables 5.2 and 5.3). Note that MSS Band 1 is not the same as TM Band 1. The TM, in relation to the MSS sensor, improved resolution, has narrower wavelength bands more specific to particular tasks, and added extra bands for new applications. Bands $1 - 6$ were chosen primarily for vegetation monitoring but Band 7 is specially for the geologist. Sampling $2.08 - 2.35\,\mu$m in the medium infrared range it gives the best means of discriminating rock types and detection of hydrothermally altered zones around ore deposits.

For each pixel the intensity of radiation in every waveband is measured separately and the analogue output of each detector is converted to digital format for either storage on magnetic tape on the satellite or for direct telemetering to ground receiving stations. Variations in the amount of reflected radiation are recorded on scales that usually range from 0 to 63 or 0 to 255 depending upon the sophistication of the sensors. Each pixel for each wavelength has a digital number (DN) or brightness value (BV) assigned to it within these scales. The MSS instrument registers information from 3240 pixels with every sweep of the mirror. With 2340 lines

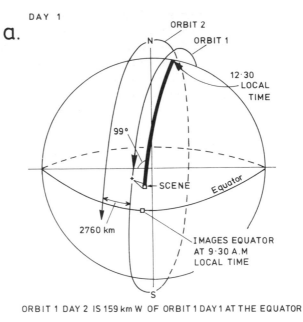

Landsat orbit characteristics.

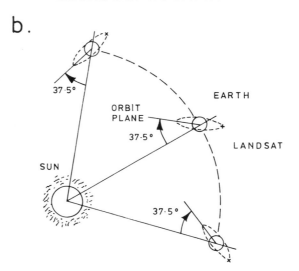

Landsat sun synchronous orbit.

Fig. 5.15 (a) The polar orbit system is shown for the early Landsat satellites. Orbit cycles, flight heights and other parameters have varied, but the basic orbit style is similar. (b) The orientation of the orbit plane is at a constant angle to the sun — earth join to give a sun synchronous orbit. The early Landsats always crossed the equator at 09.30 local time. Other satellites have different times but all maintain a sun-synchronous pattern

scanned for every 185×185 km scene, about 30 million observations are made in the 25 seconds devoted to each scene. The rate of data generation is quite daunting especially when it is considered that a TM scene of the same size requires nine times as many digital numbers. (*Note*: This is the world of acronyms, TM, MSS, DN, plus lots more!)

Data from each waveband sensor can be displayed individually in black and white represented as a scale of grey tones either on a computer terminal or as a photographic print (Figure 5.14). In colour the standard approach is to create a false-colour composite where the colours have the same meaning as those in studies of near-infrared photography. The actual procedure in making a false-colour composite is somewhat intricate. Once MSS data have been converted to black-and-white images for each wavelength band, the separate images are exposed through different colour filters on to film. For Landsats $1-3$, Band 7 (infrared) is printed using a red filter, Band 5 (red) through a green filter and Band 4 (green) is passed through a blue filter.

Because multispectral scanner data are held in digital form it can be manipulated by computer techniques. This feature is useful in both of the two modes of interpretation used to study the data. Firstly, we could interpret the information visually using the same approach as in air photography. Here computer techniques help to produce geometrically correct images, remove detector and transmission errors and expand contrast of originals with limited tonal response. The second method, digital interpretation, fully utilizes the capacity of the digital data to be modified and juggled numerically leading to the creation of entirely artificial images. An example of this style is the generation of geological maps for particular rock types or types of alteration. In an area of diamond-bearing pipes with a distinctive host rock it is possible to use a training area to determine the spectral characteristics of these hosts to economic deposits. The data banks for the rest of the scene are then searched for all similar responses and a map can be produced to show their distribution. These areas become the prime, but not exclusive, targets for the next phase of exploration. The DN or BV in one wavelength band may be used as the best discriminant for a search, but it may turn out that the ratio of Bands $6-7$ is better; such ratioing is a good illustration of the suitability of computer manipulation. Digital interpretation is now a major field of study and, with powerful minicomputers, handling single Landsat or SPOT scenes can be done interactively at a terminal at reasonable cost. Soon personal computers will be doing the job and even small exploration companies will be able to afford their own system.

Another example involves the automatic production of maps showing the traces of linear features on images (lineaments). Most of these are traces of steeply dipping fractures and many geologists believe these are significant in the origin of some types of ore deposits. Machine handling of the data can go further to produce a contour map of the density of lineament intersections per unit area. Some geologists would say that the higher the density, the greater the chance of finding a structurally localized ore deposit. Such approaches are just a hint of the massive range of applications of satellite-based remote sensing.

The advantages of space-borne multispectral remote sensing are: (1) ready availability world-wide; (2) lack of political and security restrictions (and originally free of copyright); (3) low cost; (4) repetitive coverage; (5) multispectral capability; (6) only small amounts of image distortion. Advantage (2) will be appreciated by all those who have struggled through the red tape of overly sensitive governments in search of a vertical air photograph of interest. The fourth advantage was not fully appreciated in the early days of geological analysis because it was felt that, even in tectonically very active areas, the geology was fairly static. However, other variables have a dramatic effect on the appearance of the geology in a single area, particularly seasonal variations in sun angle and vegetation growth.

The early Landsats were in near-polar orbit, making fourteen orbits per day (Figure 5.15). They were also sun-synchronous and each time they crossed the equator it was always 9.30 a.m. local time. In keeping pace with the sun's westward progress as the earth rotates, this configuration means that Landsats 1, 2 and 3 were overhead at the same point every eighteen days. Different repeat cycles have been used on Landsats 4 and 5, and for SPOT. It is thus possible to obtain images of scenes at different times of the year which will change the lengths and orientations of shadows. Linear features (fractures, strike ridges, fold traces) will be emphasized very differently by such changes. Seasonal variations can be pronounced in regions with distinct dry and wet seasons. In some areas dry-season images may be best for highlighting cryptic faults, but generally rock- and soil-type discrimination is enhanced just at the end of the wet period. Experimentation is always recommended as there are many variables to contend with.

A relative disadvantage to some observers has been the fairly coarse resolution of satellite images. Some would argue that this was useful to give the overview and to avoid getting lost in the detail. However, the early products from TM and SPOT are demonstrating that the much improved resolution of these sensors is very welcome and will allow geological interpretation to expand even further. Structural interpretation was also limited on the Landsat series because overlap between scenes was not a priority of the system. Sidelap of adjacent paths gives some stereoscopic coverage but this is a function of latitude, and at the equator only 15 per cent of one scene appears on its neighbour, increasing to 80 per cent near the poles. Stereo viewing has had only limited applicability. One of the significant advances of the SPOT system is that the optics can be pointed $\pm 27°$ from the vertical so that one scene at the equator may be viewed seven times during the 26 days orbital cycle from different positions. At 45° N and S the revisit can be completed eleven times in the cycle. In addition to the benefit of easily available stereoscopic coverage, this revisit frequency will allow natural disasters such as major floods or bushfires to be closely monitored. At this point it is worth mentioning that several of SPOT's enhancements (improved radiometric and geometric accuracy and higher spatial resolution) are in large part due to the push-broom arrangement of the sensors. Long linear arrays of cells record the reflected radiation from each pixel on the ground directly without the use of an oscillating mirror. The optics in the SPOT system are to focus the radiation on the receptors and to control the off-vertical viewing hence the process does not involve line-scanning. For the moment push-broom systems do not have the radiometric range of the latest oscillating mirror scanners but improvements are being made.

5.5   ACTIVE REMOTE SENSING SYSTEMS

All the systems we have examined to date are passive because the original source of the monitored radiation (energy) is external to the system. Mostly we have been concerned with solar EM radiation reflected by materials at the earth's surface. Radar remote sensing is different in that the source of EM radiation is generated on board the aircraft or satellite—an active system. Reflected microwaves are collected by sensors and, as with other systems, analysed in terms of the variations in energy of the reflections. The data may be both recorded electronically on magnetic tape for later manipulation and/or photographic film. Interactions between microwaves and materials is largely dependent upon the geometry of surfaces (variability of relief, angularity or curvilinearity of form, particle size of sediments, etc.). These controls are fundamentally different to short-wavelength interactions with objects and hence radar images are very different to photographic or MSS images. In particular shadows are often very dramatic and no information is received from these areas.

Radar images are very good for structural analysis because the shape of the landforms is emphasized. Fracture patterns may be starkly displayed given appropriate viewing directions though one region may look very different when recorded on flight paths with variable orientation. A major advantage of radar is cloud penetration and, despite its cost when flown in an aircraft, may be the only system that can operate in persistently cloudy tropical areas.

Seasat was an unmanned satellite that ran for three months in 1978 obtaining radar images of the ocean surfaces. A by-product of this programme was the study of departures in the satellite orbit to create regional gravity maps which have proved very useful in tectonic scale interpretation, of oceans, large seas and adjacent land masses. A similar style radar system for land observation (Shuttle Imaging Radar—SIR-A, SIR-B) has been carried on two Shuttle missions on an experimental basis. To date only a very limited number of runs have been completed but the images give a very different view from other systems and appear very useful for structural analysis (see Figure 11.11).

5.6 THE FUTURE

Each year *Geotimes* devotes an article to advances in remote sensing and there are always a great many activities to report. For me to write down the products of my crystal-ball gazing is a very dangerous occupation. It could be depressing to look back at this hard copy in years to come when all will be different. New systems are being planned all the time, push-broom sensors are getting bigger, geometric fidelity together with radiometric resolution and range are slowly improving, etc. Perhaps the most significant proposal in the offing is the International Geosphere—Biosphere Program to be largely funded by the USA. The aim is an integrated study of the earth with a permanent network of satellites in orbit and instruments on the ground. The whole earth approach will involve simultaneous study of climate, oceans, biosphere, dynamics of the solid earth and biogeochemical nutrient cycles; particular emphasis is placed on interconnections. Earth sciences figure prominently in NASA's plans for the mid-1990s to build an advanced instrument platform. This will be a major observatory in polar orbit and regularly serviced and upgraded by space Shuttle flights. Known as the Earth Observing System (EOS) it will contain three groupings of instruments. One multispectral scanner group, a radar grouping and a third group to study the composition and dynamics of the atmosphere.

6 THE FOURTH DIMENSION— CHRONOLOGY

6.1 STRATIGRAPHY/HISTORICAL GEOLOGY

This chapter should be regarded as being no more than a preface to further stratigraphic study. Here we deal with the generalities of classifying rock strata and have no time to go in detail into the myriads of properties (lithology, fossil content, time of formation, geophysical properties, etc.) used in stratigraphic studies. In dealing with strata the distintive aspect of classification involves the vertical and lateral arrangement of the layers. Besides the geometrical consideration, time relations—our fourth dimension—are crucial in our attempt to understand geological histories from maps. However, not all university courses follow a logical progression (for staffing, timetable, historical, etc., reasons) and some readers may well be in a system that completes the map analysis section before any stratigraphy is given. This chapter is mainly aimed at such unfortunates, though the more stratigraphically enlightened may benefit from an alternative presentation. The latter perhaps might like to test their grasp of stratigraphic principles by following the arguments (scientific discussion) about Cambrian stratigraphy in part of the USA (*Geology*, **13** (9), 663 – 8, 1985). First-timers in stratigraphy might like to try the same test having read this chapter.

Even a cursory survey of the present-day earth reveals a vast array of tectonic settings: ongoing continent/continent collision along the Alpine – Himalayan chain, active subduction of oceanic lithosphere at continental margins (Central and South America), island arcs marking the zone of convergence between oceanic plates, conservative plate boundaries where the elements of the plate mosaic slide past one another (San Andreas Fault system), rifting at various stages from incipient (East African Rift Valleys) to mature (Atlantic Ocean) and many more. At our present level of understanding of earth dynamics, we are at the same time starting to see some order in the pattern of geological processes within a plate tectonic framework and beginning to appreciate the complexities of the system we are investigating. Each tectonic setting is characterized by different processes and products. Particular sedimentary, structural, metamorphic, and igneous styles typify the different regimes, but significantly it is the contemporaneous variation within each of these styles that is most diagnostic. In many island arcs, the lavas and intrusives vary in composition according to distance from the trench. Also in the same subduction zones, heat flow and hence metamorphic effects are very different in the trenches and in the volcanic arcs. Sedimentary environments within single tectonic settings show the greatest variety, but their lateral and vertical arrangements are diagnostic for each tectonic process. A rifted continental margin at any one time displays a regular progression from coastal plain through littoral environments across the shelf to slope and rise conditions.

Most tectonic settings also show characteristic evolutionary sequences. Again using continental margins as the example, the first signs of break-up are rift valleys that contain fluvial and lacustrine sediments controlled by active faulting. In addition most rift valley systems have a chemically distinctive (peralkaline) volcanic contribution to the rift infill. Minor marine incursions may flood the rift zone as pull-apart proceeds, but the major change takes place at eventual separation and generation of new oceanic crust. By this time the edge of the continent is considerably thinned and, as it moves away from the zone of sea-floor spreading, it cools and subsides. The result is a marine transgression along the entire edge of the continent cutting across the earlier rift structures and sediments which are superimposed by a kilometres-thick succession of shelf, slope and rise sediments. For one region, on a continental margin several thousands of kilometres long, the local sequence of events — a relative chronology — is established using some very basic principles that can be applied in all situations to delineate a geological history. Along the whole margin the major events in the evolutionary process occur in the same order but from the relative chronology approach we cannot say if the stages took place at the same time. In fact the generation of one continental margin involves very different timings from place to place for rift initiation, first formation of oceanic crust and marine transgresstion, thus providing vital information on the nature of the tectonic machinery.

To understand and document this tremendous tectonic diversity throughout earth history requires an ability to make time correlations. An ideal but unobtainable goal of **historical geology** would be to chart continuously through time the distribution of continents and oceans, and all their varied tectonic settings. There are sizeable problems in achieving this aim for even the last 100 Ma, and further back in time we rapidly have to learn to live with limited success. Only tiny fragments of ancient oceans are preserved and inferred plate distributions of around 250 million years (250 Ma) ago are speculative, degenerating to wild guesses for around 600 Ma ago, and in the Archaean (>2500 Ma) we are not even sure what tectonic processes were operating. In dealing with the material that survives tectonic recycling, sedimentary successions provide the most complete record of the earth's history. Other results of tectonic processes are generally more discrete in their time-span; in the birth, growth and death of an ocean there are long periods, in most parts of the system, without deformation, metamorphism or igneous activity, whereas sedimentation is much more continuous. It is, therefore, not surprising that studies of strata — **stratigraphy** — have dominated historical geology. The purists would say that stratigraphy is solely the study of the geometry and time relations of bedded sedimentary rocks and even exclude consideration of sequences of lava flows. The literal translation of stratigraphy, writing about strata, would include sedimentary petrography and sedimentation but these now have a separate identify as the discipline of sedimentology. From an outsiders' point of view it appears that common usage has just about closed the gap between stratigraphy and historical geology acknowledging the very strong bonds created by much common ground. Codes of stratigraphic nomenclature issued by many countries always include sections on non-stratified plutonic and metamorphic terrains showing a willingness to embrace the wider history-of-the-rock-record approach. The basic principles of stratigraphy expounded in **every** text include some methods that can only be applied in non-stratified rock, further illustrating the close interrelation between stratigraphy and historical geology. Most undergraduate courses labelled stratigraphy deal with earth history. Whilst accepting that historical geology and stratigraphy are not strictly synonymous to some authors, they will be used interchangeably in the following discussion. At first glance the definition of stratigraphy quoted below is restricted to layered rocks but modern interpretations make sweeping reference to grand scale layering of the crust and thus include virtually all comers in their brief.

> **Stratigraphy** — Definition. The science of rock strata. It is concerned not only with the original succession and age relations of rock strata but also with their form, distribution, lithologic composition, fossil content, geophysical and geochemical properties — indeed with all characters and attributes of rocks as strata. . . . *Glossary of Geology* . (Bates, R. L. and Jackson, J. A. 1980. Second edition. American Geological Institute.)

6.2 GEOCHRONOLOGY

The time factor is the most distinctive feature of geology and it is at the heart of our 4-D understanding of geological maps. Our analysis of geological time will follow similar lines to the historical development of the subject. At first we shall deal with methods that lead to the construction of local sequences of events and then move on to the thorny problem of establishing time equivalence. Both topics come under the umbrella of **geochronology**, the science of dating and determining the time sequence of events in earth history. There are many methods that can be used to establish the section or slice of geological time to which a particular rock or event belongs. Subdivisions of geological time were identified and labelled long before the duration of the units in years could be measured but this did not invalidate the recognition of time equivalence or sequence. The quantitative measurement of geological time, expressed in years, is **geochronometry** and unfortunately careless usage in geological circles has led to only this specific activity being referred to as geochronology. Other facets of geochronology seemingly have been swamped by a massive explosion of mass spectrometry work on isotope systems to give dates for rocks and geological events.

The basic rules of relative chronology were developed very early. The most basic is the **principle of superposition**, first published in 1669 by Steno, which states that older rocks are found beneath younger rocks (Figure 6.1a). For a tectonically undisturbed region, this may seem self-evident to us, but in an age dominated by non-scientific ideas about earth processes it was a major advance. Because of the cryptic nature of some 'disturbed' regions, we now very rarely rely only on the basic principle to determine which is the oldest and youngest part of a sequence of layers. Many small-scale sedimentary structures tell us which is geological top and bottom and give way-up or younging (Figure 6.1a). Collectively such structures are referred to as geopetal. Younging is still extremely important data in analysing the sequence of units and internal geometry, and can be key information in determining tectonic displacement directions. Presentation of younging data on maps is very varied (see Appendix 2, Symbology) but is usually related to the dip and strike symbol of layering. I suspect that there are still many areas in the external parts of orogenic belts where the layers are assumed to be right-way-up which are in fact upside-down, suggesting that insufficient attention has been paid to recording younging in the past. Palaeontological data are also used to determine the younging direction.

The **principle of inclusion** is sometimes useful in confirming the younging of a layered sequence (Figure 6.1b). If a bed includes fragments of another bed, then the

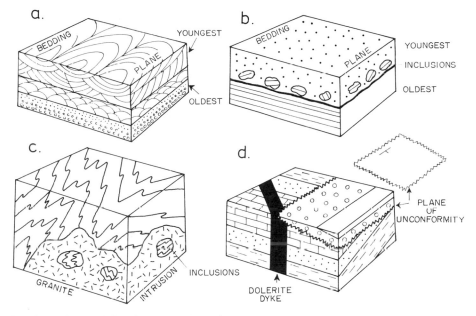

Fig. 6.1 (a) Principle of superposition for an untectonized region where older rocks lie beneath younger rocks. Because some deformation events are cryptic it is wise to check the younging direction indicated by sedimentary structures (e.g. graded bedding, truncated cross-lamination) or other features (pillow lavas). Many Geological Survey maps and besides giving dip and strike of layering give younging data (see Appendix 2). (b) Principle of inclusion — the unit containing the inclusions is the younger. (c) The principle of inclusion may also be applied to intrusive igneous bodies. (d) Principle of cross-cutting relations. The dolerite dyke is younger than the lower sedimentary succession which it truncates — both are truncated by the unconformity which is an erosional surface

unit containing the inclusions is the younger. By implication an erosive contact must exist between the two beds which may or may not represent a significant time-gap. The same principle has more general applications in establishing sequences of events. If an igneous (or sedimentary) intrusion breaks off fragments of country rock (Figure 6.1c), then again the rock with the inclusions is the younger. A third principle, that of **cross-cutting relationships**, is commonly applied to events other than sediment deposition. Its statement is that a rock body (or structure) that cuts across another rock body (or structure) is the younger of the two (Figures 6.1c, d). This approach is classically applied to igneous intrusions. Irregular masses of granite or tabular sheets (dykes) cutting layering in the country rock sediments are both younger than the host. Two dykes generate cross-cutting relationships, the second cutting the first, as do fault structures (fractures). These are very simple notions but very effective in building up the geological history of an area from a published map. Applying the above principles in the field may be another matter where you have to contend with patchy exposure, the vagaries of the weathering process, metamorphic and deformational overprints, hydrothermal alteration and lots more.

Until now the discussion has centred on successions of parallel sedimentary layers deposited without major interruptions. A noticeable gap in the geological record is referred to as an **unconformity** and the clearest example is where there is a difference in attitude between the older and younger layers (Figure 6.1d). The surface of unconformity is then a cross-cutting feature establishing the order of evens. Some unconformities divide younger and older beds with parallel layering, either being marked by a zone of minor erosive activity or by a perfectly planar contact. Unconformities play such an important role in historical geology that they are the subject of a whole chapter once folding and faulting have been discussed.

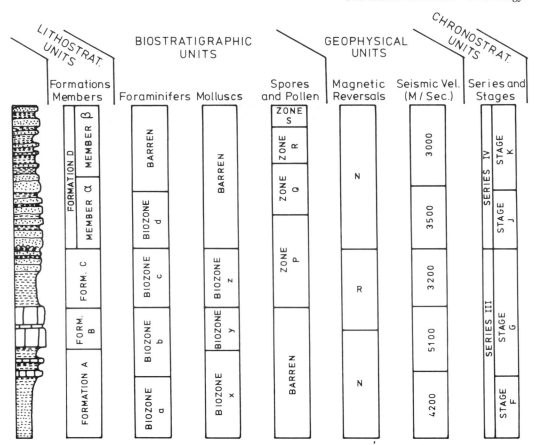

Fig. 6.2 A vertical stack of sedimentary layers may be subdivided according to a variety of characteristics. Lithostratigraphy classifies on lithology. Biostratigraphy classifies on fossil content and schemes will be different depending on the group of organism used. Geophysical units can be based on properties like seismic velocity or magnetic polarity (normal or reversed). Chronostratigraphy is a time-rock classifiation

6.3 STRATIGRAPHIC PROCEDURES

Stratigraphic procedures can readily be considered under a number of distinct headings which are dictated by the various properties of strata that are useful for classification (Figure 6.2). The major subdisciplines are based on lithology (**lithostratigraphy**), fossil content (**biostratigraphy**) and time of origin of an interval of strata (**chronostratigraphy**). Magnetic polarity, geophysical characteristics, chemical composition and mineralogy, have also been used for classification in a stratigraphic context. Of all the bases for classification, lithology is probably the most objective but not completely so as the following discussion explains.

The primary aim in geological mapping is to document the distribution of different rock types. Clearly individual beds cannot be represented on maps with scales smaller than about one to one hundred, and, even at this detailed scale, laminations in a siltstone or shale could only be schematically portrayed. It follows that virtually all maps involve a compromise in portraying the distribution of lithologies. Scale dictates the amount of compromise, and the first stage in a mapping campaign is to decide what groupings of layers are (1) distinctive enough to be recognized from place to place, and (2) can be represented at the scale of eventual publication. The results of these assessments define **formations**, the basic lithostratigraphic unit. That formations are so dependent upon scale has led to criticisms of their worth, but a reasonably objective description of the lithologies in the geological column is a fundamental first step in any stratigraphic study. Figure 6.3 is an example of the sort of choices that have to be made when first studying a region at the start of a mapping campaign. Air photographs are the medium of illustration in this case, but the same process is employed where vertical photography is unavailable or of limited use because of extensive soil and/or vegetation cover. Ten formations have been delineated (S_1 to S_{10}) as natural groupings of the strata. For example, S_9 consists of nearly 20 layers that can be identified at the scale of photography yet the overall coherence of the package of layers readily defines a formation. Formation S_7 is a thin, very well laminated unit but again is a good candidate for an easily recognizable, and thus mappable, unit. It is worth pointing out that some formations show significant variations in their expression, particularly S_3 from the north side of the fold to the south. In part, such variable units are recognized by their stratigraphic position relative to other more constant units.

Stratigraphic procedure requires that each formation be defined at a type locality as a reference point where its boundaries (top and bottom), and characteristics, are well displayed. Formation names and details are usually approved and recorded by a central committee in each country typically run by a national geological society or geological survey; the names are then regarded as being formal. Without such coordination the same name might be given to different lithologies from different places and the confusion would be unbearable. The smallest lithostratigraphic unit is a **bed** (e.g. a thin but widespread air-fall tuff layer) and, between this and a formation, a **member** may usefully be recognized under some circumstances (e.g. lenses of one lithology surrounded by another). Several formations with some connection are linked in a **group** and at the top of the hierarchy allied groups form a **supergroup** (Table 6.1). As an example of this process in operation, a tectonic element in Western Australia called the Pilbara Craton contains the Pilbara Supergroup which comprises the Warrawoona Group, the Gorge Creek Group, and the Whim Creek Group. Eighteen formations have been recognized within the Supergroup including the Duffer Formation and the Honeyeater Basalt. The example is mainly presented to show how the nomenclature is used and to indicate the style of capitalization. Lithostratigraphic unit names must contain a part derived from a geographic feature local to the type area together with a term appropriate to its rank (group, formation member, bed) (e.g. Newark Group) or a term describing the dominant rock type (e.g. Compton Schist) or both (e.g. Burlington Limestone Formation). The unit term (formation, group, etc.) of a formally defined lithostratigraphic unit should be capitalized, a practice which is extended to all parts of formal names (e.g. Spiti Shale). The above conventions are sufficient for introductory work but I would recommend readers, as soon as they can, to study, Owen, D. E. 1987; Commentary: Usage of stratigraphic terminology in papers, illustrations and talks. *Journal of Sedimentary Petrology*, **57**, 363 – 72.

6.4 STRATIGRAPHIC CORRELATION

We now move on to the more intricate conceptual aspects of traditional stratigraphy mainly centred on time-correlation. Disputation is so rife in this domain that the

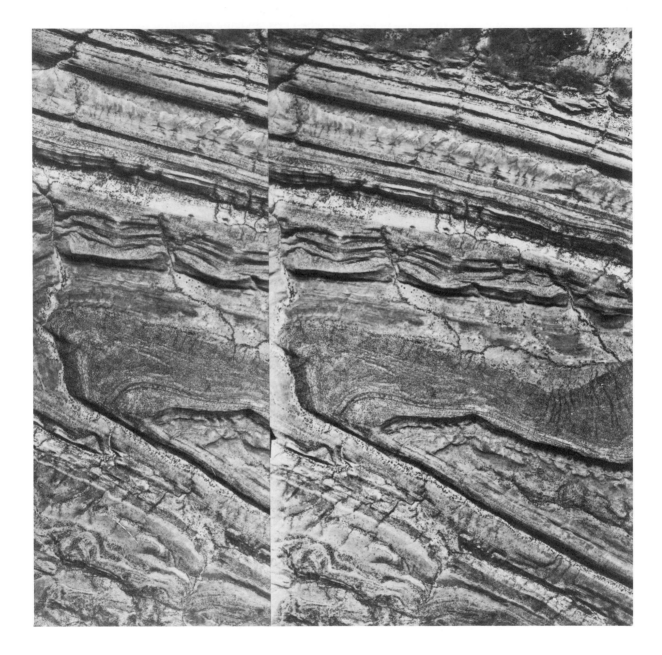

Fig. 6.3 A stereopair of vertical air photographs of a folded sedimentary sucession. The first question in mapping is 'what are the mappable units?' — the formations — the basic units of lithostratigraphy. Formations have some unifying aspect that outlines a distinct package of layers. The interpretation shows the formations that most practising geologists would adopt though, in a detailed study, smaller units might be employed. A large fold structure has caused a repetition of several of the formations and different dips on either side of the fold largely account for the variation in outcrop width (in map view)

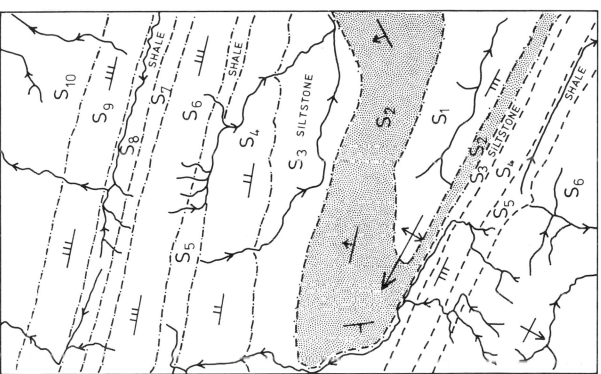

b.

FORMATION BOUNDARIES

VERY GENTLE (1-3°)

GENTLE (3-10°)

MODERATE (10-30°)

STEEP (30-60°)

VERY STEEP (60-90°)

S₁ TO S₁₀ LITHOSTRATIGRAPHIC UNITS

WATER COURSES

PLUNGING ANTIFORM

N

1 Km

Table 6.1 Conventional hierarchy of formal lithostratigraphic terms

LITHOSTRATIGRAPHY
Supergroup — two or more groups
Group — two or more formations
Formation — primary unit of lithostratigraphy
Member — named lithological entity within a formation
Bed — named distinctive layer in a member or formation

fundamentally different concepts of biostratigraphy and chronostratigraphy are to some authors so similar they should not be distinguished separately. Much stratigraphic practice has evolved from studies of rocks formed in the last 600 Ma where the attempts to establish time equivalence have heavily relied on palaeobiological criteria (biostratigraphy). The method is totally dependent on the fact that species, genera, and groups, evolve, reach peaks, and become extinct. The trend of these patterns may be followed in single sections to determine younging and comparison of the patterns from place to place leads to palaeontological-correlation and indirectly to time-correlation. By the time reliable radiometric ages (geochronometry) became available, the latest 600 Ma of earth history had been elaborately subdivided (Figure 6.4), on the basis of the fossil content of the rocks, into **geochronologic units** (Table 6.2) of varying lengths of time (eras >periods >epochs > ages). Without radiometric techniques the actual length in years of these subdivisions could only be estimated by fairly crude methods (e.g. sedimentation rates). However, the advent of geochronometry had no influence on the palaeontologically established subdivisions of time, and even today there is much debate about the absolute age of many of the biostratigraphically defined units though the range of estimates is generally narrowing. The greatest uncertainty relates to the base of this part of the column, that is, the time of explosive evolution that marks the start of the Cambrian Period placed variously between 530 and 580 Ma ago.

The biostratigraphic method of correlation says that two sections of sedimentary rock contain equivalent fossils. To make the jump and claim that this means the same age is an assumption that can never be proved. Under favourable circumstances the assertion may be subject to little uncertainty, but the more steps and complications in the correlation process the more dubious the correlation becomes. Biostratigraphic correlation is usually based on the overlap of the ranges of several taxa (species, genera, families, etc.) forming a concurrent range-zone named after two or more diagnostic components. The pessimistic view of such a comparison highlights the problems. A new species evolving in one area may take considerable time to migrate to other regions. At the present there are many biogeographic provinces defined by latitudinal constraints and physical barriers. Taxa distribution is commonly strongly influenced by warm or cold ocean currents, rain belts or rain shadows, all of which lessen the latitudinal controls. Also within each province, many elements of the flora and fauna are restricted to particular environments. Intercontinental correlations in the ancient record normally require several steps each between somewhat isolated biotas such that at every alternate step there may be no species or genera in common across the intervening province. The uncertainties of biostratigraphic correlation, which extremists claim to be as much as a quarter of a period, validate the current chronostratigraphic approach (Table 6.2). The base of each major chronostratigraphic unit (systems) will eventually be defined by international agreement at a particular place in a selected sequence of well-exposed sedimentary rock with as continuous as possible a record of sedimentation. The definition of

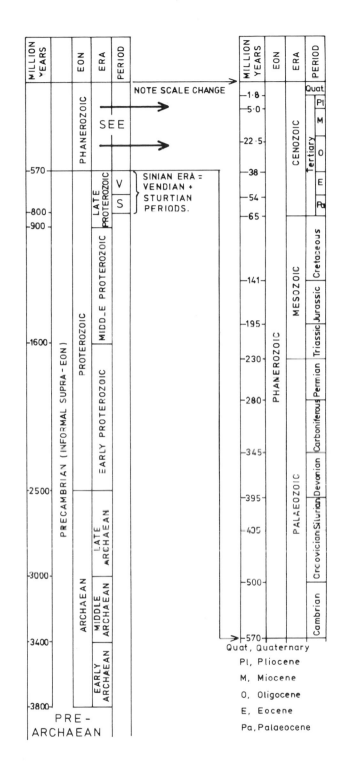

Fig. 6.4 The geological time-scale showing the major divisions. Not many geologists know all the smaller subdivisions into ages, but if you work for some time on rocks of a particular period the details should be known. The left-hand column shows earth history on a linear scale. The Phanerozoic is expanded on the right to represent all the major units in this more favoured part of the column

Table 6.2 Conventional hierarchy of chronostratigraphic and geochronological terms (Hedberg, 1976). Names in brackets are examples of each category

Chronostratigraphic (time-rock units)	Geochronologic (geological time units)
Eonothem (not often used)	Eon (Phanerozoic)
Erathem (not often used)	Era (Mesozoic)
System (Cretaceous)	Period (Cretaceous)
Series (Upper Cretaceous)	Epoch (Late Cretaceous)
Stage (Campanian)	Age (Campanian)

Remember — **Geochronometry** gives ages in years.
Note: Upper + Lower relate to rock units; Late + Early relate to time units.

Cambrian and younger systems will be on biostratigraphic grounds at specified biozones but recorded in terms of an actual sedimentary layer. The task then is, on a world-wide basis, to identify this same time horizon as closely as possible. We will never know, for example, how closely the chosen base for the Silurian in China corresponds to that in Wales. They may well be somewhat diachronous because of errors in the biostratigraphic correlation. The only known point is the defined base in the rock at Moffat in the Southern Uplands of Scotland. This then is the **time-rock** link that is crucial to chronostratigraphy.

The process has been criticized by some authors who state that chronostratigraphy is only as good as the biostratigraphy employed in its definition and they question the necessity of introducing another category of stratigraphic divisions. Chronostratigraphy is necessary because of the uncertainty in all of the methods for establishing time equivalence which generates the need for reference/anchor points around which arguments involving assertions and assumptions can rage.

Care is needed in applying stratigraphic terms and many practising stratigraphers are horrified by the lax use amongst the general geological population. The following passage demonstrates the nuances of nomenclature that have to be observed. The Cretaceous Period is a well-defined span of time in the relative time-scale (geochronologic unit). During this time rocks of the Cretaceous System were formed. Thus it would be correct to state that the Gingin Chalk (a lithostratigraphic term) belongs to the Cretaceous System but that pterodactyls flew (or glided) during the Cretaceous Period. The primary aim of the nomenclature is communication. Because there are myriads of local lithostratigraphic names each geologist will only retain a tiny subset in memory; however, any geologist will understand a colleague who says he is working on the Cambrian of an area.

In the above discussion on time-correlation very little was said about the detailed methodology of biostratigraphy which organizes strata into units based on their fossil content. A biostratigraphic unit (biozone) is a thickness of sedimentary rock characterized by one or more diagnostic fossils. Many types of biozones may be defined and there has been some discussion as to which is best for correlation. The simplest type of zone is based on the **range** of a single taxon (Figure 6.5a) both vertically in a stratigraphic sequence and horizontally (lateral extent). A concurrent-range-zone (Figures 6.5b, c) is formed by the overlap of the range-zones of two or more selected taxons. Oppel- and lineage-zones are more subjective styles of range-zone and little used in correlation. An acme-zone is a body of strata representing the maximum development of a taxon, but as this may be environmentally controlled its time significance is questionable. Of the many remaining biozones the assemblage-zone defined by a natural association of taxa is frequently discussed in

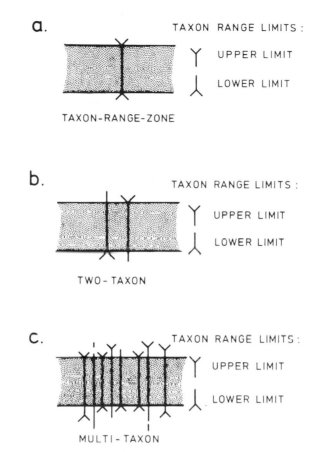

Fig. 6.5 There is a great variety in the type of biozones recognized. The simplest is based on the incoming and disappearance of one taxon. Greater definition is achieved by including more than one taxon

biostratigraphic theory, but is again more likely to reflect environment than time and is not used for correlation. Whichever zonal scheme is chosen the selected fossils need to have most of the following characteristics to be useful: (1) existence as a clearly identifiable taxon for a short time-span; (2) occurrence in several lithologies; (3) widespread distribution; (4) abundance and easy preservation.

　　Biostratigraphy is now being applied with some success in the latest Precambrian (approximately the last 200 Ma) in which chronostratigraphic units may be defined in the near future (Figure 6.4). Apart from this, the Russians have claimed that stromatolites are useful for much of the Proterozoic in defining fairly large biozones but this has been hotly disputed, mainly by the Canadians. The negative case is largely based on evidence for a strong environmental influence on the morphology of stromatolites such that particular forms more reflect factors like tidal range, frequency of storms, degree of protection from open sea, etc., rather than an evolutionary period. Also stromatolites are restricted to intertidal and shallow subtidal carbonates making them very limited in occurrence.

　　The question remains of what to do with the bulk of geological time, the 80 per cent of earth history from 800 Ma age to the creation of the earth at about 4600 Ma (Figure 6.4). Despite the overall international coordinating body recommending a chronostratigraphic solution for all of geological time, the Subcommission on Precambrian Stratigraphy has adopted subdivisions based on geochronometry and the time-rock concept does not figure in their deliberations (Plumb, K. A. and James,

H. L. 1986, Subdivision of Precambrian time: recommendations and suggestions by the Subcommission on Precambrian Stratigraphy. *Precambrian Research*, **32**, 65 – 92). This departure from the well-tried methods of the Phanerozoic has some bizarre consequences. When the Great Dyke of Zimbabwe was studied in 1977 its radiometric age of 2514 ± 16 Ma placed it in the Archaean as defined geochronometrically. However, a revision of the decay constants changed the date to 2461 ± 16 Ma and this major intrusion lurched into the Proterozoic. The oscillation of rock units from eon to eon could have been prevented if the time-rock or in-the-rock philosophy had been applied. By nominating the crystallization (cooling) age of the Great Dyke as being the Archaean/Proterozoic boundary would anchor this major divide-in-time in-the-rocks. Critics of chronometric subdivision claim that it is not based on stratigraphy but the subcommission in a spirited defence states that it is taking the broad view of stratigraphy by using the best attribute for classifying Precambrian strata and including all petrographic styles (sedimentary, igneous and metamorphic).

6.5   LITHOLOGIES THAT TRANSGRESS TIME (DIACHRONOUS)

A fundamental concept to be grasped is that lithological boundaries are rarely parallel to surfaces representing equal time horizons. Almost all formation boundaries are, therefore, diachronous. In conformable sequences where depositional slopes are low the angular differences between the two surfaces (time and lithology) are typically very small but when traced regionally they became significant. Continental shelves and abyssal plains, the most extensive depositional areas, have average slopes of 1 in 500 and 1 in 1000 respectively. The narrow continental slope has an appreciable gradient of 1 in 15 (4°) and the apron at the foot of the slope, the continental rise, has inclinations around 1 in 40. Locally more pronounced topography creates abrupt lateral changes in lithology particularly in submarine canyons and barrier reefs.

A brief consideration of almost any small region on the present-day earth will emphasize the difference between time surfaces and rock unit boundaries. Across many shorelines there is a wide range in the types of contemporaneous environments, perhaps from fluvial and aeolian through beach, lagoonal and barrier islands to offshore conditions in open seas (Figure 6.6a). The deposits formed in each of these environments are different and the term **facies** is used by many geologists to refer to the coexistence in time of many disparate lithologies. To such workers a facies boundary is a junction or transitional zone beween different rock types that were formed at the same time; to be specific this should be referred to as a **lateral facies change**. Recently the emphasis on the definition of facies has changed and qualifiers are now recommended to specify the exact meaning intended. Examples include lithofacies, biofacies, mineralogic facies, volcanic facies, marine facies, etc. Lithofacies in this usage is simply a particular lithology without the implication that it had to have formed at the same time as another unit.

To create a sheet of sand (the commonest geometry) from, for example, the barrier island depositional process (Figure 6.6a), we need a substantial period of either transgression or regression. Transgression is illustrated in Figure 6.6a where for some time either sea-level has risen (glacially or tectonically driven) or the land has fallen for tectonic reasons. Sediment supply may be an important control on which process happens as large amounts of input into a subsiding basin could bring about a

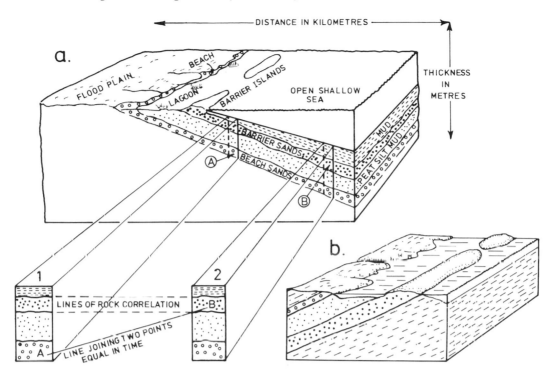

Fig. 6.6 (a) A marine transgression leaves a characteristic vertical stack of lithologies/lithofacies. At any one time the flood plain, beach, lagoon, barrier island and open-sea deposits coexist such that lithological boundaries cannot be equal time horizons. Vertical drill-holes at localities 1 and 2 will have similar sequences, but the beach rock at 1 formed at the same time as the barrier sands at 2. The vertical axis has been greatly exaggerated to make the point and angles between time planes and lithological boundaries are typically less than a degree or so. (b) Regression reverses the vertical stack of lithologies/lithofacies

regression. During a transgression the position of the beach moves towards points fixed within the hinterland and the belts of lagoons, barrier islands and offshore conditions follow suit. At any one time all of these environments coexist such that in vertical drilling at two places different distances out to sea (1 and 2, Figure 6.6a), the equivalent lithologies could not have formed at the same time. Time equivalence is oblique to the lithological contacts with the beach sands (A) at site 1 having been deposited at the same time as the barrier sands (B) at site 2.

Because of the normal low depositional slopes, the diagram (Figure 6.6a) has had its vertical dimension grossly exaggerated (several hundred to one) to clearly show the discordance between time surfaces and lithological/formation boundaries. From the time of deposition of the sediments, the angular discordance in fact is reduced by compaction which induces a 70 per cent vertical shortening in shales and around 25 per cent in sands. Compaction decreases the vertical thickness of the sediments without changing the horizontal dimensions of the layers and this deformation changes shapes and modifies angular relations (see Chapter 7). Any surfaces, except vertical ones, inclined to the horizontal, have their dip reduced during compaction, hence time surfaces and lithological contacts converge. Another example of a massive vertical stretch for cartographic clarity is shown in Figure 6.7. Such exaggeration is typical of lateral facies relationship diagrams yet it is invariably forgotten or not fully appreciated by observers (and ?authors). Even in regions with dramatic initial topographies such as carbonate reefs, an exaggeration of several hundred to one is needed to highlight the lateral facies variations (Figure

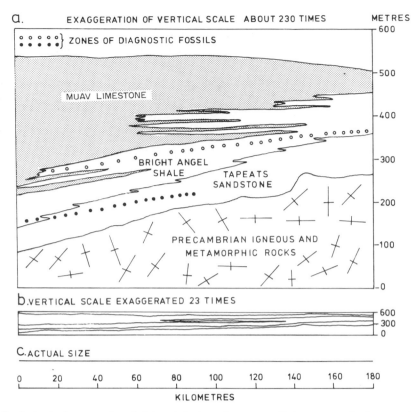

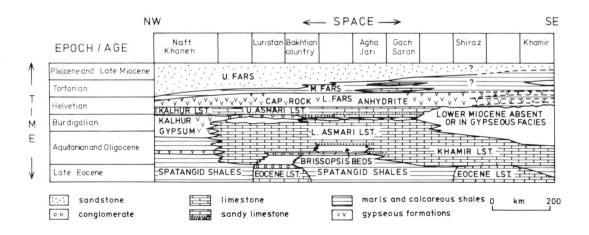

Fig. 6.7 A typical example of lateral facies changes which are gradual over tens of kilometres such that gross vertical exaggeration (a) is required for clear display. Even a large exaggeration (×23 in (b)) obscures the lateral relationships. When drawn to true scale (c) nothing can be resolved because the thickness (600 m) is so much less than the horizontal distance (180 km)

Fig. 6.8 A time-space plot draws attention to lateral facies changes, but remember the vertical axis has nothing to do with stratigraphic thickness. Gaps in the rock-record are also highlighted by these diagrams. This example shows relations between units that developed in a shelf setting in what is now the Zagros area of Iran

6.8). In this Tertiary succession palaeontological control is good and biostratigraphy provides well-defined time horizons through the different lithologies.

The lateral migration of different environments through time creates a stacking of different lithologies and vertical facies changes. Sedimentological analysis makes considerable use of these vertical varitions in interpreting palaeoenvironments. At its simplest this approach can distinguish between transgressions and regressions. Compare the vertical facies changes of Figures 6.6a and b. In a regression (Figure 6.6b) the finer offshore muds are overlain by the barrier islands sands, lagoon muds, beach sands and perhaps by continental deposits. The reverse vertical stack is generated by transgression (Figure 6.6a) with the fine offshore muds on top.

Having emphasized the generality of diachronous lithological contacts, it is now appropriate to mention the isochronous exceptions. Major catastrophic eruptions like that of Toba, 75 000 years BP, pour vast quantities of pyroclastic material (up to 2000 km^3) over large areas in a geological instant (days or weeks). The Toban event produced 400 times more ash than the better-known Krakatoan eruption with presumably devastation approximately in proportion. The Toba Tuff was deposited on land, and in shallow shelf waters over the whole range of environments in such regions. This tuff and the many equivalents in the ancient record provide geological time surfaces and are very useful in constraining biostratigraphic correlations in some areas. Instantaneous units of this nature are commonly referred to as **key beds**, though again the term has several meanings and to some workers key beds are distinctive layers not necessarily deposited everywhere at the same time.

6.6 STRATIGRAPHICAL INFORMATION ON MAPS

The standard geological survey product is a coloured map. The key to the colour scheme is on a lithostratigraphical basis (formations, groups, etc.) with geo-chronological (time) information (eons, periods, epochs, etc.). A wide range of conventions exists and even within one survey, through the years, the style of presentation will have evolved. The simplest key is a set of coloured boxes, labelled with formation names, arranged in order from oldest at the bottom to youngest at the top (Figure 6.9a). Sedimentary and volcanic rocks are always treated in this way but intrusives may appear (1) grouped together at the bottom of the key (2) in their correct stratigraphic position or (3) kept separate from the stratified rocks. If style (1) or (3) is adopted for intrusives, within this grouping, they are still arranged from oldest at the bottom to youngest at the top. In addition to colour coding, most keys provide a letter code which is helpful when there are so many units that colour differentiation may become difficult. In making a sketch-map of part of the published map to illustrate a report it is useful to follow the letter code.

The next most informative approach is a set of boxes with brief lithological descriptions for each formation (Figure 6.8b). A slight variation on the above two methods is to join up the boxes in a vertical stack which creates a schematic stratigraphic column — our example of this style (Figure 6.9d) was published in combination with a more detailed representation of the stratigraphy. The latter is achieved as a columnar portrayal of the formations showing thicknesses to scale with variable amounts of commentary on the lithology (Figure 6.9c). The thicknesses may be averages for the map area or those seen in one small part of

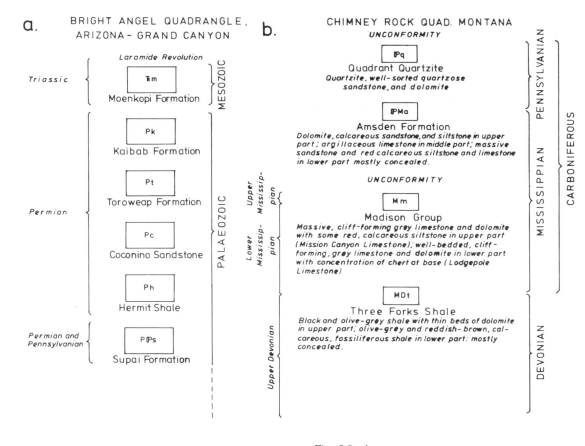

Fig. 6.9 a,b

the area. This style allows lenticular units to be identified and also for graphic presentation of major facies boundaries or angular unconformities. A thumb through one issue of a geological journal revealed a bewildering array of names given to the columnar style of stratigraphic representation, viz. schematic columnar sections, stratigraphic sections, columnar section and generalized stratigraphy and stratigraphic columns. The British Survey somewhat misleadingly calls them generalized vertical sections.

Detailed information on lithologies is seen on only a few maps. Beginning students are often unaware of the variability in lithology that is sometimes lumped together and called a mapping unit—a formation (see e.g. Figure 6.3). Bearing in mind the requirements of scale and mappability, some formations are very mixed lithologically. The style of Figure 6.9e allows us to see that the single colour on the map for the Mauch Chunk Formation from Pennsylvania is a mixture of shale, sandstone and limestone, with no one dominant lithology. Further details (Figure 6.10), in terms of a graphic, measured or sedimentary log, is rarely seen on published maps. Even Figure 6.9e does not show all beds, whereas graphic logs typically can represent beds down to a few tens of centimetres though scales again may vary to suit the aim of the project and publication constraints. Graphic logs usually give information on sedimentary structures and fauna in addition to lithology. The horizontal scale on modern examples relates to grain size, though older examples give a more fuzzy weathering profile with less information.

Several variations on the above themes are possible. Maps of a large region may cover very different domains that evolved together, for example, an island arc and

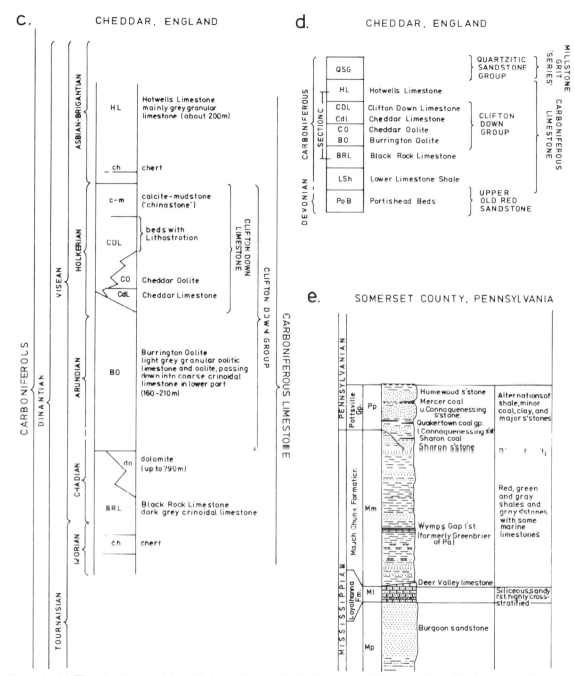

Fig. 6.9 (a) The simplest and least informative method of representing a stratigraphic key/legend on a geological map. Formations are named, colour coded and lettered but no lithological data are given. Geological time units are also given. (b) A move in the right direction in that a reasonable amount of lithological information is provided in addition to (a). (c) A stratigraphic column with some lithological commentary and a good graphic portrayal of lenticular units (lateral facies). At a glance this style gives relative thickness which may be averages for the map area so variations will still have to be analysed on the map. (d) This standardized representation is similar to (a)+(b) and the amount of lithological information given varies considerably. As a stand-alone version this would rate poorly in data transfer, but it was published as a summary diagram in combination with a more extensive stratigraphic column which is partly shown in (c). (e) It is rare to see on published Geological Survey maps this amount of detail. From the pictorial style of the stratigraphic column a good idea can be gained of sand/shale, etc., ratios within each formation. Such information is useful in an analysis of depositional environments, but often such crucial information is in limited supply

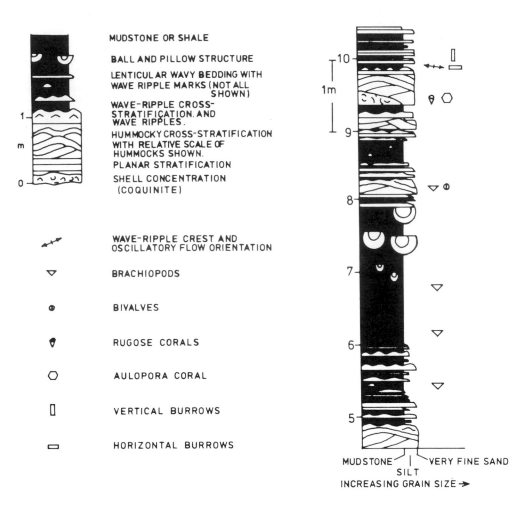

Fig. 6.10 This is a graphic, measured or sedimentary log which at its most detailed records every bed, gives grain size, fossil content, sedimentary structures and palaeocurrent direction. The same style can also be very effective on a smaller scale to give overall trend in stratigraphic columns kilometres thick. This very detailed style is normally found in reports, memoirs or published papers, whereas Figure 6.9e is a small-scale example which is not as rigorous about representing grain size

an adjacent marine region (see Figure 10.17). The pronounced lateral facies contrasts may require two separate stratigraphic columns placed side by side to show interpreted correlations. If the two disparate regions are then linked by a later more widespread unit, this may be shown diagramatically. On tectonic maps, a time-space plot may be used in a similar fashion but containing more information on structural and metamorphic events, and tectonic style. A welcome modern trend is to provide rock relationship diagrams (Figure 6.11) which schematically give the stratigraphic relations between the mapping units. In contrast to a cross-section, this diagram involves all units not just those found on the section line. These diagrams give a very quick way of assessing the structural style, lateral variations in lithology, sequence of events, nature of unconformities, intrusive styles, etc.

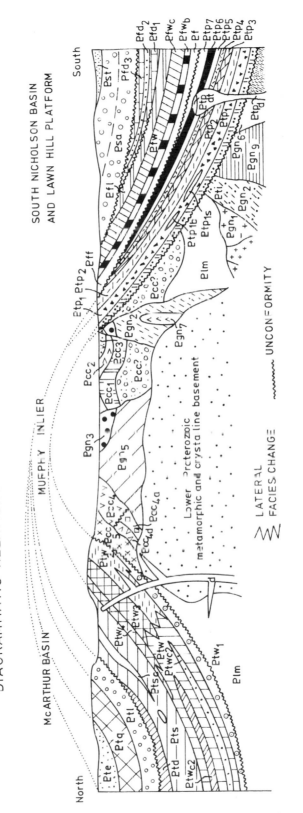

DIAGRAMMATIC RELATIONSHIP OF MAJOR PROTEROZOIC ROCK UNITS

Fig. 6.11 Rock-relationship diagrams are a powerful means of conveying the essential interrelations of most of the units on a map. Key overprinting, intrusive or lateral changes are brought to your attention and then the r expression on the map can be more easily detected. Modern maps typically have several styles of summary diagrams (e.g. block diagrams) to speed the process of data transfer

6.7   FURTHER READING AND IMPORTANT SOURCES OF INFORMATION

Eicher, D. L., 1976, *Geologic Time* (2nd edn) Prentice-Hall, New Jersey, 150 pp.

Hedberg, H. D., 1976, *A Guide to Stratigraphic Classification, Terminology and Procedure.* International Subcommission on Stratigraphic Classification of the International Union of Geological Sciences Commission on Stratigraphy, John Wiley, New York, 200 pp.

7 DEFORMATION BASICS

7.1 GENERAL COMMENTS

Stratigraphic methodology established the fundamental principles of superposition, original continuity, horizontal deposition, etc., but the many departures (exceptions) demanded the birth of a new discipline — **structural geology**. Fine-grained sediments dipping at high angles were clearly not in their original state and **displacement** had to be invoked. Displacement in this context simply means a change in position. In fact it is the analysis of displacement that is the distinctive aspect which distinguishes structural geology from other branches of geoscience. If displacement has occurred, then structural geologists are interested and the following questions are asked: What is the nature (type) of displacement? What is the amount (quantification)? What caused it? What mechanisms allowed it to occur? A good definition of structural geology is:

> 'The branch of geology concerned with the description and mechanism of displacement.'

In geological terms displacement and deformation are synonymous though the *Concise Oxford Dictionary* defines the latter as 'disfigurement, a change for the worse'. Most stratigraphers would sympathize with these sentiments because original features which they like to study are modified or obscured and complications arise, but many deformation structures are strikingly aesthetic, i.e. they can be admired despite the problems (challenges) they pose.

It is fairly common for geologists to consider structural geology to be small-scale tectonics. I do not share this view because tectonics is a much more wide-ranging subject as is hopefully conveyed in the following definition:

> 'Tectonics is the study of the thermal and mechanical history of the lithosphere.'

Tectonics is very much concerned with the consequences of the mechanical behaviour of the crust and immediately underlying mantle. For example, large-scale differential movements of the lithosphere commonly create sedimentary basins; the sedimentation pattern and history become important in the tectonic analysis, but this is a far cry from the structural heartland of studying displacement.

We will now briefly consider the causes of displacement and the nature of the processes within rocks that allow deformation to occur. The earth is a heat engine and this creates an active planet. At depth, heat production may be fairly uniform, but transfer is by the markedly heterogeneous mechanism of convection and therefore concentrated by cells into zones of upwelling and sinking. Important by-products of this activity are variations in the magnitude and orientation of forces being applied to the outer skin of the earth. Geologically it is usual to work in terms of forces normalized per unit area to facilitate comparisons from place to place and time to time. After normalization the forces are referred to as **stresses**, and despite the voluminous literature on this topic we shall only deal with the subject at the simplest level. The heat engine typically generates unequal forces (**stress difference**) and it is these inequalities that give rise to virtually all geologically significant displacements. Stress acting equally in all directions is known as **hydrostatic** and is an important characteristic of fluids. Hydrostatic stresses play a critical role in deformation but rather than being primary forces they normally develop as an internal response to externally imposed stress difference.

In all examples of stress difference the highest and lowest values are readily identifiable and are referred to as the **maximum principal stress**, (σ_1, sigma one) and **minimum principal stress** (σ_3). Stresses are vectors and these two principal directions are always perpendicular. Two special stress field conditions occur when the planes at right angles to σ_1 or σ_3 contain isotropic stress levels (Figure 7.1b and c) but more usually stresses vary in intensity in these planes (Figure 7.1d) such that there is an intermediate principal stress (σ_2) mutually orthogonal to σ_1 and σ_3. The latter stress state is the most general where $\sigma_1 \geqslant \sigma_2 \geqslant \sigma_3$ but $\sigma_1 = \sigma_2 > \sigma_3$ and $\sigma_1 > \sigma_2 = \sigma_3$ do occur. It should be noted that geologists and engineers have opposite sign conventions; compression is positive for us and tension is negative. If we examine a plane containing two unequal principal stresses (Figure 7.1d), we find that there is a systematic variation in stress levels in directions intermediate between the two main axes. A vector diagram of such a plane (Figure 7.1d) shows an elliptical distribution where stress values in particular orientations are represented by lengths proportional to intensity. By extension into 3-D, the complete state of stress can be portrayed by **ellipsoids** (Figure 7.1bii, cii, dii). Here $\sigma_1 = \sigma_2 > \sigma_3$ is a discus (Smartie) shape (oblate).

A unique feature of any principal stress is that the plane at right angles (containing the other two principal stresses) has no couple or moment of forces (shear stress) acting along it that could cause shear/slip displacements. All other directions contain resolved rotational forces (shear stresses) which may give rise to blocks sliding past one another.

The response of rocks to differential stress (studied in a subject called **rheology**) is dominantly controlled by the environmental parameters of temperature and confining pressure which are largely dictated by depth of burial. Several other factors are important including rate of deformation, material properties, and the forces exerted by fluids within the rock. Under near-surface conditions, low-temperature materials at low confining pressures normally fracture when the stress difference exceeds critical values which are essentially controlled by the cohesion of the rock. This is **brittle behaviour** which occurs when rocks rupture whilst acting as elastic solids, that is, during shape changes that are proportional to the applied load (Hooke's law). At higher temperatures and pressures solid rocks respond to differential forces by cohesive flow and behave as fluid in that they would eventually (over $10^5 - 10^8$ years) take up the shape of a container. This style of deformation is known as **ductile** and shape changes (**distortion**) occur without rupture or loss of cohesion. **Viscosity** is a measure of the resistance to flow under applied stress difference. Low viscosity means ready flow whereas high viscosity equates with great reluctance to change shape. Because continuity of the material is maintained during ductile deformation, displacement amounts vary gradually from place to place and there are no abrupt jumps in the displacement field; this is **continuous deformation** (Figure 7.2c). In contrast, if relative displacement occurs across fractures generated by brittle deformation, these surfaces represent sudden jumps in the amount of displacement hence this is **discontinuous deformation** (Figure 7.2b). Many deformed terrains contain both styles (Figure 7.2d) reflecting either superposed events generated at different stages in a burial and exhumation cycle or synchronous

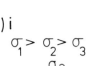

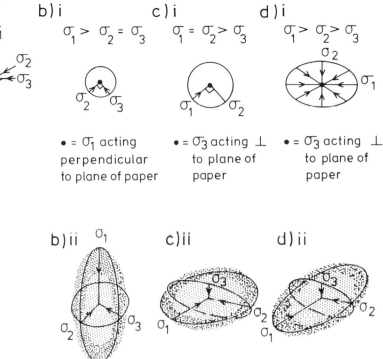

Fig. 7.1 (a) The principal stresses — maximum (σ_1), intermediate (σ_2) and minimum (σ_3) — are always mutually orthogonal. (b) (i) A view of the $\sigma_1 > \sigma_2 = \sigma_3$ stress state looking perpendicular to σ_1 on to the plane where stresses are equal in any direction; (ii) 3-D representation of (i). (c) (i) The $\sigma_1 = \sigma_2 > \sigma_3$ stress state seen on the σ_1/σ_2 plane perpendicular to σ_3; (ii) 3-D representation of (i). (d) (i) The most common stress state is $\sigma_1 > \sigma_2 > \sigma_3$ here viewed on the σ_2/σ_1 plane where stress level varies systematically with orientation to define an ellipse; (ii) 3-D representation of (i)

development. Because mechanical properties of rocks are strongly dependent upon lithology it is possible to find adjacent layers behaving quite differently. A thick quartz sandstone may well fracture whilst neighbouring laminated silts and muds undergo bending. Such differential responses are more obvious in the **transitional zone** between the brittle and ductile regimes; this is certainly not an abrupt boundary. Experimental work and field observations have confirmed the styles of deformation characteristic of transitional situations. Typically narrow zones of ductile deformation known as ductile shear zones are separated by zones of virtually negligible distortion (Figure 7.3d). Towards the more brittle end of the spectrum, the equivalent would be a concentrated zone of fractures, with relative displacements between slices of rock, again separated by zones relatively free of such structures (Figure 7.3a).

The effects on rocks of brittle deformation are clearly conveyed by terms such as fracture and rupture. However, unless you have had exposure to metallurgy, the scope of everyday activities gives little insight into the mechanisms that allow solid rocks to flow like fluids. During ductile deformation, processes at the scale of individual grains integrate to allow large amounts of cohesive flow (i.e. shape change). One type of response involves sliding along atomic lattice planes to change the shape of each particle within the rock (**crystal – plastic deformation**). Another common response involves rock components going into solution at highly stressed

a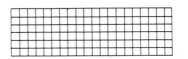

b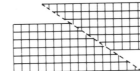

c

d

Fig. 7.2 (a) An undeformed grid with horizontal lines equivalent to bedding in sedimentary rocks. The vertical lines are to permit distortion to be monitored. All parts of this figure are 2-D and changes perpendicular to the page are not considered though they would have to be in a full analysis. (b) Brittle deformation creates fractures which can allow blocks to slide past one another. Displacement fields from block to block show abrupt jumps at the fractures generating discontinuous deformation — faulting. (c) Cohesive flow of rock produces gradual changes in displacement hence the name continuous deformation — folding. (d) Most deformed areas are mixtures of continuous and discontinuous deformation

grain contacts and being transported varying distances to be precipitated in a variety of sites (**mass transfer**). Shape changes of the whole rock can also be achieved by grain – grain rotations or by distributed microfracturing (**cataclasis**). The levels of stress difference required to generate ductile flow are remarkably low being of the order of 10 – 50 MPa. In contrast confining pressures at 15 km depth of burial (where most rocks are ductile) are about 500 MPa (1 kbar = 100 MPa).

7.2   DISPLACEMENT CATEGORIES

Four different styles or categories of displacement are recognized. By far the best description involves quantification, but for a variety of geological reasons this is a

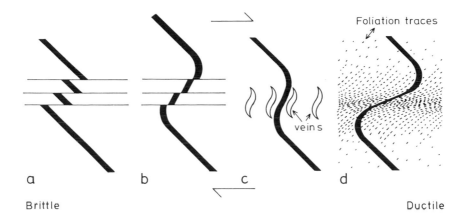

Fig. 7.3 Shear zones are narrow concentrations of deformation. (a) Several closely spaced faults create a brittle shear zone; (b) a limited amount of ductile deformation is shown by the folded layer but brittle effects still dominate; (c) in a progression towards a more ductile response, the layer is now continuous and fracturing is restricted to short vein structures within the shear zone; (d) deformation in this shear zone is totally ductile though the two blocks of country rock on either side have acted in a rigid fashion

difficult goal to achieve on a regular basis and we are often left with qualitative accounts. Scale is an important consideration in this discussion. Beyond the scope of this text are the approaches used in describing motion at the scale of lithospheric plates: these rely on spherical geometry and require a detailed study of their own. The exact upper scale limit of our discussions will not be defined but normally we shall be considering areas of tens to hundreds of square kilometres. One theme of structural geology is that many patterns analysed and understood on a small to medium scale can be extrapolated to very large structures. The somewhat loosely defined terms **microscopic, mesoscopic** and **macroscopic**, are used to denote a progression of scales. Microscopic naturally refers to structures that can only be resolved under the microscope. Mesoscopic covers structures that can be observed from one vantage-point ranging from hand-lens investigations to features exposed on a mountainside. Macroscopic involves structures too large or too poorly exposed to be observed in their entirety from one position. Perhaps the plate-tectonic scale should be called megascopic?

7.2.1 TRANSLATION

If all points within a body move in the same direction by the same amount then translation has occurred and the displacement vector is constant throughout the body. Figure 7.4a illustrates the effects of translation in 2-D and whilst it is dangerous to limit our thinking in this way our first steps will be tentative. The heavy lines of Figure 7.4 represent bedding whilst the thin lines create an imaginary grid which allows us to monitor shape changes. After translation there is no internal distortion and thus the recognition of these effects in the field can be difficult. Large translated displacements are known from all orogenic belts and values of over 10 km are often quoted in the literature. Slabs of rock (thrust blocks), hundreds of metres thick and tens of square kilometres in area, undergo approximately translatory movement achieving overall shortenings of 50 per cent or more in belts hundreds of kilometres wide (see Fig. 9.17).

DISPLACEMENT CATEGORIES

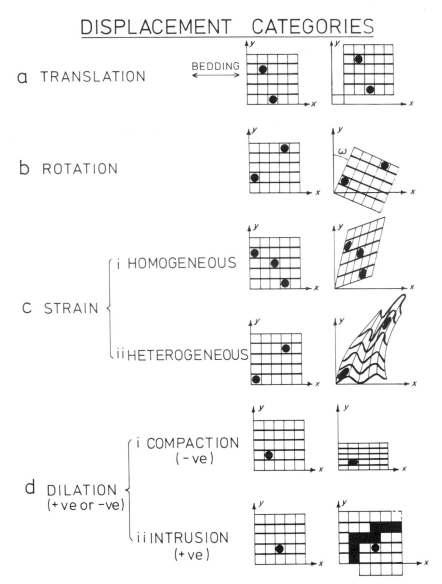

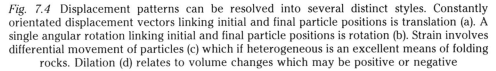

Fig. 7.4 Displacement patterns can be resolved into several distinct styles. Constantly orientated displacement vectors linking initial and final particle positions is translation (a). A single angular rotation linking initial and final particle positions is rotation (b). Strain involves differential movement of particles (c) which if heterogeneous is an excellent means of folding rocks. Dilation (d) relates to volume changes which may be positive or negative

7.2.2 ROTATION

In this displacement pattern (Figure 7.4b), the original and final positions of points are linked by the same angle of rotation, ω. Sediments that have only undergone rotation allow simple quantification because layering was generally horizontal pre-deformation. It is important to note that this style of displacement again does not involve distortion and the circular marker within the grid of Figure 7.1b remains circular. By inducing variable amounts of body rotation a planar surface becomes curved, i.e. folded.

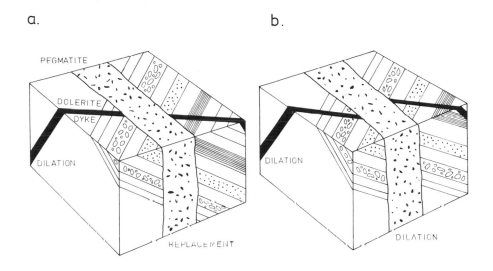

Fig. 7.5 At high metamorphic temperatures a volume of rock may be replaced volume for volume by another lithology which typically has the appearance of an igneous intrusive. (a) After replacement pre-existing features can be projected through the new body (pegmatite in (a)) with no offset. (b) Similar materials can also be generated by a dilational event. Shown here is the simplest dilation where a fracture is pulled apart perpendicular to the fracture walls and there is no shear-type displacement parallel to the fracture. Pre-existing features appear offset and 'tie-lines' are perpendicular to the boundary of the intrusive. Note on the top surface of (b) that differently oriented pre-dilational features can have opposite senses of offset even though the dilational direction was constantly orientated. Dilation is further discussed in Chapter 10

7.2.3 STRAIN

This behaviour involves particles moving relatively with respect to each other, thus establishing an internal deformation that changes shapes and angular relations (Figure 7.4c). Strain and distortion are synonymous and their effects are best introduced in the case of homogeneous strain (Figure 7.4ci). Here, each element of rock has undergone the same amount of shape changes as recorded by the originally rectangular grid and the transformation of the circles to ellipses. Quantification of strain in such a 2-D slice is fairly simple, it is the axial ratio of the now elliptical but once circular sections. Extreme examples have inferred ratios of hundreds to one but values of two or three to one induce noticeable shape changes. A key feature of homogeneous strain is that originally straight lines remain straight and parallel lines remain parallel — confirm this by looking at the bedding traces and grid lines of Figure 7.4ci. On this diagram also note that the bedding traces are now inclined to the horizontal, i.e. a rotation has occurred. The style of rotation is, however, different to that of Figure 7.4b and being strain induced it is not considered to be a rigid body rotation.

Natural processes of ductile deformation invariably lead to heterogeneous strain (Figure 7.4cii) where the amount of strain varies from place to place. The variability is monitored in Figure 7.4cii by the grid distortion and the circular markers, and is expressed by straight lines becoming curved. Any originally parallel lines lose this geometry and complex shape changes occur. 3-D heterogeneous strain clearly provides an effective method of producing folded layering. A complete description (i.e. quantification) of the displacements during an episode of heterogeneous strain is naturally a complex operation. Basically the end-product is broken down for analysis into approximately homogeneous areas (domains); the individual results are integrated to provide an account of the sum.

7.2.4   DILATION

We are now considering volume changes which can be either positive or negative. The simplest case is that of a sequence of sediments being buried where the ever increasing weight of overburden causes shortening (compaction) in a direction controlled by the gravitational field. For undisturbed horizontally bedded sequences this is perpendicular to layering (Figure 7.4di) and is achieved by expulsion of fluid from pore space associated with the resultant collapse of the particle structure of the sediment. In the standard case, no dimensional or shape changes occur within the bedding plane and only the vertical dimension is reduced. In shales compaction values of around 80 per cent are common, whereas coarse sands may only record 10 or 20 per cent shortening.

Positive dilation is best illustrated by igneous intrusions and vein structures (Figure 7.4dii) (see also Chapter 10). Veining processes have considerable economic significance as many metals are either mainly found in veins or have important vein style occurrences (e.g. gold in quartz veins, tin and tantalum in pegmatite veins). Under near-surface brittle conditions it is common to find sheets of igneous material that have forced their way into and opened up fractures. Many regions of high-grade crystalline rocks (cratons) are criss-crossed by numerous tabular igneous injections recording extensions of 10 to 20 per cent. The most dramatic increase in volume is demonstrated by the sheeted dyke complexes of ophiolites where only dykes are present — infinite extension related to sea-floor spreading and creation of oceanic crust.

At lower crustal levels within the ductile regime it is possible that replacement processes could mimic a dilational injection. Given sufficient information, however, it is possible to discriminate between dilation and replacement (Figure 7.5). Intermediate situations exist where igneous materials ascending from their point of generation either partially assimilate the host rocks or raft off fragments. Incorporation of wall rock into a magma modifies the chemistry and knowing the proportions of assimilation and shouldering aside is an important consideration in petrogenetic studies.

7.2.5   GENERAL CASE

Typically a volume of metamorphic rocks that has undergone ductile deformation will in terms of displacement be a composite of perhaps all four categories. Strain, translation, rotation and volume change, may well all have combined to create the final disposition of the rock mass. To provide a good description of the cumulative effects of these displacements, the contribution of each category has to be isolated and measured; a very difficult task because the rocks rarely provide enough information for this to be done satisfactorily.

8 CONTINUOUS DEFORMATION

Folds, the products of continuous deformation, display a myriad of shapes. Such diversity of often quite complex variably curved shapes requires careful documentation if we are to successfully convey to others an accurate picture of the fold(s) being described. In this chapter folds are firstly considered in their own right as 3-D entities and methods will be established for their analysis and description: then we shall progress to examine the style of outcrop patterns that can be generated by folded rocks. Considering the potential variety of topography and fold morphology, the results of their interaction are seemingly endless: hopefully the principles dealt with in the latter half of this chapter will equip readers with the means to tackle most situations. One of the most difficult conceptual problems in this type of analysis is the appreciation of the cut-effect. Taking the one fold structure and slicing it in several different directions produces markedly different cross-sectional shapes. A graphic, though simplified and non-geological, example is shown in Figure 8.1. An inclined concentrically laminated cylinder displays its simplest cross-section (a) at right angles to the axis: a circular section with constant curvature. A vertical slice (b) produces in 2-D an elliptical section with varying curvature whilst that of the solid object clearly does not change. When sliced horizontally (equivalent to a map view) the cross-section (c) is again elliptical but of different axial ratio, and hence curvature, to that of the vertical cut. The elliptical sections each have two well-defined points of maximum curvature but these are artefacts of the cut-effect, illustrating the importance of orientation when studying fold structures. Virtually all folds show varying curvature in most slices, and, because the changes in curvature are different from slice to slice, a standard viewing direction, with respect to the geometry of the fold, has to be established. It is rare for this direction to be the same as the one we use to view maps, that is, perpendicular to the earth's surface, and hence most maps of folds bear as little relevance to the 3-D shape of the folds as the elliptical sections do to the cylinder. Folds are a true test of ability in thinking in 3-D!

8.1 FOLD MORPHOLOGY AND DESCRIPTION

The rigorous definition of a fold is as follows.

A **fold** is a continuous curved surface convex in a single sense.

The normal implication is that the folded surface was once planar and through deformation achieved its present shape, but as the mechanism is not specified the process is not important: the definition is purely geometric (an ideal not always attained in structural geology). The curved surface is referred to as the **form**

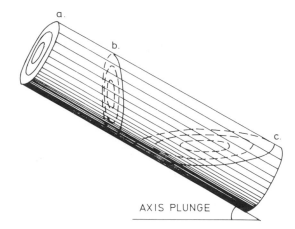

AXIS PLUNGE

Fig. 8.1 Differently oriented slices through the same 3-D object give very different impressions. Here the constant curvature of the cylinder is represented as variable curvature in vertical (b) and horizontal (c) slices. The only 'true' section is perpendicular to the cylinder axis and is equivalent to a fold profile

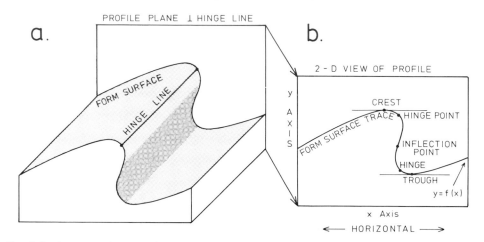

Fig. 8.2 A curved form surface allows a fold to be recognized and on this surface a line joining all points of maximum curvature, for a single fold, is a hinge line. Perpendicular to this line is the profile plane (b) in which many aspects of fold geometry are defined. On the profile the trace of the form surface outlines a curve $y = F(x)$ and, where the gradient is zero, crest and trough points occur. The hinge point is the maximum curvature and the inflection point marks the change from one fold to another

surface, this is the feature that allows the fold to be recognized (Figure 8.2). Most commonly this surface is sedimentary layering but many other structures fulfil this role, particularly tectonically generated planar fabrics such as slaty cleavage, gneissic foliation, etc. Fold form surfaces are so varied that they include the margins of once tabular igneous intrusions and surfaces of unconformity.

When a single folded surface such as that shown in Figure 8.2 is examined in 3-D it is possible to recognize a line joining all the points of maximum curvature — the **hinge line**. The example shown has a rectilinear hinge line, but in nature such regularity is usually destined to be short-lived as illustrated by Figure 8.3. With other than slight curvature of the hinge line the analysis of a single fold as a whole becomes geometrically very complicated. The standard procedure when confronted with significant hinge curvature is to subdivide the structure into segments (**domains**) with approximately rectilinear hinge lines: a very time-consuming practice, rarely fully implemented, but necessary if a detailed description is required. In the following

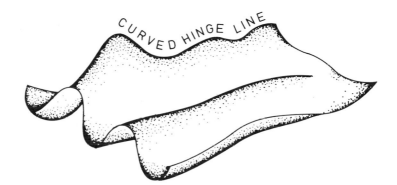

Fig. 8.3 Hinge lines may be approximately rectilinear for kilometres but equally they can be markedly curved on a small scale

sections assume a straight hinge segment is being referred to unless otherwise stated. The geometry of the folds in Figure 8.2 is very straightforward in type. They are characterized by the ability of a line parallel to the hinge line to always rest on the form surface when moved around the fold which is, therefore, known as **cylindrical** (Figure 8.4a). This moving line is the **generatrix** which, as a result has the orientation (plunge and bearing) of the hinge line but unlike the latter does not have a fixed position. **Fold axis** and generatrix are synonymous. The other regular type of geometry occurs where the form surface takes on the shape of a partial **cone**. In this style the shape of the fold is described by the movement of the generatrix through space with one end fixed (at the cone apex) and the other end free to move (Figure 8.4b). Cones with small half-apical angles are fairly distinct but as the angle increases they become more and more cylindrical. In fact cylindrical folds can be regarded as cones with half-apical angles of 90°. Cones can only be properly recognized by stereographic or numerical analysis and hence many have been mis-identified as cylindrical. However, cones are still in the minority most commonly occurring at the tapering ends of cylindrical folds (Figure 8.4c) and in zones of complex geometry where refolding of previously folded surfaces has taken place.

Stereographic analysis can recognize zones of cylindrical folds with approximately straight hinge lines and zones of conical folds: it cannot handle more complex geometries such as domes and basins, or folds with markedly curved hinge lines. Qualitative means for describing these complexities will be given as we develop our technique.

Following our discussion of the sections through the inclined cylinder you will appreciate that, for variably oriented slices, the point of maximum curvature on 2-D views may not correspond to the 3-D equivalent. For this reason we can only sensibly describe a fold's variation in curvature (and some other properties) in a section that is established with respect to the 3-D shape of the fold. In the case of cylindrical folds this section is at right angles to the hinge line and is known as the **profile plane** which geometrically has several very special relationships to the fold shape (Figure 8.2). Of great importance to the stereographic technique is the fact that normals or poles to the form surface are everywhere parallel to (contained within) the profile plane orientation (Figure 8.5). Many folds are too large to be measured directly and the pole/profile plane geometry allows the plunge and bearing of the fold axis to be calculated (Appendix 1). Geometrically the profile plane orientation may be established either by using the hinge line or the form surface normals. The latter gives rise to the term normal-plane-curvature which refers to descriptions of fold shape in the reference profile plane.

If we deal mathematically with the profile shape of a single folded surface ($y = f(x)$),

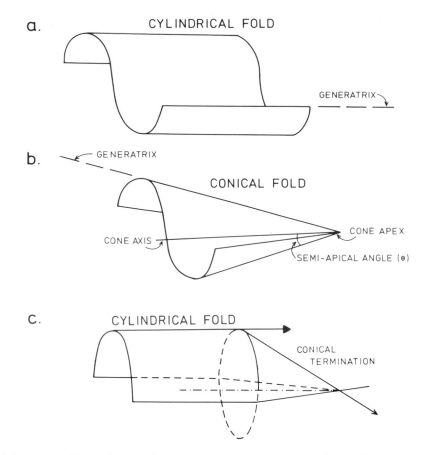

Fig. 8.4 (a) A cylindrical fold where the generatrix moves parallel to itself and stays in contact with the form surface; (b) now the generatrix is fixed at a point and defines a conical shape — always incomplete; (c) many cylindrical folds terminate as partial cones

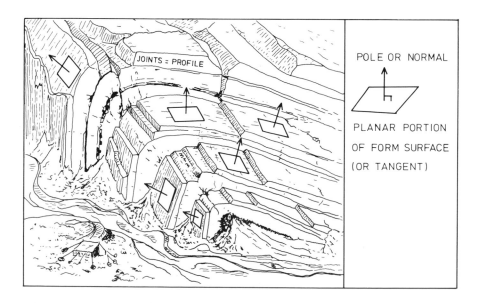

Fig. 8.5 A unique property of cylindrical folds is that normals/poles to the form surface are all parallel to the profile plane. This relationship is the basis of stereographic analysis of folds

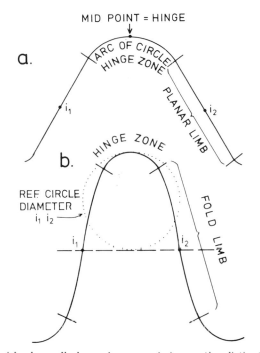

Fig. 8.6 In a fold with planar limbs and a curved closure the distinction between the hinge zone and limbs is simple (a). Specific definition of fold limbs and hinge zones may not be straightforward with a continuously curved surface (b) and depends upon the relative curvature of a circle of diameter i_1i_2 (the inflection points join) and the fold closure

Figure 8.2) we can define several important features. The curvature is the rate of change of the slope (gradient) of the surface

$$\frac{(\mathrm{d}^2 y)}{\mathrm{d}x^2}$$

and where this quantity is zero we have the **inflection points** which mark the end of one fold and the beginning of another (Figure 8.2b is showing **two** incomplete folds). At the position of maximum curvature we have the **hinge points**. Both inflection and hinge points are defined relative to curvature. A fundamentally different reference is used for crest and trough points as these are defined relative to a horizontal datum. They occur when the slope of the curve ($\mathrm{d}y/\mathrm{d}x$) is zero. The **crest point** is the highest position on the folded surface. The **trough point** is the lowest point on the form surface. It is possible for hinge points to coincide with crest and trough points depending on fold shape and attitude as will be demonstrated later.

Further variations on the fold shape theme are illustrated in Figure 8.6. Many folds have planar segments known as **limbs** separated by curved sections known as **hinge zones**. If the latter have variable curvature then it is easy to locate the hinge point; however, if the hinge zone has constant curvature (fairly common) then by convention the hinge point is placed half-way through the hinge zone (Figure 8.6a). Where a fold has no planar limbs the definition of limbs and hinge zones is more involved. The distinction is drawn by comparing the curvature of the form surface with that of a circle whose diameter is the length of the join between the inflection points (i_1, i_2, Figure 8.6b). Curvature less than the circle defines the fold limb and greater curvature outlines the hinge zone. If a fold is all straight limbs then the hinge

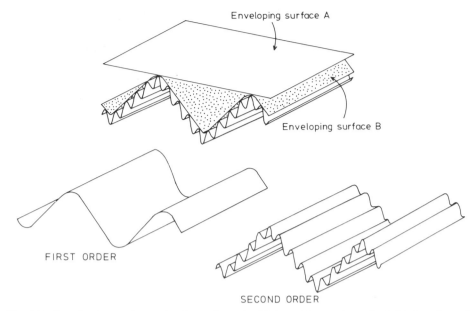

Fig. 8.7 Enveloping surfaces simply enclose the high and low points of a folded surface. The enveloping surface of the folded layers is itself folded defining first-order structures

zone is reduced to a point, and if the fold is a circular arc then the definition of limbs seems meaningless in this context.

When dealing with a series of folds it is sometimes useful to delineate the **enveloping surface** to the waveform. This surface or sheet dip simply joins successive trough or crest lines to enclose the overall wavetrain (Figure 8.7). In the example illustrated the regular periodic nature of the folds means that the enveloping surfaces are smoothly curved surfaces or planes. More variability to the fold shapes and sizes along a fold train produces irregularly shaped enveloping surfaces. Two scales or **orders** of folding are shown in Figure 8.7. Enveloping surface B defines a form surface to a larger order set of folds which are regarded as first order and enclosed by enveloping surface A. Within surface B we have the second-order folds and there is the possibility for smaller third-order folds to be present. The folds within surface B are **parasitic** to the first-order folds. If an enveloping surface (or significant part of one) is planar then its attitude can be specified by using dip and strike which gives the average attitude of a folded layer or sequence of layers. Such an average attitude cannot be gained from individual dip and strike readings of the form surface. If Figure 8.8a is taken to be the profile of a fold with a horizontal hinge line then the majority of form surface readings taken around the structure would have steep dips, whereas the average attitude of the folded sheet is horizontal.

In terms of fold shape, symmetry is a very important but widely misunderstood parameter. To demonstrate symmetry/asymmetry, the minimum requirement is that the analysis is carried out on one complete fold and on each immediate neighbour up to the hinge point; part of one fold is **not** enough. **Symmetrical folds** have identical (or mirror image) curve shapes between adjacent hinges; all other folds are **asymmetrical**. Symmetry or the lack of it has nothing to do with limb dips. Symmetrical folds can have different limb dips and asymmetrical folds can, with the appropriate orientation, have the same dips (usually, though not necessarily in the opposite directions) (see Figure 8.8). In Figure 8.8c a series of symmetrical folds alternate from flat-lying limbs to vertically dipping. It has recently been proposed that folds with limbs of the same dip be referred to as equal folds, the opposite being

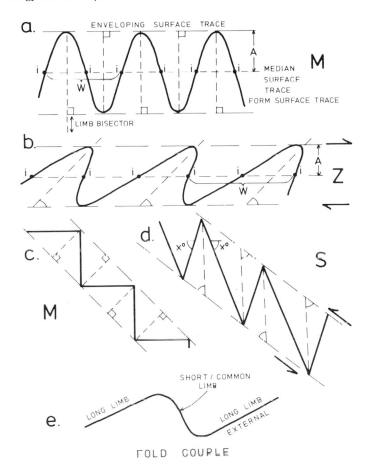

Fig 8.8 Symmetry/asymmetry requires careful definition. (a) Symmetrical folds have identical or mirror image shapes between adjacent hinge points. Also the axial surface trace is at right angles to the enveloping surface; A = fold amplitude, W = wavelength. M represents the symmetrical shape. (b) Asymmetrical folds with axial surface traces oblique to the enveloping surface. To convert symmetrical folds to this shape requires a clockwise rotation and an overall Z shape results. (c) Symmetrical folds with unequal limb dips. (d) Asymmetrical folds with equal limb dips. (e) A fold couple — an antiform and a synform sharing a short common limb. This is an asymmetric case where the external limbs are longer than the common limb

unequal folds. With asymmetrical folds one limb is longer than the other, and two adjacent folds (Figure 8.8e) make a fold couple that share a common/short limb.

Symmetry or the lack of it can be recognized in another way. Take a line through a hinge point bisecting the angle between the fold limbs (Figure 8.8). If this line is perpendicular to the enveloping surface the folds are symmetrical, any other angular relation means asymmetry (compare Figures 8.8a, b). For the asymmetric folds the sense of rotational forces required to transform a perpendicular relationship into the one observed (Figures 8.8b, d) is a key feature. A shorthand notation has been applied to describe fold symmetry/asymmetry, the symmetrical folds of Figure 8.8a being referred to as M folds. For Figure 8.8b the best approximation letter is Z (not N) and for Figure 8.8d S. Note that to define S, Z and M, shapes more than one fold needs to be examined. Unfortunately this style of description is affected by the viewing direction as can easily be demonstrated by making a tracing of Figure 8.8b and reversing the overlay to look from the other side. The Z is transformed into an S (see Figure 8.9). To overcome this problem the concept of vergence has been introduced

Fig. 8.9   A cartoon to emphasize that Zs looked at from the opposite direction (behind the sheet of paper) become Ss

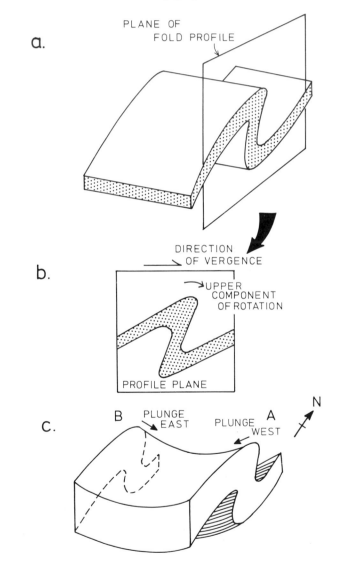

Fig. 8.10   Fold vergence is defined in the profile plane (a) as the direction of the upper component of rotation necessary to convert symmetric folds into those observed (b). This approach means that at both localities A and B (c) the vergence is in the same direction (to the north) despite the plunge change. Viewing each of A and B down the local plunge, A would be recorded as Z-shaped and B as S-shaped

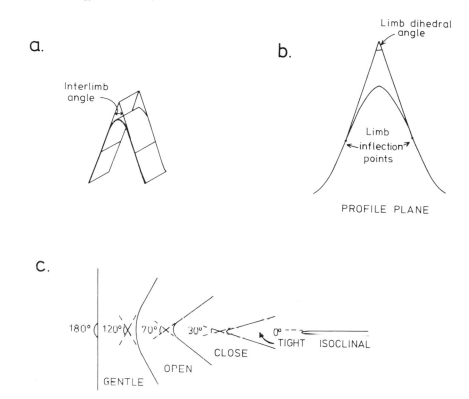

Fig. 8.11 (a) A 3-D view of the dihedral angle; (b) profile plane view of dihedral angle; (c) descriptive names for broad ranges of dihedral angles

to describe the sense of deflection from the symmetrical case; this parameter can be used in the analysis of larger-scale structures. Vergence (Figure 8.10) is defined as:

'The horizontal direction, within the fold profile plane, towards which upper component of rotation is directed.'

Vergence is, therefore, a direction at right angles to the fold hinge line and consistently points to the next highest order upwards closing fold (antiform) (see Figure 8.19). The map representation of vergence will be discussed after fold/cleavage geometries have been outlined (see Figure 8.25). The usefulness of vergence is conveyed by Figure 8.10c where the asymmetric fold has a gently curving hinge line with opposite directions of plunge at either end of the exposed block. One imprecise convention would record the asymmetry in terms of S and Z as seen looking down the plunge of the hinge line. However, in this case the same fold would be S-shaped when viewed by an observer in the west looking east (location A) and Z-shaped to someone looking west from the east side (location B). Using vergence, the whole structure has a single vergence which is directed northwards.

Fold size is a very important feature to be described. For reasonably regular (periodic) folds this is best done using amplitude and wavelength (Figures 8.8a, b). Clearly readers need to know whether the folds are metres or kilometres in scale.

Angular relations between fold limbs strongly influenced the style of the fold and give a rough estimate of the amount of shortening involved. Fold tightness is measured by the angle between the limbs (**dihedral angle**) as seen on the profile plane (Figure 8.11). Again the cut-effect is very important in that differently oriented

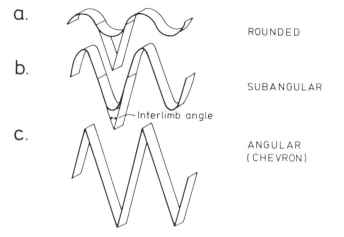

Fig. 8.12 Three folds of the same dihedral angle but varying angularity

sections through a single fold may give a wide range of 'apparent dihedral angles'. If a fold does not have planar limbs then the angle used is that between planes tangential to the form surface at the inflection points. A series of terms applied to bands of dihedral angle is in common use (Figure 8.11).

Apparent fold tightness is controlled by the relative proportions of planar limbs versus hinge zones and is best described as **angularity**. The three folds of Figure 8. 12 have the same dihedral angle but have very different shapes. The folds of Figure 8.12a have constant curvature and hence have no limbs making them rounded, whereas in contrast Figure 8.12c is all limbs and is angular. Intermediate situations are subangular or subrounded though specification can be more rigorous. The classical examples of angular folds are kinks (Figure 8.13) and chevrons (Figure 8.12c). What is normally referred to as a kink fold or kink band is in fact two coupled angular folds with two parallel external limbs. The common limb represents a segment rotated from the orientation of the long limb (through angle z Figure 8.13a). Until now we have been concerned with folded single layers, but to best display kinks they are shown in their natural environment affecting a multilayer of well-laminated rocks. Very commonly the form surface to small-scale (centimetric) kink bands is a slaty-type cleavage. Separating the rotated zone of a kink from the long limbs is an imaginary surface known as a **kink-band boundary** (KBB) (which may be a fracture in outcrop because the layers broke at the sharp deflection of trend). The KBB is also a **hinge surface** because it contains the hinge lines of each folded layer (Figure 8.14a). In Figure 8.13a the angle x and y give the angular relations of the KBB to the long limb and the short limb respectively. These angles need not be equal and they largely depend on the amount of rotation of the common limb. Values of around 60° for x and y are frequently developed for mechanical reasons. Large-scale kink-like folds are commonly referred to as **monoclines**. Kink folds are often arranged in pairs with opposite senses of rotation affecting the same zone of constantly oriented layering (that is, external limbs are parallel—Figure 8.13b). This arrangement is a **conjugate** one where both sets developed synchronously. At the zone of convergence of conjugate kinks, **box folds** or chevron folds are generated (Figure 8.13c).

Now that we are into considering multilayers it is essential to study layer thickness variations around folds. Figure 8.15 shows some aspects of this theme; in Figure 8.15a the layer (stratigraphic) thickness is maintained (**parallel fold**) whereas in (b) and (c) considerable changes take place. In Figure 8.15b the variations are symmetrically arranged about the hinge surface trace, but if the layering is measured

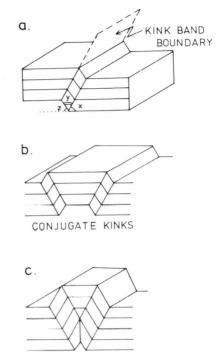

Fig. 8.13 (a) A kink fold which in fact is two angular folds with unrotated fold limbs separated by a rotation zone (common limb) — angle of rotation $z°$. The surface containing the hinge lines is the KBB. Large examples are monoclines. (b) Kinks often occur as inclined pairs with opposite sense of rotation — conjugate kinks. (c) At the convergence of two conjugate kinks a box fold is generated

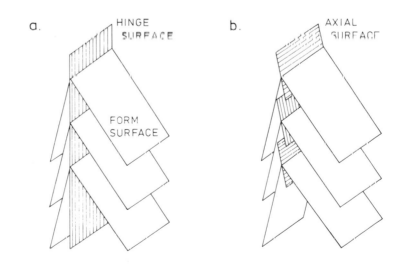

Fig. 8.14 (a) A hinge surface is the locus of all hinge lines. (b) An axial surface bisects the limbs and with unequal stratigraphic thickness on the two limbs gives rise to an *en échelon* arrangement

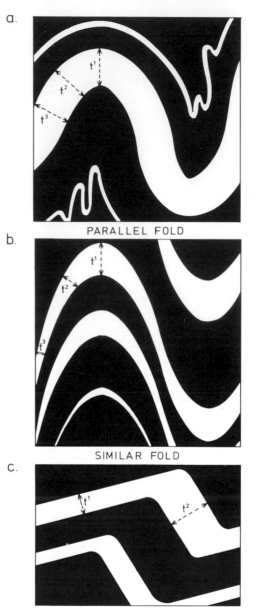

Fig. 8.15 Profile views showing stratigraphic thickness variations around folds. (a) Constant stratigraphic thickness — parallel fold; (b) thickness parallel to the hinge surface trace is constant — similar fold; (c) the short common limb is thicker than the external limbs in an example of asymmetric thickness variation about the hinge surface

parallel to this trace a constant value is recorded defining a **similar fold**. Stratigraphic thickness on the closure is four times greater than on the limbs as a result of high strain and clearly these variations have nothing to do with facies changes or any other pre-deformation factors. Figure 8.15c illustrates a common limb of an asymmetric fold has undergone thickening by ductile deformation; to assess changes (if any) to the long limbs relative to original stratigraphic thickness would require detailed knowledge of the folding mechanisms. Many other layer thickness variations are possible including both fold limbs thicker than the closure. It is very common for fold layers to be intermediate between parallel and similar in their layer thickness geometry, particularly in materials like sandstones.

For the similar fold each layer has an identical shape to its neighbours which

Fig. 8.16 Parallel folds eventually run out of room in trying to squeeze in layers of constant thickness and this requires a change in style. Either fractures develop in the closure or folds have to die out rapidly along a detachment horizon

means that this same geometry can affect an infinite number of layers. In marked contrast a stack of layers with constant thickness will soon cause space problems in a fold because the layers can no longer be accommodated (Figure 8.16) and either the fold dies out rapidly or the closure becomes a fracture zone with significant movement. The space problem will also necessitate a change in dihedral angle in a single fold (Figure 8.17). Layers of different composition within a single fold will usually have different styles of thickness variation. Full documentation requires detailed work and more advanced courses give methods for measuring and analysing these patterns in detail.

The term **axial surface** suffers from a lack of agreement as to its meaning. Herein it is 'the plane that bisects the angle between the limbs' (Figure 8.14b). For folds with symmetrical layer thickness variation about the hinge surface, the axial surface coincides with the hinge surface. However, the need for two terms is demonstrated by folds like Figures 8.15c and 8.14 with different limb thickness; in both of these cases the hinge surface is easily defined as a continuous plane. However, limb bisectors, though parallel, form a series of discontinuous planes in different positions and to join these would be a fruitless exercise in view of the definition. Hinge surfaces may be quite curved, but where they are reasonably planar (**plane folds**) they may be measured and their attitude specified by dip and strike as can be done for axial surfaces. Using a combination of hinge line plunge and bearing, and hinge surface dip and strike, fold attitude can be described (Figure 8.18). Table 8.1 gives the recommended nomenclature.

For plane folds it is worth remembering that the hinge line is an apparent dip of the hinge surface. In the upright plunging and inclined plunging folds of Figure 8.18 the hinge lines are pitching at about 30° within the hinge surface. By definition for a reclined fold the pitch is 90°, whereas in horizontal folds the hinge line is parallel to the strike of the hinge surface, that is, zero pitch.

In dealing with a single fold it is very important to describe the sense of closure which is defined relative to the horizontal. If the form surface is arched upwards the fold is an **antiform** and the converse is a **synform** (Figure 8.19a, b). The term **neutral** fold covers the case of a sideways-closing structure whether it is recumbent

Fig. 8.17 A field photograph of a parallel fold in 3.3 billion years (Archaean) iron formation. Note how the fold shape varies up and down one hinge surface trace

or vertically plunging and upright. With such folds an orientation should be quoted to give the sense of closure.

The above style of notation is purely geometrical/structural in that it deals only with the shape of the layers. We now move on to a set of terms that are defined relative to a stratigraphic datum, that is, the sense of younging of a sequence of sediments and/or volcanics. An **anticline** is a fold structure which has the oldest rocks in the core. Anticlines occur in both parts of Figure 8.20 though with opposite

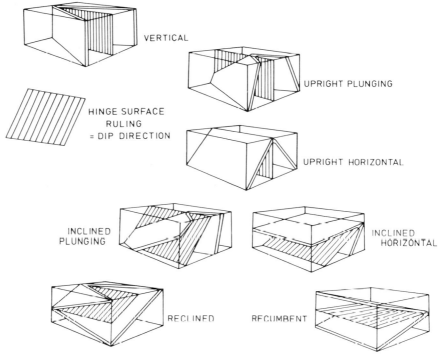

Fig. 8.18 Combinations of axial surface and hinge-line attitudes are used to describe fold attitude in broad categories (see Table 8.1). This diagram emphasizes the apparent dip nature of the hinge line relative to the axial surface

Table 8.1 Terms describing the attitude of folds

Dip of hinge surface (°)	Terms	
0	Horizontal	⎫ recumbent fold
1 – 10	Subhorizontal	⎬
10 – 30	Gently inclined fold	
30 – 60	Moderately inclined fold	
60 – 80	Steeply inclined fold	
80 – 89	Subvertical	
90	Vertical	⎰ upright fold

Plunge of hinge line (°)	Terms	
0	Horizontal (horizontal fold)	
1 – 10	Subhorizontal (subhorizontally plunging fold)	
10 – 30	Gentle (gently plunging fold)	
30 – 60	Moderate (etc.)	
60 – 80	Steep (etc.)	
80 – 89	Subvertical	
	Vertical	⎰ vertical fold

Note: Terms are used in combination as indicated in Figure 8.18.

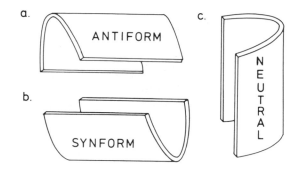

Fig. 8.19 (a) An upwards-closing antiform; (b) a downwards-closing synform; (c) neutral, neither closing up or down

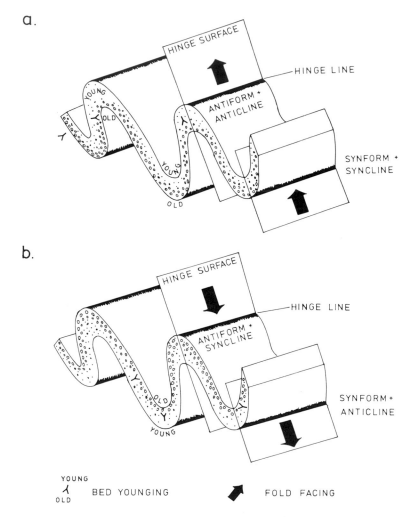

Fig. 8.20 Stratigraphic relations of folding. An anticline has the oldest rocks in its core and a syncline is cored by the youngest rocks. Combining fold shape and stratigraphy it is possible to have upwards-facing antiformal anticlines and downwards-facing antiformal synclines (same shape different younging). The inverted Y is one method of showing the younging of layers. Facing is a feature of a whole fold or sequence of folds

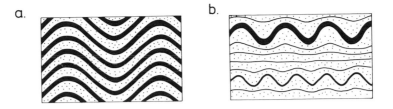

Fig. 8.21 (a) Harmonic folds where antiforms are stacked on antiforms and synforms on synforms; (b) disharmonic folds where waveforms are not in phase

Table 8.2 Cylindrical fold description. Check list

Size — amplitude, wavelength
Sense of closing — antiform, synform
Stratigraphic relations — anticline, syncline — **Facing**
Attitude — hinge line, hinge surface
Dihedral angle
Angularity
Symmetry, vergence
Enveloping surface attitude
Layer thickness variations
Harmony/disharmony

senses of closure. In Figure 8.20a the anticlines coincide with the antiforms and hence are antiformal anticlines, whereas in Figure 8.20b it is the synforms that have the oldest rocks in their cores making them synformal anticlines. A **syncline** is a fold which has the youngest rocks at the centre of the structure. Figure 8.20a shows synformal synclines in contrast to the antiformal synclines of Figure 8.20b. If the stratigraphy is known or younging is given by small-scale features (graded-bedding, etc.) then folds must be considered in terms of fold facing. A fold **faces** in a direction normal to its hinge line within the hinge surface, and towards the younger beds (Figure 8.20). All the folds of Figure 8.20a are upwards facing which is where antiforms are anticlines and synforms are synclines. The converse arrangement is brought about by downwards-facing folds (Figure 8.20b).

Fold size is strongly controlled by layer thickness and the physical properties of the materials. Generally the thicker the layer the larger the fold. In a multilayer of different thicknesses and variable lithologies a wide range of wavelengths and fold shapes is to be expected. **Harmonic folds** occur where layers share the same wavelength and are in phase, that is, antiforms and synforms are stacked along planar hinge surfaces (Figure 8.21a). **Disharmonic** folds are created by different wavelengths (Figure 8.21b) or offset relationships between closures of the same sense.

Having dealt with many of the 'nuts and bolts' aspects of folding it is now possible to further develop the study of more complex shapes though still limiting ourselves to plane folds. The spectrum shown in Figure 8.22 is effectively between cylindrical folds and domes. Departures from cylindricity are measured by two parameters. Firstly the hinge angle which records how far the hinge line is from being rectilinear (Figure 8.22a). The hinge angle is defined within the hinge surface by tangents to the curved hinge line at its points of inflection. The second parameter is the degree of domicity. Ideal domes, as shown in Figure 8.22b, occur where the form surface has

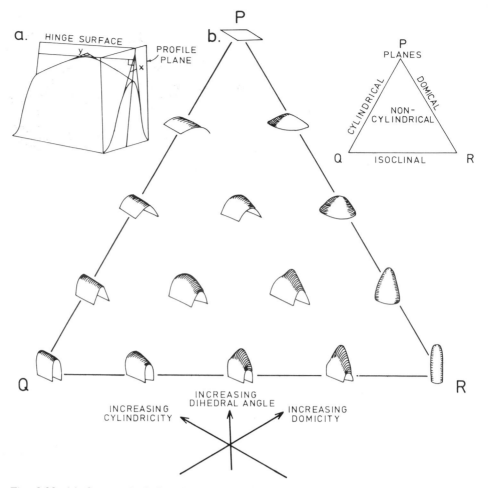

Fig. 8.22 (a) Geometrical description of a non-cylindrical plane fold; x=dihedral angle, y=hinge angle. (b) The range of shapes from cylindrical to domal which can be specified by PQR ratios

equal inclinations in all directions away from a reference axis. A full specification of domes has yet to be drawn up. Dictionaries typically define domes in terms of equal dips in all directions away from the core of a structure but, if the domes of Figure 8.22b were titled, their shape would not change but dips relative to the horizontal would. Some authors define domes as fold structures with an outcrop length less than three times their width — a not very specific approach.

8.2   MINOR STRUCTURES ASSOCIATED WITH FOLDING

During a deformation event that leads to folding many other structures may be generated. Synchronous development produces geometrical relationships between the structures which contribute substantially to the techniques of analysing folded areas. If several orders of folding are present it is usually only the smaller scales that can be observed in exposures especially in poorly exposed terrains. Also large folds,

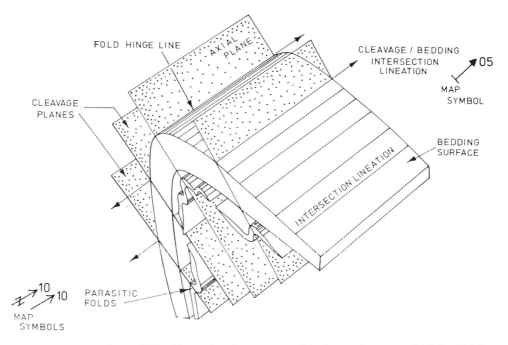

Fig. 8.23 A cylindrical fold with, at the closure, an axial planar cleavage which is slightly fanned on the limbs. The lineation on the form surface (sedimentary layering) is generated by the cleavage intersecting the bedding: for a cylindrical fold this lineation is parallel to the hinge line. In regular examples the hinge lines of the parasitic (minor) folds (symmetric and asymmetric) are subparallel to the major fold hinge line. The map symbols are illustrations of the style used; there are many variations around the world

up to the kilometric scale, may occur within thick, featureless, sedimentary sequences that cannot be subdivided into easily recognizable formations. In both these circumstances minor structures can help to locate and map out larger-scale folds. A standard procedure in the field is to measure the plunge and bearing of mesoscopic-fold hinge lines. Figure 8.23 shows the subparallel relationship between the parastic fold hinge lines and that of the major fold (**Pumpelly's rule**). This is the typical case but the rule does not always hold. With pronounced disharmony the more rigid dominant layers behave independently and both plunge and/or bearing of hinge lines may become quite variable. These departures are favoured by appreciable separation of the dominant layers by weak rocks such as shale. If minor fold hinges are reasonably consistent in attitude their mean value will be a good indication of the major hinge direction and plunge.

Under metamorphic conditions cohesive flow of rocks commonly produces planar fabrics. During strain-type transformations, platy particles are organized into preferred orientations, equant particles become deformed oriented tablets and planar zones of different composition are created by mass transfer; all of these elements are subparallel and give rise to a direction of weak cohesion in the rock. This tectonic fissility is **cleavage** or **foliation**. In a region that has had a simple deformation history (a **slate belt**) the geometric relation between cleavage and folding is usually straight-forward (Figure 8.23). The common situation is for the cleavage to be subparallel to the fold axial plane (generally itself close to the hinge surface in attitude). Taking any one small part of the fold form surface (bedding) we may calculate the attitude of the line of intersection between bedding and cleavage (intersection of two planes). For an approximately cylindrical fold this line is close to the orientation of the major fold's hinge line wherever it is measured on the form surface.

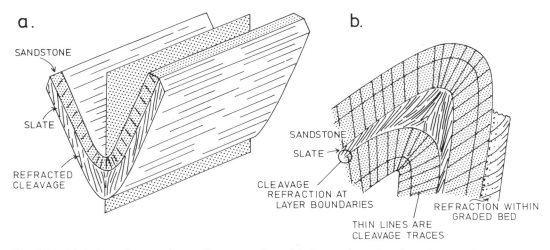

Fig. 8.24 (a) A slaty cleavage is usually nearer the axial plane orientation than cleavage in competent beds like sandstone where cleavage fans may be pronounced. The change in cleavage attitude from slate to sandstone is refraction. If the fans are symmetrical about the axial plane, then cleavage/bedding intersection lineations will still parallel the major hinge line. (b) Profile view of cleavage fans and refraction. In continuously graded beds (grain size variation), the cleavage will show a sympathetic curvature (refraction) such that the slaty part will be closest to the axial plane attitude

The proof of this relationship follows from a good understanding of fold geometry. A cylindrical fold is defined by a fold axis moving through space but always resting on the form (bedding) surface. It follows that any one planar segment of, or tangent to, the form surface contains a line parallel to the fold axis. The axial surface also contains the fold axis and the intersection of an axial planar cleavage, and any part of the form surface must, therefore, be this line they share. As with the definition of a fold axis, the intersection lineation is primarily concerned with orientation not position because in a cylindrical fold all such lineations will be parallel. Cleavage is commonly spaced at the millimetric or smaller scale and the cleavage/bedding intersection lineation in the field is defined by surface irregularity where weathering has etched out the weak cleavage planes. If the intersection lineation is reasonably well oriented over part of a map then it can be taken to approximate the macroscopic hinge line.

Several cautionary words have to be added. Even in simple areas with one dominant deformation event, quite an array of fold/cleavage geometries have been recorded. To have a planar cleavage cutting uniformly through all layers either requires very high strain or a sequence of very similar lithologies. In strongly contrasting multilayers, such as sandstone and shale alternations, cleavage will vary in attitude controlled by lithology (Figure 8.24) in a structure known as a **cleavage fan**, The illustrated case typifies the common situation where cleavage in the incompetent rock (e.g. slate) is subparallel to the axial plane and hence intersection lineations still give the overall hinge orientation. The fan is in the competent layers, diverging downwards in a synform and converging downwards in an antiform. The change in angle of cleavage from bed to bed is known as refraction. In a graded bed the change of cleavage orientation occurs within the single bed and is directly attributable to the change of composition. If a cleavage fan is symmetric about the axial plane, the cleavage/bedding intersections will parallel the hinge line of the fold. Still considering our simply deformed terrain, perhaps as many as 20 per cent of the folds have significantly non-axial planar cleavages and the folds are said to be **transected**. This gives rise to widely varying cleavage/bedding intersection lineations which hopefully will be obvious on a well-made map. Several other

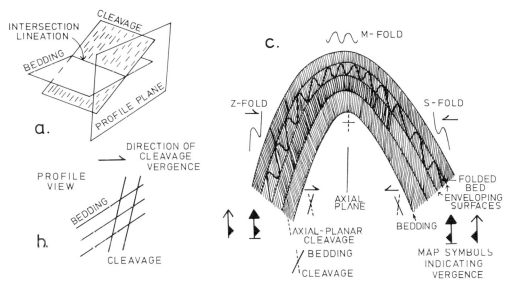

Fig. 8.25 (a) Cleavage vergence is defined in the plane (=profile plane) at right angles to the intersection lineation. (b) Cleavage is the younger fabric and the vergence is the horizontal direction that this fabric has to be rotated to become parallel to the older fabric (bedding). It has the same meaning as fold vergence in locating antiformal closures. (c) A large antiform with near-axial plane cleavage and parasitic folds. Cleavage/bedding angles vary around the fold which can indicate how close a small exposure is to the fold closure. Cleavage and fold vergence are opposite on the two limbs

variations on the fold/cleavage theme give rise to complex patterns. Be aware that the simple state shown in Figure 8.23 does not hold in every case. To prove whether or not a cleavage is axial planar requires that you know the attitude of the limb bisector. This is not as easy as it seems as will be discussed later.

With a cleavage that is approximately axial planar the angular relations between bedding and cleavage may be used to determine if an exposure is from a fold limb or the closure. A near 90° angle between bedding and cleavage is found at the hinge, whereas the lowest dihedral angle (measured in profile plane) represents a reading on the fold limb (Figure 8.25c). Some complex cleavage patterns have extreme cleavage fans that wrap around fold closures giving low bedding/cleavage angles at the hinge. Fortunately these are fairly rare. Returning to the axial planar situation, bedding/cleavage relations are used in a similar way to fold vergence to determine which fold limb is being examined. **Cleavage vergence** is defined by studying the cleavage/bedding relations in the profile plane, that is, perpendicular to the intersection lineation (Figure 8.25a). The vergence in this plane is the horizontal direction towards which the cleavage (the younger fabric) needs to be rotated so that it becomes parallel to the bedding (the older fabric) (Figure 8.25). The rotation should be through the acute angle not the obtuse. A logical outcome of the definition is that the vergence direction is at right angles to the trend of the intersection lineation or fold hinge line. Fold and cleavage vergence both point towards the nearest antiform and a change of vergence means that you have crossed either an antiformal or synformal closure. For a complete fold the directions of cleavage vergence are opposite for the two limbs and a map recording vergence should readily distinguish outcrops belonging together on one limb.

The primary advantage of recording vergence is that for one locality the vergence direction is constant no matter which way you look along the fold hinge line or cleavage/bedding intersection lineation. Instead of cleavage vergence some authors

124 Continuous deformation

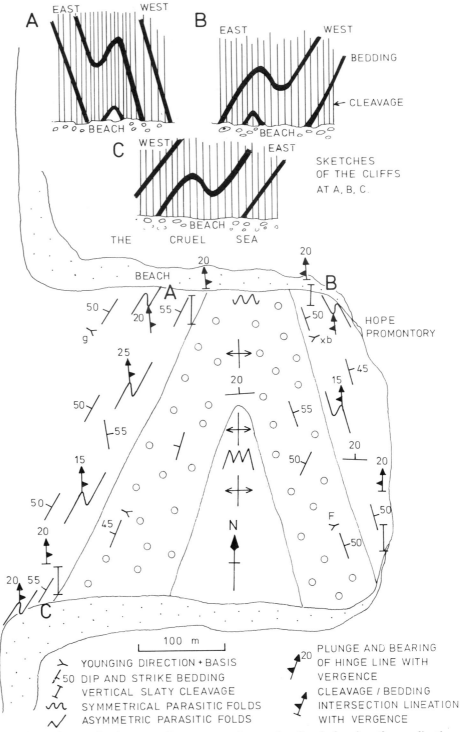

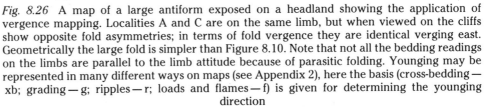

Fig. 8.26 A map of a large antiform exposed on a headland showing the application of vergence mapping. Localities A and C are on the same limb, but when viewed on the cliffs show opposite fold asymmetries; in terms of fold vergence they are identical verging east. Geometrically the large fold is simpler than Figure 8.10. Note that not all the bedding readings on the limbs are parallel to the limb attitude because of parasitic folding. Younging may be represented in many different ways on maps (see Appendix 2), here the basis (cross-bedding — xb; grading — g; ripples — r; loads and flames — f) is given for determining the younging direction

recommend that the relations be recorded in terms of the bedding being clockwise (left limb, Figure 8.25c) or anticlockwise (right limb, Figure 8.25c) from the cleavage. This suffers the same ambiguity as noting fold asymmetry in terms of S or Z because it depends on viewing direction and, even if down plunge viewing is adopted, regional variations in plunge would produce a complicated map pattern from just a single limb (cf. Figure 8.10). A large north-plunging antiform on a headland shows the usefulness of vergence (Figure 8.26). Minor folds sketched on the cliffs at locality A have an S shape but the equivalent folds on the same limb observed on the cliffs at locality C have Z shape. Both fold vergences, however, are to the east and this is shown on the map by a mark on the symbol for the hinge line plunge and bearing. Likewise cleavage vergence for the most common bedding attitude at A and B is also to the east. The eastern limb has opposite vergence clearly defining the closure of the major fold.

8.3 OUTCROP PATTERNS OF FOLDS

The basic principles relating to fold outcrop patterns will be introduced using simplified examples. Initially we shall deal with maps of plane horizontal surfaces to avoid the complications derived from the variable shape of the earth's surface. It should be noted that the same effect is gained from studying structure contours where the scale of the map is appropriate for this style of construction. Secondly, by working with angular to subangular folds, we shall temporarily postpone much of our encounter with the problems caused by random slices through curved surfaces, i.e. map views of rounded to subrounded folds. Maps give only profile views where the folds are vertically plunging. All other hinge-line attitudes make maps produce distorted views of folds. The degree of distortion is a function of the angular difference between the map and the profile plane orientations, reaching a maximum for non-plunging folds. On a horizontal surface, such a fold merely consists of parallel straight outcrop boundaries (Figure 8.27a). All horizontal folds have this style of outcrop pattern no matter what the inclination of their hinge surface. In contrast plunging folds create patterns of converging and diverging outcrops (Figure 8.27b). A **very important rule** is that outcrop patterns converge in the direction of plunge of an antiform and diverge in the direction of plunge of a synform. These patterns are related only to the sense of closing (geometry) of the layers and are not a function of stratigraphic relations. Once the stratigraphy is known it is possible to determine fold facing and label the folds in terms of anticline and syncline (Figure 8.27b, c). An upwards-facing antiform has the oldest rocks in the core and is an anticline, whereas an associated synform would have the youngest rocks in the core making it a syncline. The converse applies to downwards-facing folds.

The simplest way to generate downwards-facing folds is to refold an earlier set of folds. One example is given in Figure 8.28 where a recumbent fold has been refolded by a set of upright structures. The first-phase recumbent fold has caused a repetition of the stratigraphy and is an anticline because it has the oldest rocks in its core. Second generation folds above the first fold hinge surface are upwards facing, but where erosion has cut below this surface downwards-facing folds are exposed. Even though we are dealing with a relatively simple deformation history,

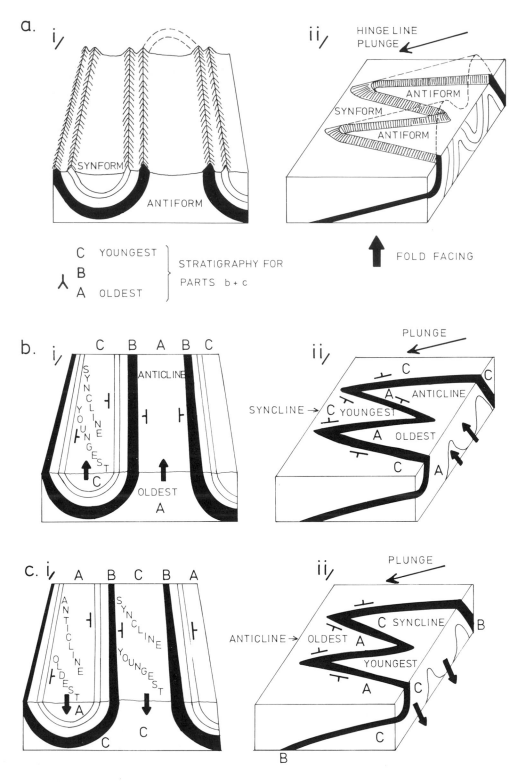

Fig. 8.27 (a) (i) Outcrop pattern of non-plunging folds with parallel strikes to the limbs; (ii) outcrop patterns of plunging folds — convergence in the direction of plunge for an antiform and divergence for a synform. (b) Same geometry as (a) but with upwards-facing folds such that the antiforms are anticlines. (c) Same geometry as (b) but with downwards-facing folds such that antiforms are synclines. (Reproduced by permission of Harper Row Inc)

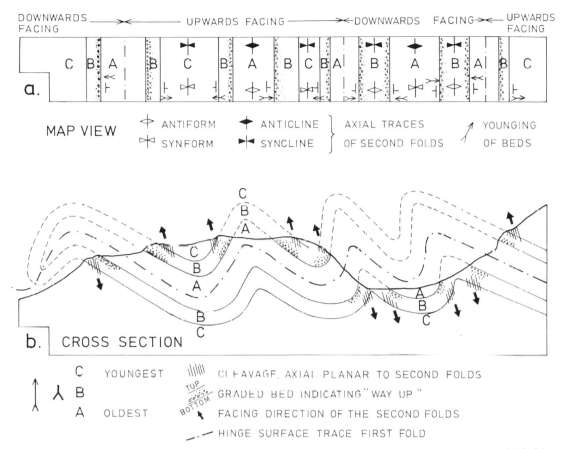

Fig. 8.28 Cross-section (b) and map (a) of a recumbent anticline (oldest rocks in the core) refolded by upright folds. The later folds are upwards facing above the hinge surface of the recumbent fold, but become downwards facing once this surface is crossed. Single synforms can be traced down from synclines to anticlines. The map view can be fairly involved but careful analysis of fold facing is the basic technique. Note the different symbol for bed younging (cf. Figure 8.26)

the map symbols required are quite complex. We have to distinguish between folds of different generations and between upwards- and downwards-facing folds within each generation. The symbols shown represent but one of the many published schemes. The terms antiformal anticline and synformal syncline are cumbersome and some shorthand is allowed. If your analysis of a map shows that all folds are upwards facing then this should be stated in the introduction to the section on fold description and readers will then know that reference to an anticline implies it is also an antiform.

For a constantly oriented fold with well-developed planar limbs it is a simple matter to calculate the orientation of the hinge line. Because we are initially confining ourselves to plateau-like areas of topography we need to construct structure contours using dip data from the fold limbs. In Figure 8.29 structure contours are drawn for the top surface of the black formation. The 500 m contours (the plateau height) are extended from the strike of the two limbs and, where they intersect (H_1), an imaginary hinge point is defined. This point does not lie on the form surface because the fold is not completely angular. Similar hinge points are constructed for the 400 and 300 m structure contours and H_1—H_3 is 165 m; in this distance the 'hinge line' falls 200 m hence the plunge is 50°. The bearing is towards 320°, thus the hinge-line attitude is 50 → 320. If a varied topography had been

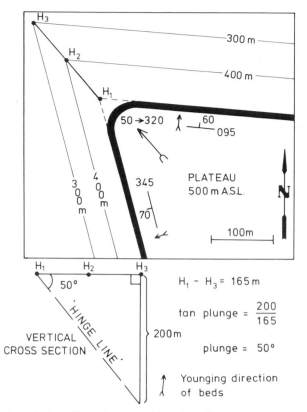

Fig. 8.29 Calculation of hinge line plunge and bearing. Construct structure contours for the two limbs and, where like values cross, a hinge point is defined (e.g. H_1, H_2, H_3). The join H_1 to H_3 is the vertical projection of the hinge line and a vertical cross-section in this direction (or trigonometry) gives the plunge angle. This fold has an 85° dihedral angle yet the angle between the map traces of the fold limbs is 70°. The fold is an antiformal anticline, both limbs are right-way-up and the axial surface has a dip and strike of 84/132

present then the outcrop pattern would have provided several structure contours for each limb and a similar construction could have been made.

The fold of Figure 8.29 is a fairly simple antiform with limbs dipping away from each other. If the fold was also upwards facing it would be an anticline and both limbs would be stratigraphically right way up. This arrangement is largely dictated by the steep dip of the axial surface (85°). By reducing the dip of the axial surface, eventually a situation is reached where one limb becomes vertical and, with more rotation, is inverted. At this point the two limbs have opposite senses of younging, one is stratigraphically right-way-up and the other is younging downwards. In Figure 8.30 the fold of Figure 8.29 has been rotated about the strike of its axial surface to give a dip to the latter of 35°. This has also achieved, in what is still an antiformal anticline, limbs with opposed younging; the limb striking 048° is right-way-up but the 168° striking limb is inverted. Had the fold been tighter the limbs would have been dipping in the same general direction. Once one limb has been rotated through the vertical, structure contour construction for finding the hinge line is more awkward looking (Figure 8.30). The structure contours now cross over one another but the principle of finding imaginary hinge points is exactly the same.

Once some part of the form surface becomes vertical (which includes all folds with one limb rotated through the vertical) there is a rapid method of estimating hinge-line attitude. Given some rounding of the fold and enough dip and strike readings from the hinge zone, the attitude of vertical beds and those striking at right angles are used

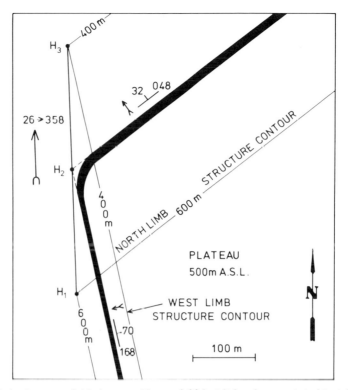

Fig. 8.30 This is the same fold shape as Figure 8.29 but it has been rotated to take one limb through the vertical to become stratigraphically inverted: the fold still faces upwards. The hinge line is calculated as before but the structure contours cross over giving an apparently more complicated construction

in combination. Vertical or near-vertical beds strike in the direction of the hinge bearing. Beds with a perpendicular strike have a dip that is equal to the plunge of the hinge line (Figure 8.31). A cursory examination of Figure 8.31 leads many to suggest that the fold (an antiform) is plunging towards the north-west whereas its hinge line actually is on a bearing of 016°. This is a demonstration of the apparent dip relation between hinge surface and hinge line.

Any variation in hinge-line attitude can be monitored by studying pairs of vertical and orthogonal beds remembering that this applies only to folds with overturned limbs. Also any change in the outcrop pattern (independent of topography) of a folded surface will show the folds to be non-cylindrical. In a detailed description, each change of pattern should be measured to quantify the variation in hinge-line attitude. For a more general description, estimates may suffice. Figure 8.32 shows a series of culminations and depressions in a group of non-cylindrical folds and how these affect the outcrop pattern. Note how the rules derived in Figure 8.27 may be applied to individual segments that approximate cylindricity to relate the outcrop pattern to plunge direction.

Folds do not go on for ever with the same geometry. Undulating hinge lines is one example of variability but an equally common situation is the lateral termination of a fold (block diagram, Figure 8.33). Even in the relatively simple fold structures shown on the map of Figure 8.33 there are many laterally impersistent folds. The clearest example is the synform in the Shenandoah Mountain which when traced south-westwards passes through what must be a non-folded planar zone into an antiform (note all folds face upwards in this region). Three-dimensional visualization of these situations may appear difficult at first, but you need this capacity if your mapping and

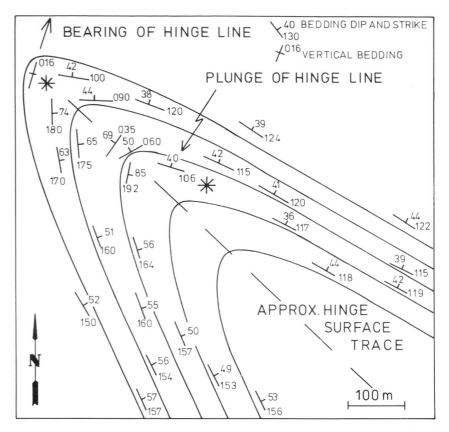

Fig. 8.31 For folds that include vertical beds, the hinge-line orientation can be rapidly assessed. The strike of the vertical bed gives the bearing of the hinge line; the dip of a bed striking at right angles is the plunge of the hinge line

map analysis is to go beyond first base. Another problem in analysing geometry as shown on maps like the Williamsville Quadrangle, is the efficient selection of data. Fold distribution is mainly defined by the outcrop pattern and, where the latter is complicated, detailed study is required to see if this is a true reflection of fold geometry or if it is largely brought about by topography. Further analysis of fold characteristics requires that limb dips be determined. On a small-scale map this is normally dependent on readings placed on the map. In lithologies with abundant parasitic folds the bedding readings may be very variable and identifying those related to the overall dip of the limb may be a difficult task. For example, even in a very simplistic exercise like Figure 8.26, several bedding readings were from common limbs of asymmetric fold couples; if this were not realized an incorrect impression of the fold geometry could result.

By comparing Figures 8.29 and 8.30, some idea may be gained of the distortion introduced by map views of plunging folds. Both of these folds have the **same dihedral angle** yet the 'apparent dihedral angles' on the maps are very different. Stereographically it is a simple matter to measure the dihedral angle between two planes (see Appendix 1) but without this technique it is more complicated. A rather naive approach to the problem is to just consider the dips of the limbs which for Figure 8.29 would give a guesstimate of 50°, much less than the true 85°. The best geometric method, in the absence of stereographic methods, is to construct a profile view of the fold which is the equivalent of looking end on to the cylinder in Figure 8.1, section (a). By tradition such sections are viewed down the plunge to

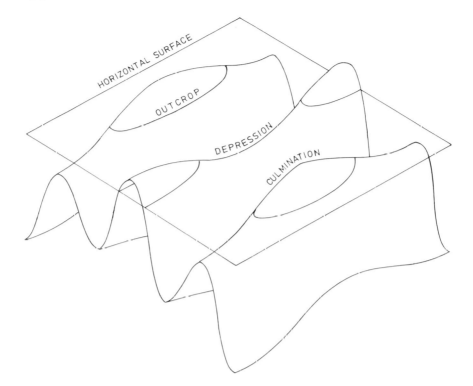

Fig. 8.32 Even in tectonically fairly simple areas fold hinge lines may be noticeably curved. A hinge-line high point is a culmination producing a domal shape and a low point is a depression giving rise to a basin. Note that outcrop patterns are closed elliptical shapes

provide a common basis for comparison. For a plane horizontal topography the construction is relatively easy. The projection direction from the map to the profile is the fold axis/hinge line (Figure 8.34). The projection method does not change dimensions perpendicular to the bearing of the hinge line but a foreshortening occurs along the hinge direction. If you sight the map down the hinge line you should get some appreciation of how the profile will turn out. It may not be necessary to produce a profile of the whole of a fold (or map area) and in the case of Figure 8.29 once part of the limbs have been plotted their planar nature means that the remainder may be drawn in easily. To construct the profile, a square grid is drawn over the map oriented with a base line perpendicular to the bearing of the hinge line (Figure 8.35). Looking down the plunge of the fold a foreshortened grid is constructed below the square grid. The amount of foreshortening in the hinge direction is by a factor sine of the plunge angle (Figure 8.34), so that all dimensions parallel to the hinge/axis orientation are reduced by this factor. Once constructed the profile is normally presented such that the strike of the profile runs horizontally across the page (Figure 8.35). Because the profile is an inclined section its attitude (perpendicular to the hinge line) should be clearly marked to indicate that it is not a standard vertical section. A square grid was recommended for this exercise but the spacing of the grid lines may be varied to suit different circumstances.

Once the profile has been constructed, the precise dihedral angle may be measured directly which for Figures 8.29/8.35 is 85°. Without such techniques students have been known to clutch at straws and even use the difference in strike between the limbs to spuriously calculate the dihedral angle. This approach may be easily falsified by considering an open non-plunging fold which has limbs of parallel

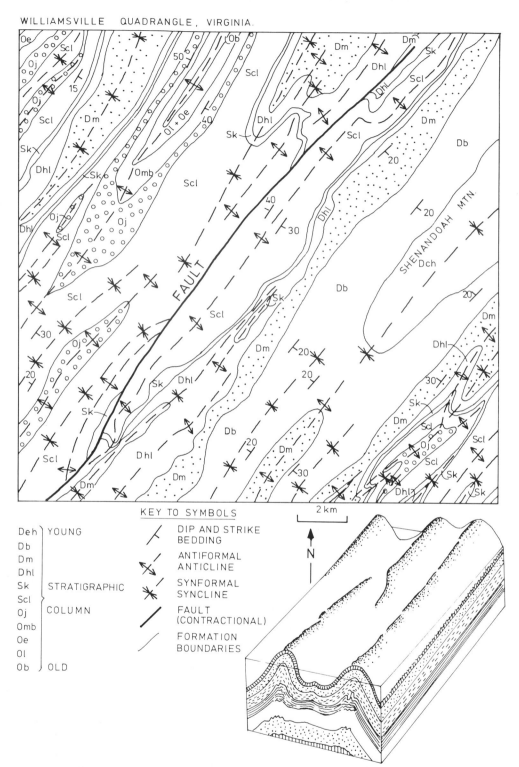

WILLIAMSVILLE QUADRANGLE, VIRGINIA.

KEY TO SYMBOLS

Deh ⎤ YOUNG
Db
Dm
Dhl
Sk   STRATIGRAPHIC
Scl
Oj   COLUMN
Omb
Oe
Ol
Ob ⎦ OLD

╱⊤ DIP AND STRIKE BEDDING

⟿ ANTIFORMAL ANTICLINE

✳ SYNFORMAL SYNCLINE

━━ FAULT (CONTRACTIONAL)

╱ FORMATION BOUNDARIES

2 km

N

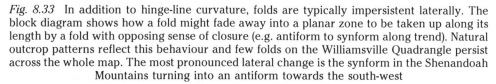

Fig. 8.33 In addition to hinge-line curvature, folds are typically impersistent laterally. The block diagram shows how a fold might fade away into a planar zone to be taken up along its length by a fold with opposing sense of closure (e.g. antiform to synform along trend). Natural outcrop patterns reflect this behaviour and few folds on the Williamsville Quadrangle persist across the whole map. The most pronounced lateral change is the synform in the Shenandoah Mountains turning into an antiform towards the south-west

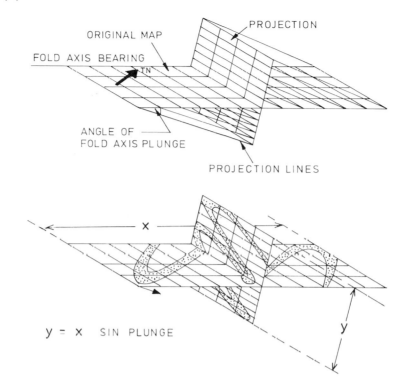

ORIGINAL MAP

FOLD AXIS BEARING

PROJECTION

ANGLE OF
FOLD AXIS PLUNGE

PROJECTION LINES

y = x SIN PLUNGE

Fig. 8.34 Map views of folds are converted into profile plane views by projecting data up and down the orientation of the fold hinge line. The map is then foreshortened when seen in the profile view by a factor sine of the plunge angle

strike. In Figure 8.35 the difference between the map and the profile may not seem great: the difference between the angle defined by the strike of the limbs (70°) and the dihedral angle (85°) is not great. However, with some geometries the profile is very different to the map. Also Figures 8.29 and 8.30 have the same dihedral angles but the angles between the limb strikes are very different.

Without the use of stereographic projection it is also troublesome to calculate the attitude of the axial surface or hinge surface. One approach is to use the fact that the hinge line rests on the hinge surface/axial surface; the hinge line is an apparent dip of the axial/hinge surface. If we can determine two apparent dips on the axial surface then its dip and strike can be calculated. For angular to subangular folds, the trace of the axial surface may be drawn on the map from an inspection of the outcrop pattern (Figure 8.36). In a plateau-like region this line on the map gives the orientation of the strike, that is, it is an apparent dip of zero, pitching at 0°. The first step in the method (Figure 8.36) is to draw on the map the axial surface trace joining points of maximum curvature on successive folded layers. **Note;** with other than angular/subangular folds these points are not on the hinge or axial surface if the folds are inclined. This trace should approximately bisect the angle between the limbs for tight folds. From the point where this trace crosses the boundary of one of the folded surfaces draw a line parallel to the bearing of the hinge line. On the projection of the hinge mark off points representing a fall of 100 m, 200 m and 300 m using the map scale and the fold plunge. Through each of these points draw structure contours parallel to the axial surface trace and by measuring the dip at right angles to these contours the attitude of the axial surface is determined. This method will give an exact answer for angular folds, it is an approximation for subangular folds, but might be very misleading for rounded structures. The

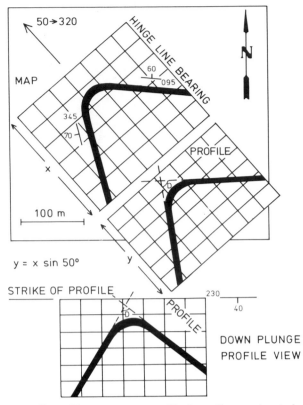

Fig. 8.35 In topographically uniform areas, a profile is easily constructed. Place a square grid over the fold with its edge oriented along the trend of the fold hinge line. Construct a foreshortened grid such that $y=x$ sine plunge and then transfer equivalent points from the map grid. On the profile the true dihedral angle may be measured and the true stratigraphic layer thickness around the fold may be studied. The profile should be clearly oriented because it is not a standard vertical section. Traditionally profiles are presented as down-plunge views

construction of Figure 8.36 gives an axial surface attitude of 44/139, whereas stereographic methods give 46/139. Very close agreement!

Having learnt profile construction we are now in a position to look at some of the problems posed by rounded and subrounded folds. On inclined folds, with any degree of rounding, the crest/trough points do not coincide with the hinge points (Figure 8.37). The departure between these two points increases with more rounding of the fold shape and a lessening of the axial surface dip (dihedral angle is a factor as well). On a map, the outcrop of a folded surface shows its points of maximum curvature at the crest or trough point (cf. the various slices through an inclined cylinder, Fig. 8.1). Hence with inclined and rounded folds there may be a considerable distance between the hinge-point outcrop and the maximum curvature of the form surface on the map (see Figure 8.38). Figure 8.38 was deliberately chosen to highlight these problems. The fold plunges 11 → 360, it is isoclinal with limb dips of 30/160. Note that even with a plunge to the fold, the strikes of the limbs are parallel, this is dictated by their parallel relationship.

All aspects of fold outcrop patterns on plane surfaces may be demonstrated with a simple experiment. Using a folded aluminium sheet (or similar material) in a basin of dyed water, the outcrop pattern for a variety of fold shapes and attitudes may be reproduced. I have improvised with a student's folder and a laboratory sink when one student had a mental block over visualizing the outcrop of a plunging angular fold.

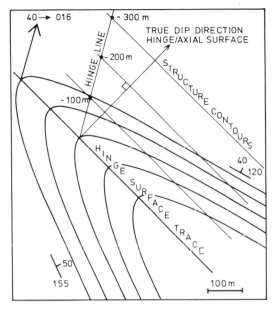

Fig. 8.36 Axial surface altitude is very easily calculated using stereographic methods but is troublesome without this technique. The alternative is a 'two apparent dips' exercise. The hinge line is one apparent dip and the other is the strike of the axial surface (a non-plunging apparent dip), which for angular folds is well defined but for rounded folds may be hard to locate (see Figure 8.38). From one hinge point on the hinge surface trace calculate the horizontal equivalent of dropping 100 m, 200 m, and 300 m down the hinge line. Then, through each of these points, draw structure contours parallel to the hinge surface trace. The dip of the axial/hinge surface may then be calculated

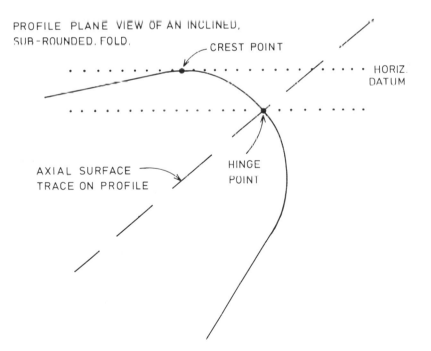

Fig. 8.37 A profile view of an inclined plunging subrounded fold. The crest point is clearly a large distance from the hinge point. Crest/trough points on the map are the location of the points of maximum curvature and thus the 'apparent axial/hinge surface' trace on the map may have nothing to do with the position and/or orientation of the true hinge surface trace

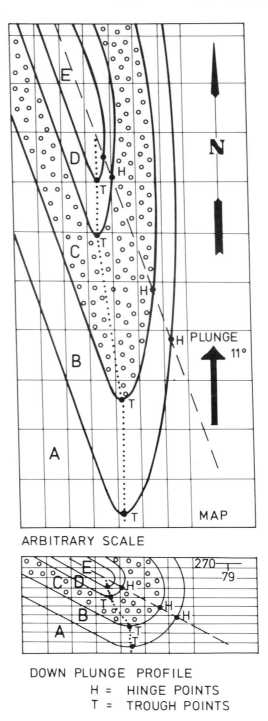

ARBITRARY SCALE

DOWN PLUNGE PROFILE
H = HINGE POINTS
T = TROUGH POINTS

Fig. 8.38 An isoclinal plunging synform where layers B and D have constant stratigraphic thickness. Here the hinge and axial surface are parallel and on the map may be some considerable distance from the points of maximum curvature of the formation boundaries defined by trough points. Also note the constant curvature on the profile view, but variable curvature on the map (cf. Figure 8.1). Hold a pencil in the hinge-line orientation $11 \rightarrow 360$ over the map and view along this direction which should give a good impression of the profile view.

The low angle plunge considerably distorts the fold when presented as a map view

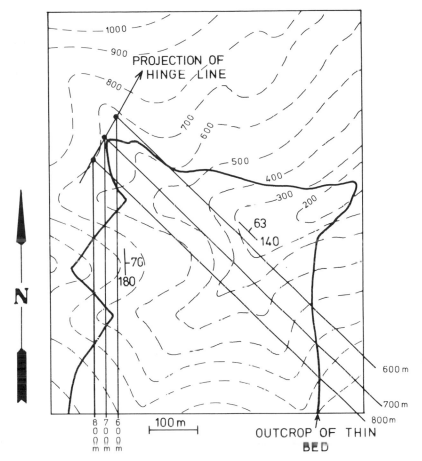

Fig. 8.39 On large-scale maps, structure contours should be constructed, in the analysis of folds, to remove the topographic influence on outcrop pattern. The structure contour pattern is equivalent to outcrop on a horizontal surface, and hinge-line attitude is easily calculated as well as sense of closing

If topography significantly influences the outcrop pattern of a fold then structure contours have to be drawn. By ignoring the topographic and outcrop information, the structure contours give the shape of the folded surface and its attitude. The major problem with abstracting this information is the initial step of recognizing fold limbs and distinguishing between fold closures and V patterns made by the interaction of planar fold segments and valleys/ridges. Once the fold style has been identified, structure contour construction should reveal coherent fold structures unless a complex deformation history has been involved. Figure 8.39 illustrates the outcrop pattern of an angular fold on a relatively large-scale map. Here the topographic variations significantly influence the outcrop of the two fold limbs which have been identified by structure contours. The same construction leads to the hinge-line attitude being determined (62 → 025).

9 DISCONTINUOUS DEFORMATION— FAULTS

9.1 ABSOLUTE BEGINNINGS

Many of us have seen the T-shirt with the logo 'The San Andreas — it's not my fault!'. This sentiment reflects the origin of the term fault referring to an imperfection in the rock mass. To the early coal-miners faults were troublesome defects but for some minerals (gold especially) localization by faults into economic deposits is common and such 'defects' are actively sought by exploration programmes. The simplest expression of a fault is a clean break (fracture) which represents an abrupt jump in the displacement fields (see Fig. 7.2b). The two blocks on either side of the fault have undergone different amounts of displacement and, in contrast to folding, a plot of displacement vectors shows sharply defined discontinuities. If a fault cuts a sequence of bedded sediments, the typical expression of this style of deformation is the disruption or offset of once continuous layers (Figure 9.1). Naturally the example chosen aims to be as graphic as possible which for a map view occurs where the fault trend is not too far from the dip direction of the tilted sediments — a **dip fault**. Though Figure 9.1 gives some 3-D feeling for the fault geometry, our observations are still constrained by a not too irregular earth's surface. Such limits to our visualization tend to focus our attention on the most obvious offset which is parallel to the strike of the fault. Perhaps because the 2-D image is concrete, its strength overpowers our capacity to see the full 3-D geometry which requires an image to be created in our minds. This polarization of observations (into the plane of the map) leads to many erroneous conclusions about fault displacement. Many observers in looking at situations like Figure 9.1 leap in and state that the differential displacements took place in the horizontal plane whereas the displacement vectors could be vertical! The difference between offset geometry and displacement directions escapes many people even after studying diagrams of the greatest clarity and receiving intensive tutoring. Clearly the subject-matter is not easy and requires some attention though the basic principles are not intractable. Fault analysis must be carried out in two very distinct phases:

1. Geometrical description of the faulted situation.
2. Attempts to determine the displacement responsible for the observed offsets.

Phase one must be done impassively as a purely descriptive exercise and any thought of displacement direction, sense of movement, etc., could prejudice the next stage. Only when a complete description has been carried out are you in a position to consider the differential displacement (the **kinematics**).

The world of fault analysis is something of a terminological jungle and it is already time to take the plunge. Fortunately, during fault movement the two blocks typically stay in contact (or close to it) which means that displacements are constrained to be

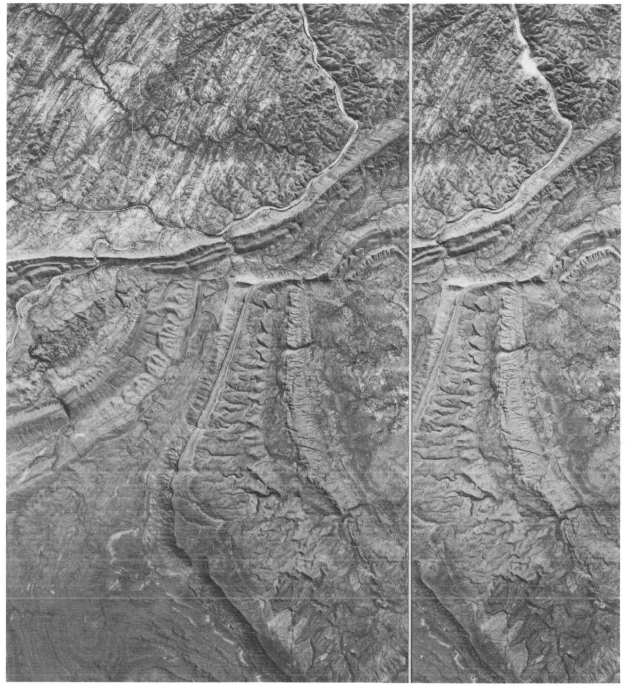

Fig. 9.1 (Caption on following page.)

oriented within the plane of the fault. This means that we are in the business of studying relative translation between adjacent blocks (Figure 9.2a), rigid body rotations (Figure 9.2b, c) where the rotation axis is the normal (perpendicular) to the fault, or a combination of the types of displacement. Translations are fairly straightforward to quantify, but rotations are a little more involved and will only be treated qualitatively in this text. Rotations are easily recognized because equivalent

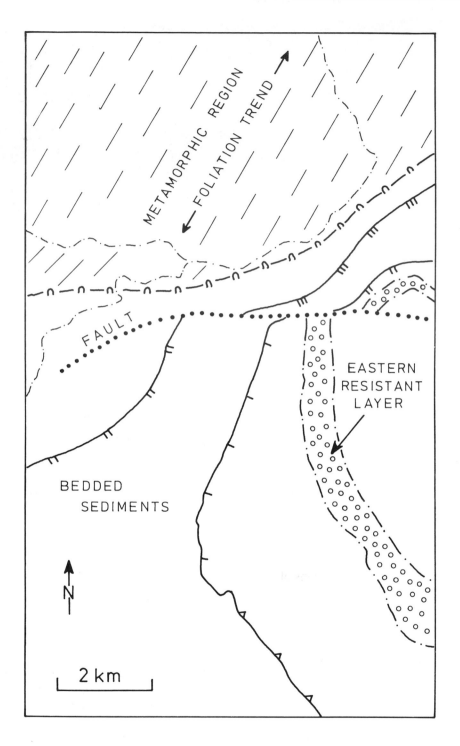

Fig. 9.1 Stereopair of vertical air photographs showing a fault trending generally down the dip of a bedded sequence. The easternmost resistant layer is clearly offset parallel to the strike of the fault. Layers lower down in the stratigraphy have responded by a mixture of bending plus fracturing (cf. Figure 7.3b). Note that the fault when traced west curves through 40°. In the northern third of the image, a metamorphic region shows a consistent north-north-east— south-south-west trend of foliation and layering. This basement trend is truncated by the bedded sequence at an angular unconformity (see Chapter 11 and Figure 6.1)

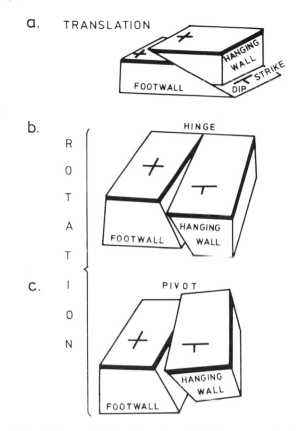

a. TRANSLATION

HANGING WALL

FOOTWALL

STRIKE

DIP

b. HINGE

ROTATION

FOOTWALL

HANGING WALL

c. PIVOT

HANGING WALL

FOOTWALL

Fig. 9.2 In translation (a) pre-faulting layering attitudes are not changed. Rotational displacements (b and c) produce differences in attitude in adjacent fault blocks in once parallel sequences. Hinge behaviour (b) occurs at the termination of a fault whereas pivotal displacement requires more complicated accommodation. Planar sections of faults can have their attitude specified in terms of dip and strike in the same way as layering. The fault block on top of the fault surface is the hanging wall and the block below is the footwall. A vertical
     drill in the hanging wall would eventually pierce the fault and then enter the footwall

portions of an offset surface in the two blocks will have different attitudes (Fig. 9.2b, c). If you are studying pre-faulting folds, make sure you compare the same fold limb in both blocks; comparisons that cross hinge surfaces in moving across the fault will lead to error. Translation does not cause a difference in attitude of a pre-faulting surface from block to block.

Many aspects of fault terminology are referred to the attitude of the fault which commonly is planar or only slightly sinuous or can be considered in approximately planar segments. Given sufficient relief or underground information, structure contours can be drawn for the fault surface and its dip and strike calculated. Exposures may also allow direct measurement of the fault attitude. For faults other than vertical the **footwall** is the block below the fault surface whereas the **hanging wall** is above (Figures 9.2 and 9.3). A bedding plane or formation boundary intersecting a fault generates a line of intersection (Figure 9.3) known as a **cutoff**. Where the surface from the footwall meets the fault the trace created is a **footwall cutoff** and if the surface comes in from above the fault it forms a **hanging wall cutoff** (Figure 9.3).

It is sometimes useful to be able to describe the geometrical relation between a fault and the sequence of rocks it affects. For an evenly inclined layered sequence:

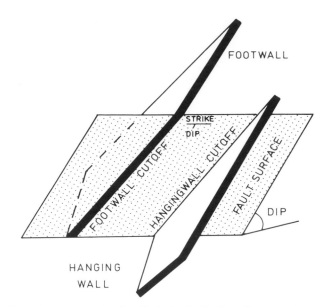

Fig. 9.3 If a sedimentary succession is offset at a fault, then the same layer on either side of the fault generates two linear intersections with the fault. The surface in the footwall below the fault creates a footwall cutoff which is the trace of the surface on the fault. From the hanging wall, a hanging wall cutoff is produced. Note that for all but thin layers, the upper and lower surfaces of a layer/formation should be dealt with separately

1. A **dip fault** has a strike close to the dip direction of the layering (approx. within ±10° of the dip direction).
2. A **strike fault** has a strike close to the strike of the layering (±10°). *Note:* some authors restrict this term to exactly strike parallel faults.
3. An **oblique fault** has strike relationships to the layering intermediate between strike and dip faults.

9.2 DESCRIPTION = SEPARATION

Offset is commonly used as a very general term for the disruption of what was once clearly a continuous surface and could equally apply to an igneous contact, an unconformity, an earlier fault, a vein or imaginary features such as hinge surfaces. **Separation** (Figure 9.4) is a more closely defined term for describing offsets being:

'the distance between the displaced parts of any recognizable geological surface which has been offset along a fault, measured in any **specified** direction'.

Separation only has meaning if a direction of measurement and a surface are specified. Note that for a unit of appreciable thickness only one bounding surface (upper or lower) should be considered — if not you will be comparing chalk and cheese!

Strike separation is measured along the fault strike (S, Figure 9.4).

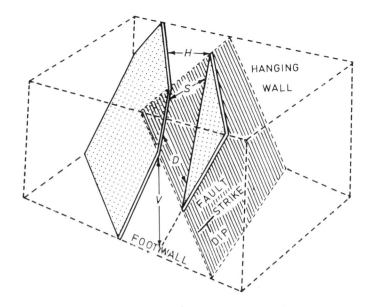

Fig. 9.4 The description of the effects of faulting on a layered sequence is based on offset (separation). A single offset surface can be described in a variety of ways depending on circumstances. The strike separation is the distance between the displaced parts measured parallel to the strike of the fault. Dip separation is measured down the dip of the fault. Horizontal separation is measured within the map plane perpendicular to the strike of the beds. Vertical separation is the distance between the same surface as seen in a vertical drill-hole or shaft. Perpendicular separation (not shown) is measured perpendicular to the offset layering. Typically both dip and strike separations are quoted and the others are referred to where appropriate. In addition to giving the amount of dip and strike separation, their 'sense' needs to be quoted. Here the hanging wall cutoff is below the footwall cutoff so the dip separation sense is normal (the opposite is reverse). Standing in one block where the layer touches the fault, and looking to the other block, the trace of the offset plane is reached by moving horizontally (along the fault strike to the left. This is left strike separation

Dip separation is measured down the fault dip (D, Figure 9.4).

Vertical separation is seen in a vertical shaft or borehole or in a vertical cross-section (V, Figure 9.4).

Horizontal separation is measured at right angles to the strike of the faulted surface within a horizontal plane (H, Figure 9.4). Some authors have this alone as their definition of offset.

Perpendicular separation is the shortest or perpendicular distance between the two parts of the faulted surface. This becomes the **stratigraphic separation** if the faulted surface is bedding.

In addition to an amount of separation, the dip and strike components are further qualified in terms of **sense** of separation.

Normal dip separation — the hanging wall cutoff is below the footwall cutoff (Figures 9.3, 9.4 and 9.5).

Reverse dip separation — the hanging wall cutoff is above the footwall cutoff (see, for example, Figure 9.15b).

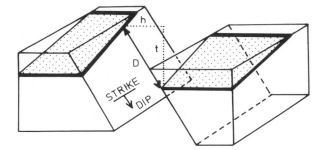

Fig. 9.5 The dip separation has vertical (*t*) and horizontal components (*h*) which can easily be calculated by trigonometry if the fault dip and the dip separation are known. Alternatively the components can be measured in a vertical section perpendicular to the fault strike (a dip section). In older texts (*t*) is referred to as the throw and (*h*) as the heave.

Left strike separation — standing at the termination of an offset marker on one block, the trace of the same surface is found to the left (Figure 9.4).

Right strike separation — the trace of the marker is found to the right across the fault (see, for example, west limb of Figure 9.7).

In each category of separation, amounts will vary if the fault is rotational or if the fault and/or offset surfaces are non-planar and such situations would require much measurement for complete specification. Before attempting detailed analysis, consider the cost/benefit equation and the requirements of the task you have been set. These might vary from writing a brief account of a regional map down to detailed underground mapping to plan an expensive drilling programme.

In terms of mine economics, offsets on faults can cause much extra development work in non-paying ground, particularly if mine levels have to be changed frequently. If the black layer in Figure 9.5 were a coal seam, the geometry after faulting involves a zone where no coal occurs. Access tunnels would have to be driven through this barren area for no income — too much of this type of activity leads to an uneconomic pit. The barren zone is equal to the horizontal component of the dip separation (*h* in Figure 9.5) sometimes referred to as the **heave**, but as this term is variably defined by different authors the more long-winded term is preferable. The vertical component of dip separation (*t* in Figure 9.5) is the **throw** of some writers.

With the above information you are in a good position to decide whether or not a fault is translational or involves some rotation, and also to describe the geometry created by differential displacement. It should be noted that the terminology is starting to pile up in potentially awkward ways. We can already contemplate a translational dip fault with 100 m of normal dip separation and 200 m of left strike separation of Formation X. Given the dip and strike of the fault and the bedding, we have a good description. The next section delves into how this disposition came about.

9.3 DISPLACEMENTS ACROSS FAULTS

Rather than dealing with the absolute displacement of each block (which is normally indeterminable) we have much more chance of calculating the relative movement.

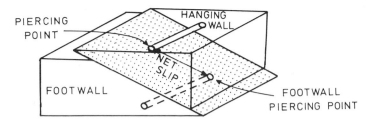

Fig. 9.6 Relative displacement between adjacent fault blocks (net slip) can only be quantified (amount and orientation) if originally adjacent points can be identified. This is best done with once continuous linear features; the disrupted parts in both the hanging wall and footwall pierce the fault at points. The join between these points measured on the fault is the net slip. The linear feature in practice is usually a fold hinge line or an intersection between two pre-faulting surfaces. If, as shown, the hanging wall piercing point is above the footwall equivalent, the dip-slip sense is reverse

This can **only** be measured if we are in a position to identify two **points** on opposite sides of the fault surface that used to be next to each other prior to faulting. '**Net slip** is the distance, measured **on** the fault surface between two originally adjacent points situated on opposite sides of the fault' (Figure 9.6).

There are a number of geological features that can define originally adjacent points, but they are not common and hence it is unusual to be able to fully specify the relative displacement vector between two fault blocks. The standard requirement is a displaced linear feature (Figure 9.6) running through both the hanging wall and the footwall. From each block the linear will pierce the fault at a point, the join between these once adjacent points is the net slip (Figures 9.6 and 9.7). Linear structures of use in this calculation include fold hinge lines (Figure 9.7), the line of intersection between a dyke and a formation boundary (Figure 9.8), two intersecting dykes, veins or a dyke and vein, and angular unconformities. It is also possible to use the intersection of an earlier fault or narrow shear zone with any of the above planar features. Theoretical discussions normally mention shoestring sands, linear ore bodies, stratigraphic pinch-out lines, but these must be considered rarities in practical terms.

The net slip is a line on the fault plane which can be measured in metres and its orientation specified either as a plunge and bearing or as a pitch on the fault plane — by definition the net slip is an apparent dip of the fault surface. If the net slip parallels the strike of the fault, relative displacement of the two fault blocks is in the horizontal plane and we have **strike slip** (Figure 9.8a). Net slip parallel to the dip of the fault is **dip slip** (Figure 9.8b) and any net slip pitching between these two is **oblique slip** (Figure 9.8c – e). In a similar fashion to separation we distinguish senses of strike and dip slip. In Figure 9.6 the hanging wall piercing point is above the footwall equivalent and hence the dip slip sense is **reverse** and the hanging wall has been upthrown relative to the footwall. The opposite **normal dip slip** sense (generated by downthrowing the hanging wall) is illustrated in Figures 9.7 and 9.8b. If you stand in one block over its piercing point (facing the other block), and the opposite block's equivalent point is to your left this is **left slip** (Figure 9.8a). The converse is **right slip**. For oblique slip situations the sense of both the strike and dip slip components must be quoted (Figure 9.8c – e) and the following combinations are possible: **Right normal slip** (Figure 9.8e) **Left normal slip** (Figure 9.8c) **Left reverse slip** (Figure 9.8d) **Right reverse slip**.

Variations on the terminology substitute sinistral or anticlockwise for left, and dextral or clockwise for right strike slip. Others talk of left or right lateral movement for strike slip and left lateral reverse (etc.) for oblique slip.

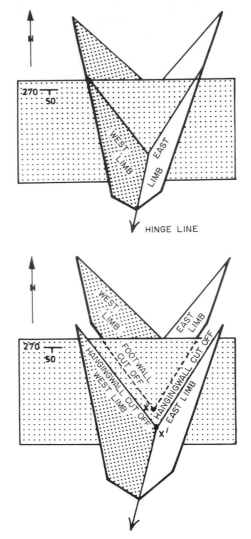

Fig. 9.7 An angular fold shows how displacement is monitored using the piercing points of the hinge line with the fault plane (X from the footwall, X′ from the hanging wall). Location of the piercing points is by construction of cuttoffs, each limb generates footwall and hanging-wall cutoffs. X to X′ is the net slip which is nearly down the dip of the fault. The sense of dip-slip is normal, i.e. hanging wall piercing point below the footwall point. For both limbs the dip separation senses are normal, but the strike separation senses are opposite (the east limb is left and west limb is right)

9.4 SLIP/SEPARATION: WHAT'S THE DIFFERENCE?

Separation describes the geometrical consequences of differential displacement (slip) across a fault. The two concepts are frequently confused perhaps because the differences are not actively thought through. Also, as we saw in Figure 9.1, incomplete visualization of the geometry leads an observer's thoughts in erroneous directions. The differences between slip and separation are highlighted by two statements:

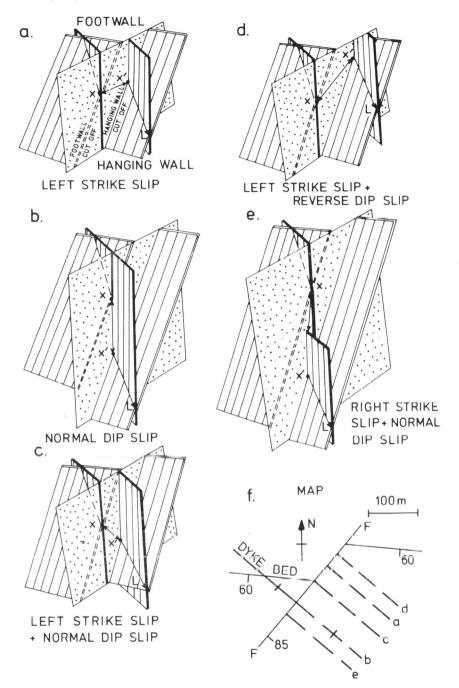

Fig. 9.8 Two intersecting planes generate lines of intersection which in turn produce piercing points (X and X′) on the fault plane (stippled). The vertical sheet is a dolerite dyke (igneous intrusion) emplaced dilationally into dipping sediments. In all diagrams the fault plane dips towards the observer who is looking from a hanging wall viewpoint. (a) Pure strike slip, i.e. the net slip is parallel to the strike of the fault and the sense is left. (b) Pure dip slip, net slip parallel to the fault dip; X′ the hanging-wall piercing point is below its footwall equivalent so the sense is normal. (c) Oblique slip, the joint X to X′ is inclined to both the fault dip and strike, hence the net slip has components along the fault dip and strike. Strike-slip is left and dip-slip is normal. (d) Oblique slip with left slip and reverse dip-slip components. (e) Oblique slip with right strike slip and normal dip-slip components (the converse of (d)). (f) Map representations of (a)–(e). Despite the varying amounts and orientations of net slip, the offset of the layering is constant! Dip and strike separations of the beds are identical in (a)–(e) hence their traces on the map are fixed in position. The varying effects of the differing displacements are best shown by the change of position of the vertical dyke — dashed lines with letter labels to refer to (a)–(e)

1. For constantly orientated surfaces, many different slips can produce the same separation geometry (Figures 9.8a—e and 9.9a); and
2. Again for surfaces of constant dip and strike, if the net slip is parallel to the hanging wall and footwall cutoff lines then no separation (no offset) is generated no matter how large the slip — this is **trace slip** (Figure 9.9b). This geometry also means that the hanging-wall and footwall cutoff lines are coincident.

If you are given, or have mapped a geometry, similar to Figure 9.9a, then it is impossible to specify the relative displacement vector responsible for what you see. This applies to single faults cutting constantly oriented beds or arrays of subparallel faults cutting beds of uniform attitude. Any one of the following combinations could have generated the final product of Figure 9.9a:

1. Left reverse slip (OA) — oblique slip;
2. Left slip (OB) — pure strike slip;
3. Left normal slip (OC) — oblique slip;
4. Normal slip (OD) — pure dip slip;
5. Right normal slip (OE) — oblique slip.

To make matters worse you cannot even tell if the hanging wall went up or down! The sense of dip separation is normal because the hanging wall cutoff is below the footwall cutoff, but if the displacement had followed slip OA this normal separation could have been generated by a reverse dip slip relative movement. **Don't give up**! — coming to grips with the last sentence will take you a long way in fault analysis. In fact, the angle between OA and OE does not cover the full range of possible slip orientations that could have produced the observed geometry; slips in virtually the entire 180° arc about point O could, given the right length, generate the offsets seen. Not much of a specification of the net slip! Of course the closer the net slip orientation is to the cutoff line, the larger the amount of slip has to be to generate the separation recorded.

The above comments are reinforced by Figure 9.8 where again several different slips have generated only one final separation geometry for the inclined bed. The introduction of a second pre-faulting surface (a vertical dyke) allows the slip to be located. Where the bed cutoff in the footwall block crosses the dyke cutoff from the same block (e.g. point X, Figure 9.8a), this defines a piercing point and its equivalent (X′) in the hanging wall block is easily found to fix the net slip vector — the X to X′ join. The different orientations and amounts of net slip (Figure 9.8a – e) have produced the same separation for the bed but have resulted in very different positions for the vertical dyke. For one fault attitude, separation is a function of:

1. The pre-faulting attitude of the surface;
2. The amount and orientation of the net slip (differential displacement).

For one fault with a single net slip it follows that surfaces of different attitudes will have different separation geometries (Figures 9.8 and 9.7). The simplest examples are folded beds (not isoclines), where a single net slip will produce different separations for the two limbs (Figure 9.7). Similar effects are noted for bed/vein,

a.

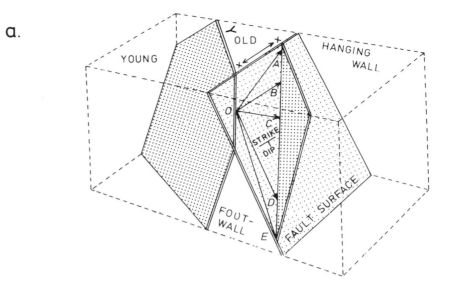

b.

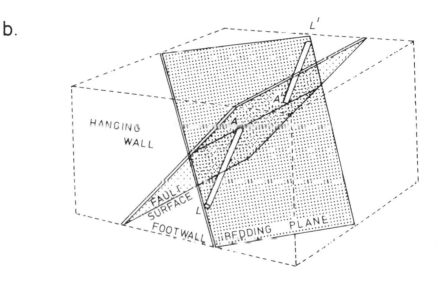

Fig. 9.9 (a) This offset (separation geometry could have been caused by point A moving from O, B from O, etc.; OA is left reverse slip, OB left strike slip, OC left normal slip, OD normal dip slip and OE is right normal slip. This gives the complete range of combinations of dip and/or strike components of slip that could have generated the observed geometry. The full range of orientations is larger than shown here being nearly a 180° arc around O including A – E. For the varying orientations, the amount of net slip has to be juggled to keep the separations constant. Between X and X on the fault plane, the youngest rocks in the hanging wall are brought against the oldest in the footwall — a relationship commonly interpreted as hanging wall down motion. However, a component of reverse dip slip could have created this situation invalidating the simplistic interpretation. For single faults cutting constantly oriented layering, a wide range of net slips could have generated the observed offset pattern. (b) A linear feature (LL') rests on a bedding plane. Net slip on the fault (A – A') is given by the piercing points but there is no separation of the bedding plane. This is trace slip where the net slip is parallel to the cutoffs of the bedding on the fault surface. No matter how large the relative displacement, this geometry never produces separation

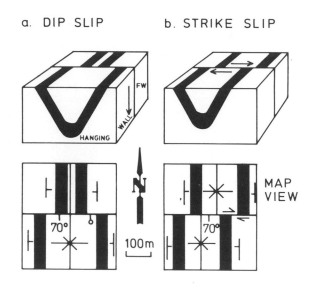

Fig. 9.10 Differently oriented planes fix the net slip. With upright folds and near-equal limb dips, dip slip (a) and strike slip (b) can be easily distinguished. Strike slip produces the same sense of strike separation (map view) whereas dip slip generates opposing senses. Dip separations are converse. In dip slip, both fold limbs have normal dip separations whereas, in strike slip, the western limb has reverse, and the eastern normal, dip separation. More inclined folds with unequal limb dips require more care for their analysis

bed/dyke, vein/vein, dyke/sill, etc., combinations; differently oriented planes will not have the same separations even when both features undergo the same relative displacement.

This apparent complication can be turned to our advantage. For steeply inclined to upright folds, with limbs of similar but opposing dips, faults at high angle to the fold trend readily show their approximate slip orientation (Figure 9.10). Pure dip slip does not offset the hinge surface trace but causes opposite senses of strike separation on the two limbs (Figure 9.10a). Strike slip in contrast offsets the hinge surface trace and creates the same sense of limb strike separations (Figure 9.10b). Oblique slip is intermediate in its effects and in the real world not quite so well-behaved fold geometries may also be met. Faulted folds also allow us to further pursue the difference between slip and separation. In the case of strike slip (Figure 9.10b), a single slip amount and orientation has produced reverse dip separation on the east fold limb but normal dip separation on the west fold limb.

If you do not have high expectations of separation then you will not be disappointed.

Separation is sometimes referred to as apparent relative displacement but this is a dangerous notion. On a map of Figure 9.9(a), the strike separation (XX) would be the most obvious feature, but to allow yourself to think it was the **actual relative displacement** would be foolhardy in the extreme. Between X and X, young beds on the hanging wall are preserved against older beds in the footwall (assuming constantly oriented and right-way-up stratigraphy). Some textbooks use this relationship to say that the block with the younger beds against older was **downthrown** but we have seen that the opposite could have been the case. This 'rule' holds for faulted flat-lying beds and for strike faults cutting dipping beds but is not generally applicable.

9.5   FAULT CALCULATIONS

Being true to the basic principles we have established, the first step in any calculation is to describe the present-day geometry and then attempt to be as specific as possible as to how it came about. Most of the information we need will come from the construction of cutoff lines. The example to be used (Figure 9.11) reduces the construction to the bones, in that topography and interactions between topography and geology are not considered. We are looking at outcrop on a level surface at a height of 0 m (it could be a wave-cut platform). Any structure contours are constructed by extrapolating the dip and strike information rather than being positioned by intersecting outcrop patterns and topographic contours. Because of the level surface, the fault and planar sections of bedding have straight-line traces on the maps; valleys and ridges would have created sinuous V outcrops.

Before plunging into the exercise, an initial assessment should be made, particularly taking 3-D visualization as far as you can. A thin bed shows inward-directed dips defining a synform and the southwards opening of the outcrop pattern shows this to be the plunge direction. A completely angular/chevron style is indicated by the absence of a curved closure. The fault is dipping due south at 45° and hence the southern block is the hanging wall leaving the north block as the footwall. Displacement is purely translational as each limb attitude is the same in both fault blocks.

Both limbs have the same sense of strike separation — stand at A with your feet on the hanging wall looking towards the footwall, the displaced part along the strike of the fault is to the **right** at B — and from C, viewed the same way, D is to the right; **but** note that the amounts are different. This is a partial expression of the relationship:

Separation = Function (net slip, surface attitude)

The two limbs had different attitudes pre-faulting so the separation must be different though as yet we have only investigated one component.

To construct cutoff lines:

1. Take one fault block and construct structure contours on the fault surface and the two limbs (Figure 9.11b) using the dip and strike data (or from outcrop/topographic contour interaction).
2. For the west limb structure contours find the points where they are at the same height as the fault contours (Figure 9.11b), i.e. the line where the two surfaces are touching (W1, W2, W3, W4). This is the west limb hanging wall cutoff projected vertically on to a horizontal (map) surface.
3. The equivalent east limb hanging wall cutoff projection is along E1, E2, E3 and E4.
4. The west limb footwall cutoff is parallel to W1 to W4 but passes through point B and the east limb equivalent is parallel to E1 – E4 passing through D (Figure 9.11c).

Remember that the cutoff traces on the map are projections and the true cutoff lines are **on** the fault surface. You should now be in a position to visualize the dip separation geometry. The west limb hanging wall cutoff is below the footwall cutoff hence the sense of dip separation is normal. The east limb hanging wall cutoff is

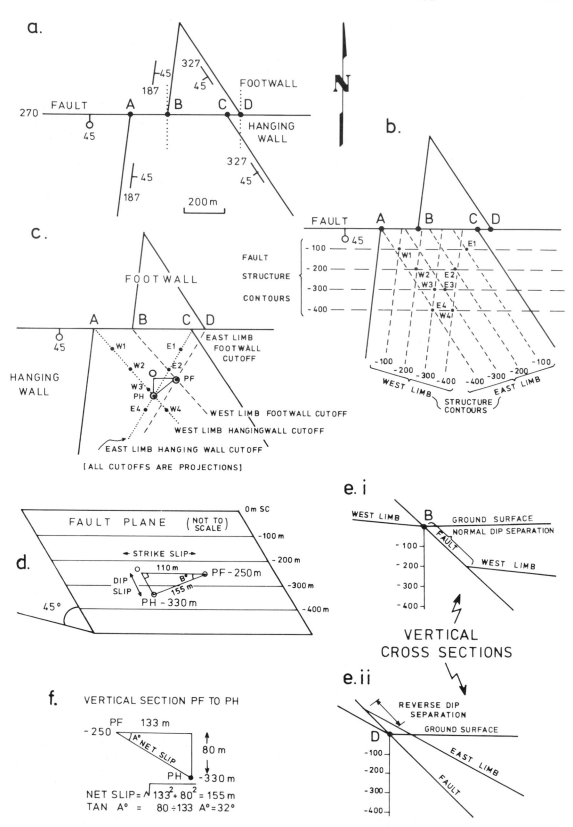

Fig. 9.11

above the footwall cutoff so here dip separation is reverse; again emphasizing that different attitudes + one slip give different separations. (Note that separations would be different even if both dip and strike separation components had the same sense but the distances were not the same.)

Any visualization of dip separations should be confirmed by construction of vertical sections striking perpendicular to the fault — dip sections with respect to the fault (Figure 9.11e). To save some work the sections are drawn through points B and D (Figure 9.11e) and because they are down the fault dip they give a direct reading of the amount and sense of dip separation.

With a fold structure we can use the hinge line to define points that were originally adjacent across the fault and hence fully specify net slip. The cross-over of the west and east limb hanging wall cutoffs is the map projection of the point where the hinge line in the uppermost block touches the fault (PH, Figure 9.11c). The hinge line meets the fault from the lower block at the intersection of the limb footwall cutoffs (PF, Figure 9.11c). On the map (Figure 9.11c) the line PH to PF is the projection of the net slip which by definition is a line on the fault — in this case the fault surface has a 45/270 attitude. The bearing of PF to PH is the bearing of the net slip, but its length and plunge have to be measured on a vertical section (Figure 9.11f) along this trend (think about the definition of plunge being the strike of a vertical surface containing the line). The hinge line from the footwall touches the fault at -250 m and from the hanging wall at -330 m, therefore, from PF to PH the net slip falls 80 m in a horizontal distance of 133 m. Simple trigonometry gives the plunge at $32°$ and Pythagoras' theorem (hypotenuse squared = sum of squares of the other two sides of a right-angled triangle) or further trigonometry gives the length. The pitch of the net slip on the fault plane is B° (Figure 9.11d) which is easily calculated once the distance O to PF and the net slip lengths are

Fig. 9.11 (on page 152) (a) A faulted fold provides two inclined planes (fold limbs) which intersect in a line (hinge line). This was once continuous and the distance between the piercing points, generated by the disrupted hinge line touching the fault, is the net slip. The displacement is translational because each limb has the same attitude in both fault blocks. The map is of a horizontal plateau surface. (b) Knowing the dip and strike of the two limbs and the fault surface, structure contours can be constructed for these three planes. Construction is shown for the hanging wall only. The height of the plateau is at datum level and structure contours are drawn with respect to this (-100 m, -200 m, etc.). E_1 is the vertical projection of the cross-over of the -100 m fault structure contour and the -100 m east limb structure contour, i.e. the point where the east limb is cut off by the fault at -100 m. $E_2 - E_4$ are more points on the projection of the east limb hanging wall cutoff. Likewise $W_1 - W_4$ are projected points on the west limb hanging wall cutoff. (c) Because the fault is translational, the footwall cutoffs will be parallel to those in the hanging wall and can be drawn from points B and D. The cross-over of the two hanging wall cutoffs PH is the projection of the hanging wall piercing point (where the hinge line in the hanging wall touches the fault plane); PF is the footwall piercing point. Part (c) is shown without the structure contour construction lines. If PH were shown on (b) it would be three-tenths of the distance from -300 m to -400 m fault structure contours and hence is at -330 m. PF is half-way between -300 m and -200 m and hence is at -250 m. PH—PF is the vertical projection of the net slip. (d) A 3-D representation of the piercing points and the net slip. (e) (i) The west limb dip separation is measured in a vertical section perpendicular to the fault strike. Construction is simplified by drawing the section through B. The hanging wall cutoff is below the footwall equivalent hence the dip separation sense is normal. (ii) A dip section, with respect to fault attitude, through D gives the dip separation for the east limb which is reverse in sense. On both (e)(i) and (ii) the vertical and horizontal components of dip separation can be directly measured. (f) A vertical section along the direction PH — PF (c) allows the true length and plunge of the net slip to be calculated either by construction or trigonometry. The net slip falls 80 m from PF to PH in a ground distance of 133 m

known. It is useful to bear in mind that the net slip is an apparent dip of the fault pitching between 0° and 90°.

Having obtained the net slip we now have to consider the components and particularly their senses. The hangng wall piercing point (PH, Figure 9.11d) is below the footwall equivalent (PF) hence the dip-slip component is **normal**—the hanging wall went down relative to the footwall. Standing over PH in the hanging wall and looking to the footwall, the point PF is to the **right** which is the sense of strike slip. So right normal slip generated right normal separation for the west limb but right reverse separation for the east limb—we clearly have to separate slip and separation!

Even if the net slip has been quantified this does not give us the displacement history. An oblique slip as seen in Figure 9.11 could have evolved as a strike-slip fault which was later reactivated with a dip-slip motion (or vice versa). It could also have been in an oblique-slip environment throughout the deformation, though even more complicated histories could be masked by the simple join of the net slip which is only the resultant displacement. Some fracture surfaces are scarred by friction grooves (striations) which give the orientation of at least part of the relative displacement. They may, however, destroy differently oriented striations of an earlier movement or perhaps one or more displacement events may not have generated grooves. In addition to grooves, some fault planes carry fibrous crystals which record part of the displacement history but again cannot be relied upon to reflect the total picture. Rare examples of curved grooves or fibres give a clue to the complexity of some fault movement patterns.

If only one limb of the fold were exposed in the above example or if the succession was of constant dip and strike, then the net slip could not have been specified. Under these circumstances our closest approach would have been to narrow the range of slip orientations to just less than 180° as in Figure 9.9a. If strike faults are present, cutoff lines may be constructed as above but they will parallel the strike, and the structure contours, of the fault. Remember in nature that perfect planes are rare, and curved faults and/or beds will result in sinuous cutoff lines.

9.6   FAULTS AND OUTCROP PATTERNS

For the moment we are staying with single faults and for the most part with bedded sequences of constant dip and strike; consideration of fault patterns comes later. **Strike** faults with the same orientation and sense of net slip have different effects on outcrop patterns depending on whether they dip in the same or opposite direction as the layering they offset (Figure 9.12). A normal dip-slip strike fault dipping against the layering brings about a **repetition of beds** (Figure 9.12a), whereas the same sense of movement on a fault dipping with the beds causes **removal** of part of a bed or several beds (Figure 9.12b). Strike faults with reverse dip slip have **opposite** effects on outcrop patterns. These are bread-and-butter issues of map reading in a faulted terrain.

Dip faults require considerable caution in their interpretation. We have already seen that a wide variety of slip orientations could produce one particular separation

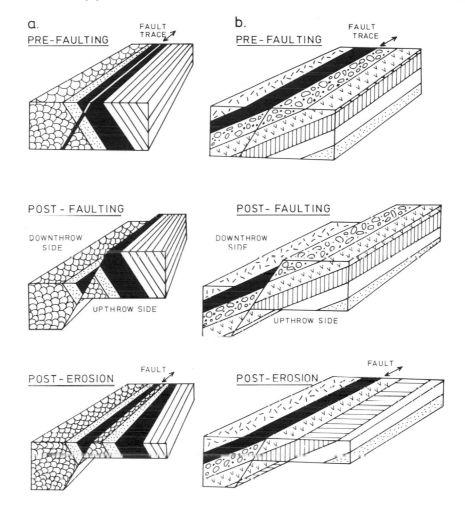

Fig. 9.12 (a) and (b) are both strike faults (fault strike parallels bed strike) and both have normal dip-slip displacement. The fault dipping against the dip of the beds repeats part of the stratigraphy (a) whereas a dip with the beds removes some of the stratigraphy

geometry. With normal dip-slip displacement and an inclined stratigraphy, it appears that the outcrop pattern has moved **against** the direction of their dip in the hanging wall (downthrown) block (see slip OD on Figure 9.9a). Pure reverse dip slip would appear to move the outcrop pattern in the direction of their dip in the hanging wall block (in this case upthrown). These apparent movements in a horizontal direction bear no relationship whatsoever to the actual relative displacement which is at right angles up and down the dip of the fault. They are, however, the sorts of effects that mislead many observers.

Some special cases are worthy of attention. If the succession is flat lying then any departure from strike slip (which would be trace slip) is very obvious as would be the sense (normal or reverse) of dip slip (e.g. Figure 9.2a). The same generalization holds for strike faults because even the smallest component of reverse slip will create reverse separation and any normal slip will create normal separation. This contrasts with the distinctly tricky dip faults where a component of reverse slip (OA, Figure 9.9a) can very readily produce normal separation!

9.7   FAULT PLANES AND FAULT ZONES

Not all faults are simple fractures and few are cohesionless surfaces because it is very common for circulating fluids to precipitate minerals that heal fault surfaces. At high crustal levels (near surface) the walls of the fault may become fractured, perhaps creating a zone of broken fault rock metres wide (in exceptional cases hundreds of metres). Coarse material of this type is fault breccia and the finer-grained equivalent is gouge, a product of intense microfracturing. Fault breccias commonly show a high degree of porosity which may be infilled by a wide variety of minerals. Many small tonnage metal mines (Pb, Zn, Cu) are located in such structures, though modern economics means that most new discoveries of this type cannot be worked unless the metals are precious (Au, Ag). A breccia that is not cohesive or bound by a mechanically weak/easily weathered mineral, will be preferentially eroded at the earth's surface and the expression of the fault line will be a linear trough. Many faults are of this style, but those breccias cemented by silica or faults occupied by quartz veins may well be positive features marked by upstanding tabular zones.

Because the change from brittle to ductile style deformation is not abrupt, there is a range of responses between the two end members. For fault rocks this range is from unfoliated non-cohesive breccias and gouges, through unfoliated but adhesive fault rocks (intensely microfractured cataclasites and melts from friction heating called pseudo-tachylytes), to cohesive foliated rocks known as mylonites. The brittle to ductile range for shear zones has already been illustrated (see Fig. 7.3). A zone of closely spaced faults (see Fig. 7.3a) may be referred to as a zone of distributive slicing. The mixture of some folding with dominant discontinuous displacement (see Fig. 7.3b) give rise to **drag** though it is usually impossible to say whether the folding or the faulting came first or if they evolved together. Note that drag reduces the separations that would have formed if the layers had remained planar. Transitional shear zones involve varying proportions of fracturing and folding with cleavages/foliations being formed towards the ductile end of the spectrum. High strains in a ductile shear zone (see Fig. 7.3d) could lead to apparent discontinuities in once continuous surfaces and because of such problems a comprehensive definition of a fault has been proposed (Wise, D. U. *et al*, 1984. Fault related rocks. *Geology*, **12**, 391 – 395.):

> Faults are relatively tabular or planar discontinuities in which the zone as a whole or any macroscopic part of it contains displacement parallel to the zone greater than 0.5 to 1 cm and displacement at least 5 to 10 times greater than the width regardless of whether the zone is marked by loss of cohesion or extreme ductile deformation.

Life was not meant to be easy! But the definition is comprehensive and is partly aimed at potential court-room battles over land zoning, earthquake hazard prediction, engineering geology assessment, etc. The distinction is also made between fractures with no discernible displacement (joints) and those with obvious movement (faults). In the field any evidence for relative movement makes a fracture a fault, though few maps could represent faults with only a centimetre of slip. Again scale has to be considered and, at the 1 : 50 000 scale, strike separations of less than 50 m could not be easily represented.

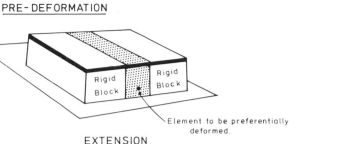

PRE-DEFORMATION

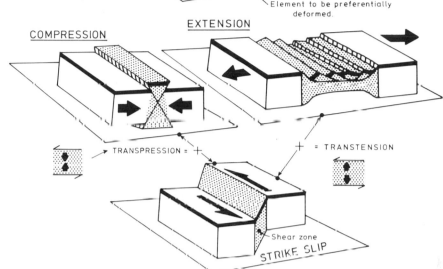

COMPRESSION

EXTENSION

TRANSPRESSION = +

+ = TRANSTENSION

Shear zone

STRIKE SLIP

Fig. 9.13 Basic deformational environments illustrated by considering a zone to undergo preferential deformation bounded by rigid blocks. Compression contracts and thickens the reference element of material, extension lengthens and thins the deforming region, strike slip distorts the element by relative horizontal displacements of the bounding blocks. A mixture of strike slip and compression is transpression; strike slip plus extension is transtension

9.8   FAULT PATTERNS AND ASSOCIATIONS

We now broaden our sights from individual faults to groups of linked or associated faults and consider specific tectonic environments. Our discussion does not use traditional tectonic classifications of passive margin, marginal basin, subduction zone and the like, but is based on the regional stress state. Taking deformation zones at the scale of orogenic belts and extended continental margins (tens to hundreds of kilometres wide) we have five fundamental situations (Figure 9.13):

1.   Compressional/contractional tectonics;
2.   Extensional tectonics;
3.   Transform tectonics (strike slip);
4.   Transpressional tectonics;
5.   Transtensional tectonics.

   Our understanding of these five environments evolved from three basic near-surface stress configurations. Theoretical considerations show that, very near the earth's surface, one of the principal stresses (σ_1, σ_2 or σ_3) must be vertical. This condition, taken with the most common pattern of synchronous fractures (conjugate

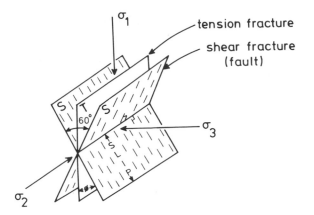

Fig. 9.14 Faults are shear fractures (S) in that blocks slide past each other. Shear fractures are commonly arranged in inclined conjugate pairs that have opposing shear senses. The shear fractures typically have dihedral angles of around 60° and they intersect in a line parallel to the intermediate principal stress. The maximum principal stress bisects the acute angle between the conjugate faults; the plane containing σ_1 and σ_2 is a tension fracture (T). The net slip orientation on the faults is perpendicular to the line of intersection of the shear fractures

shear fractures, Figure 9.14), defines the basis for much analysis of fault arrays. Experimental work and field observation have shown that inclined shear fractures (faults) frequently form at the same time. This arrangement is known as a conjugate pattern which also involves the inclined shear fractures having opposite sense of slip (Figure 9.15). Conjugate faults very commonly have an acute dihedral angle of about 60° (Figure 9.14). Tension fractures, when present, bisect this angle and all three planes intersect along a common line. Slip orientations are perpendicular to this shared line and a mechanical analysis shows the line of intersection to be σ_2, the acute angle is besected by σ_1 (the maximum compressive stress), and σ_3 is perpendicular to the tension plane. The wedges of rock within the acute angles of the faults are pushed together by the maximum compressive stress (σ_1) which generates the opposing senses of shear motion seen in Figure 9.15. Visualization of the displacements may be helped by considering the blocks in the obtuse angle segments moving against the least compressive stress (σ_3) which in many cases may be a tensional force.

In the perfect situations illustrated in Figure 9.15, the slip directions are either exactly down dip (Figure 9.15a, b) or along the strike (Figure 9.15c) of the faults. For Figure 9.15(a) the dip-slip sense is hanging wall down which is normal displacement. Because most conjugate angles are about 60° it follows that normal faults, generated near the earth's surface, should dip at about 60°; this fits observations remarkably well in many regions. Normal faults are the dominant structures in regions where the lithosphere has been stretched (Figure 9.13) as is suggested by the extension and shortening directions shown in Figure 9.15a. Considering the whole crust, only the upper to 10 – 15 km is brittle, hence fracturing is confined to these upper levels and during lithospheric stretching the lower crust should deform in transitional or ductile modes. The extensional nature is also emphasized by Figure 9.16(b) where part (i) is the pre-deformation length of a layer (l_0) and part (ii) is the length (l_f) of the same layer after a normal faulting episode. For this example, the layer has been extended by 26 per cent and all of the motions on the conjugate fault array are hanging wall down.

In Figure 9.15(b) the slip directions are again down dip but now the hanging wall movement is up, that is, reverse. Pursuing the consequences of a simple conjugate array of faults, the typical reverse fault should dip at about 30° which is a fairly common observation. Reverse faults are found in regions that have undergone

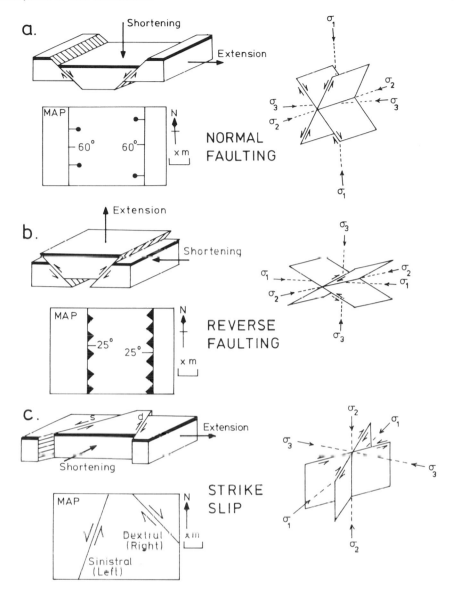

Fig. 9.15 Near the earth's surface, one of σ_1, σ_2 or σ_3, has to be vertical with the other two horizontal. This creates three basic styles of brittle behaviour. (a) Maximum principal stress vertical and faced by the acute angle between the conjugate faults. The two acute angle wedges converge under the influence of the maximum compressive stress and hence hanging walls are downthrown with respect to footwalls — normal slip. The fault blocks in the obtuse angle between the faults move against the least confining stress (σ_3) causing extension of the horizontal layering. (b) The least principal stress is now vertical so the acute angle wedges are bisected by the horizontal. Both conjugate faults have reverse dip slip and have shortened horizontal layering. The jagged teeth symbol for a reverse fault is placed on the upper block. (c) With a vertical intermediate principal stress, the slips have to be horizontal — strike slip. Again the movement of the acute angle wedges indicates the senses of movement along the faults

compressional deformation (Figure 9.13) as seen in collisional orogenic belts. Near-surface reverse faults (Figure 9.15b) result in a shortening of stratigraphic layering (Fig. 9.16a) hence the derivation of the term contractional faulting.

Where σ_2 is vertical and both σ_1 and σ_3 are parallel to the earth's surface, the two conjugate faults and their line of intersection are vertical (Figure 9.15c).

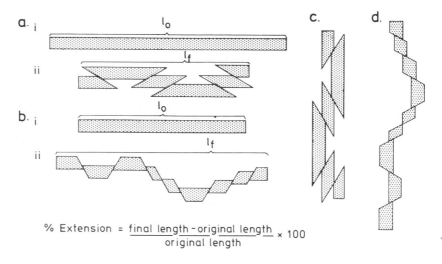

$$\% \text{ Extension} = \frac{\text{final length} - \text{original length}}{\text{original length}} \times 100$$

Fig. 9.16 Normal and reverse slip classifications are referred to the horizontal but subsequent rotations can distort perceptions. (a) Layering of initial length l_0 is contracted 28 per cent to l_f by reverse faulting. (b) Layer l_0 is extended by 26 per cent by normal faulting. (c) Rotation of (a) (ii) through 90°, by subsequent deformation, changes the reverse slip faults to normal slip but the style remains contractional. (d) Rotation of (b)(ii) through 90° converts the normal slip faults to reverse but the basic extensional style remains

Geometrically this means that the slip directions must be parallel to the strike of the faults giving the origin of the term strike-slip faulting. Synonyms are wrench, transcurrent, tear and lateral faulting. Because relative displacement is in the horizontal plane, the senses of movement are described by one of the following pairs of terms: left/right, sinistral/dextral or anticlockwise/ clockwise.

Normal, reverse or strike-slip faults that evolved in the near surface may subsequently be buried and be subjected to large rotations about horizontal axes. Later we shall see that significant rotations are expected within areas that have undergone either just tension or just compression. Such rotations can change our perceptions of faulting. Figure 9.16(c) shows a group of reverse faults rotated through 90° to create vertical bedding attitudes. As a result, all the once reverse faults are now normal (hanging wall down); the one feature that has not changed is the shortening of the layer by the fault process. There is now a move to place less emphasis on the classification based on normal and reverse apparent movements but to consider **contractional** and **extensional** effects as being more significant in any analysis that looks at the implications and environments of the fault style. There is a potential problem in using the layering as a reference. If beds are rotated to the vertical and the region is subsequently extended and normal faults are formed, these will bring about a shortening of the layering. In any analysis the tectonic history must be fully appreciated.

9.8.1   CONTRACTIONAL FAULTS

For small amounts of shortening, of the order of 10 per cent, contractional faults have geometries as illustrated in Figures 9.15b and 9.16a. Such styles are met where the original sedimentary basin is not severely modified by the deformational overprint and net slips on individual faults are not large. A very different contractional fault style is encountered in regions known as foreland fold and thrust belts characterized by high amounts of shortening (typically 50 per cent or more).

A **thrust** fault is a map scale contractional fault

Thrusts once were defined as low-angle (i.e. low-dip) reverse faults, but subsequent rotations and folding can easily change fault attitude but not disturb its basic contractional nature hence the change of emphasis. Thrusts typically involve tablet-shaped blocks of rock being essentially translated over large distances (Figure 9.17); the individual record is around 200 km. These thrust slabs may be kilometres thick down to a couple of metres, and lengths and widths also range from the metric scale up to hundreds of kilometres. For far-travelled thrusts, measurement of displacement is qualitative. The method relies on erosion cutting through the overlying thrusted block to expose portions of the lower units (Figure 9.18). An enclosed outcrop pattern of the subthrust rocks is a **fenster** (window) whereas an isolated area of the thrust block is a **klippe**. The distance from the front of the klippe to the back of the window is the **minimum** displacement. Quite accurate estimates of displacement can be made in areas with a well-defined stratigraphy and a good data base (plenty of exposure, geophysical information, oil drilling, etc.). The process depends upon unwinding the fault displacements to restore the region to its undeformed state. This is normally done in a series of vertical cross-sections through the zone, and sections that permit sensible restoration are referred to as **balanced**. Balancing of complicated areas is an advanced technique, but even in the early days of drawing sections of deformed terrains you should ask yourself 'Is the section restorable?' If not, then your section is not a valid representation of the region.

Thrust faults are the dominant structures within orogenic belts, but their most obvious expression is on the marginal foreland fold and thrust belts. The clarity of definition is a function of well-defined stratigraphy which has not been strongly overprinted by metamorphism or distortional strain effects. Within the core hinterlands of orogenic belts, ductile deformation means that any attempt at restoration requires large amounts of information about shape changes and hence balancing cross-sections is difficult. Thrust belts commonly have strike lengths of hundreds to thousands of kilometres and overall displacement directions perpendicular to the general trend. Not all parts of a thrust belt advance at the same rate and one way of accommodating the differential displacements is to develop vertical strike-slip faults (Figure 9.17) that parallel the displacement direction. Such a combination of different fault styles operating in tandem is known as a **linked** system. Other information on direction of movement might come from striations on thrust surfaces.

> **Basic rules**: (i) thrusts place older strata over younger;
> (ii) thrusts cut up section.

Both rules are illustrated in Figures 9.19 and 9.20. Thrust planes commonly have long **flat** sections parallel or subparallel to sedimentary layering and short **ramp** sections where they cut bedding at around $20 - 30°$. Moving a thrust block over this topography and keeping the blocks in contact must lead to the geometry shown in Figure 9.19. Formations 2 and 3 are duplicated and layer 2 rests on layer 3 in the doubled-up portion giving older on younger. Ramps occur where stratigraphy is **cut off**, hence in a thrust with a single climb up stratigraphy (Figure 9.19) there are two ramps — a **footwall ramp** and a **hanging wall ramp**. Another consequence of the ramp flat geometry are the folds (Figure 9.19). These are generated by the thrust block responding to bends in the fault surface and are known as **fault-bend folds** or to some workers as 'snake's head folds'. The folds are named according to their relation to the ramps, so we have (Figure 9.19) **hanging wall ramp monoclines** and

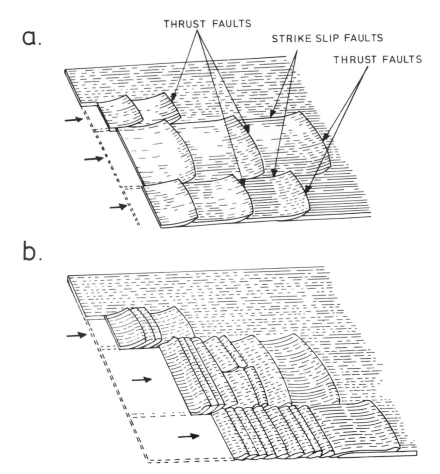

Fig. 9.17 Regions affected by significant contraction normally respond by generating large tabular thrust blocks on gently inclined thrust faults. Slicing of the blocks can locally generate steeply dipping faults. Differential movements between blocks is accommodated by strike-slip faults in a linked fault system. (b) is a more advanced state of contraction than (a). Reproduced by permission of the Geological Society from G. Mandl and G. K. Shippam, Mechanical model of thrust sheet gliding and imbrication, in *Special Publication No. 9*, pp. 79–88, 1981.)

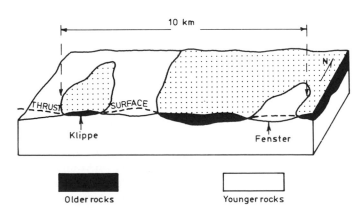

Fig. 9.18 Estimates of thrust displacement is often imprecise. An eroded thrust block may produce isolated outcrops of the overlying block (klippe) or windows into the underlying block (fenster). The distance from the front of the most distant klippe to the back of the last fenster is the minimum slip

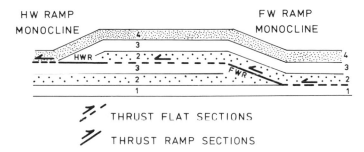

Fig. 9.19 Thrust faults are commonly long flats subparallel to bedding and short ramps at 25° – 30° to bedding. Where layering in the footwall is cut off there is a footwall ramp; there is an equivalent ramp in the hanging wall. As the thrusted block travels past a ramp, geometrically necessary folds are generated if the blocks are to maintain contact; these are monoclinal in the simplest case

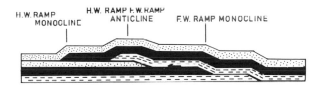

THRUST PROFILE – STAIRCASE TRAJECTORY

Fig. 9.20 Most thrust surfaces contain many ramp/flat alternations (staircase trajectory) climbing up stratigraphy in the direction of displacement. Each additional ramp creates extra folds. A major control on fold geometry is the ramp length and spacing

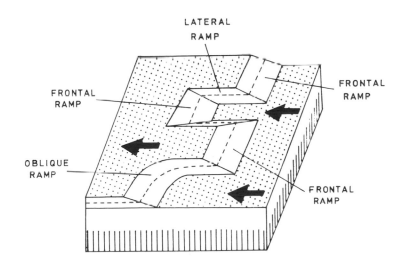

Fig. 9.21 Thrust surfaces in 3-D have complex shapes. Flats are stippled to distinguish them from ramps. Ramps do not always face the direction of transport (frontal) nor do they maintain their geometry for ever. Lateral ramps strike parallel to the displacement direction and oblique ramps are across this orientation

footwall ramp monoclines. The typical thrust cuts up across layering several times giving rise to a **staircase trajectory** and results in the production of more geometrically necessary folds (Figure 9.20). Thrust plane topography is rarely simple and the climb to higher stratigraphic levels typically takes place on differently oriented ramps (Figure 9.21). Frontal ramps face the displacement direction head on, lateral ramps strike parallel to this direction and oblique ramps have intermediate orientations. If one part of a thrust belt has a larger number of frontal and oblique ramps (which must be accommodated by lateral ramps) than the adjacent regions then a bulge or **culmination** occurs. Within the culmination lower stratigraphic layers are taken to higher levels relative to neighbouring sections. Little thought is needed to realize that distortions on oblique ramps can be quite complicated.

The majority of thrusts develop in an orderly fashion which greatly assists their analysis and restoration. In foreland fold and thrust belts, displacement is from the hinterland of the orogenic zone towards the foreland. The hinterland is deformed first and the orogenic effects move outwards towards the edges so that inner thrusts form before those in the outer zones. Thrust propagation is in the direction of movement, and though this thrust sequence (Figure 9.22) is the most common it is not the exclusive order of formation. Once a thrust has climbed up a ramp, the next thrust to form will be in the footwall of the older thrust (Figure 9.22, stages 1 and 2). Movement on the lower thrust bends (folds) both the layering within the thrust block (**a horse**) and the overlying roof thrust in a series of geometrically necessary fault-bend folds. If the slip (displacement) is not large relative to the length of the horse, the next thrust, of similar dimensions and slip, will generate stage 2 of Figure 9.22 causing some unfolding of horse 1 and the overlying part of the roof thrust. However, displacement of horse 2 leads to further folding of the roof thrust above the newly formed horse. Complex rotations can accumulate or be cancelled out depending on horse length and displacement amount. Of interest is that seemingly simple planar sections may have been rotated one way only to have the rotation removed later. Another aspect of this sequence of thrust propagation is that the older, higher thrust blocks are carried **piggy-back** fashion by the younger thrusts. Horse 3 (Figure 9.22) piggy-backs horses 1 and 2 as it moves and also leads to more bending and unbending of the roof thrust.At the final stage (3), the thrust slices have upper and lower thrust boundaries and together they form a **duplex**. The complete evolution from initial to final stage is referred to as footwall collapse. Duplexes can be much more involved than the one shown here.

Brief mention should be made of exceptions to the basic rules of thrust terrains. Contractional faults may not result in old rocks being placed over young if they are imposed on already folded rocks. Another breakdown is associated with out-of-sequence thrusts where thrusts cut through previously thrusted blocks instead of propagating by footwall collapse in already unthrusted materials. Breached thrusts (Figure 9.23) occur where the new thrust (e.g. after stage 1 of Figure 9.22) cuts up through the roof thrust instead of joining it; these clearly disturb the typical geometry of old over young at a thrust contact. Several other variations on the theme (e.g. footwall plucking) are best left to advanced courses, but be aware that these pages are not a complete account of contemporary knowledge.

The final section on thrusts makes the important connection between folds and faults. On maps of foreland fold and thrust belts, the folds are commonly more evident than the thrusts and the latter require more attention before recognition. Many fold patterns follow the fault-bend style (Figures 9.19, 9.20 and 9.24b) but a significant proportion are different and are generated during the propagation of the thrust (Figure 9.24a). Thrusts, like all faults, do not go on for ever and likewise their final extent is not created at one time. They nucleate, then grow in area and eventually

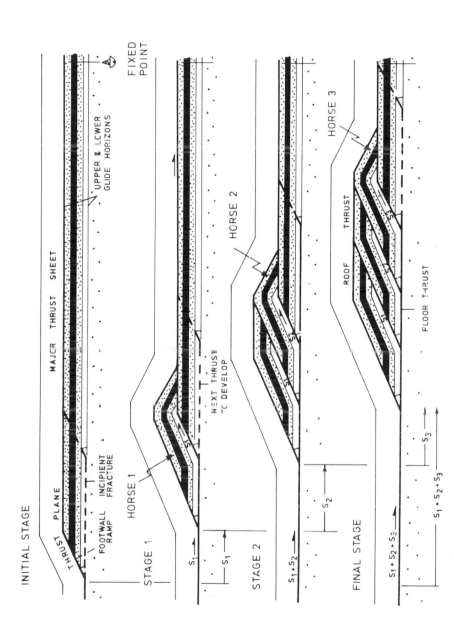

Fig. 9.22 Duplex formation. Beneath a major thrust sheet, the footwall progressively collapses in the direction of thrust displacement. Horse 1 (an individual fault bounded block) forms first, then horse 2, then horse 3. The resulting package of fault-bounded horses is a duplex. Geometrically necessary folds form with the movement of each horse, and planar sections of layering and roof thrust may have been folded then unfolded in the formation of the duplex

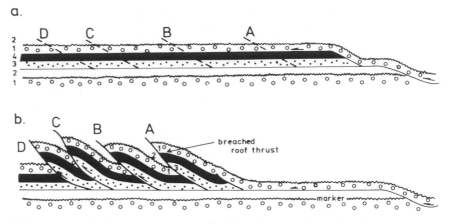

Fig. 9.23 A basic rule in thrust tectonics is that thrust planes place old rocks over younger (a). One of the exceptions to this rule is shown here (b). If in footwall collapse, the next thrust to form cuts across the roof thrust instead of rejoining it, the roof is breached. Along thrust A, formation 4 is placed over formation 2 which is the correct stratigraphic order, though formation 3 is only partly over formation 2

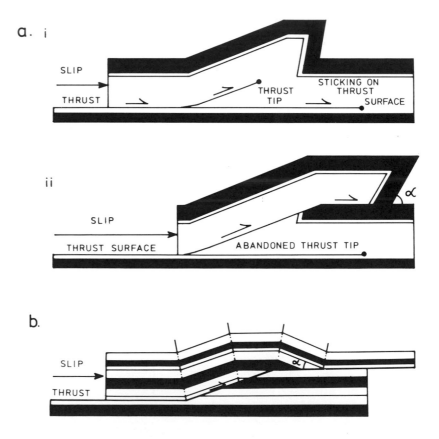

Fig. 9.24 There are two basic fold styles associated with thrust faults, fault-propagation (a) and fault-bend (b). The folds of Figures 9.19, 9.20 and 9.22 are fault-bend in type where the thrust block displaces and keeps in contact with the thrust plane and folding is dictated by the ramp/flat transitions. In fault-propagation folding (ai), if a thrust fault stops increasing its length but the thrust block continues to move, the displacement can be taken up by folding. Typically such a thrust fault will climb a ramp to avoid the area of sticking and will eventually cut through the fold and carry it off along a flat(aii)

stop moving. All this leads to boundary accommodation problems which are normally taken up by a zone around the fault of more dispersed brittle behaviour or some combination of ductile and brittle deformation. One aspect of thrust propagation is that a flat may extend into a zone of irregular sedimentary structures which could arrest its development (Figure 9.24ai). Stress difference is still being applied to the system and this might be alleviated by the thrust climbing to a higher (easier slip) stratigraphic level. During the climb, the rocks ahead of the propagating fault may take up the shortening by folding. Once the fault reaches its new level, the fold that developed ahead of it will be cut through and carried off along the flat (Figure 9.24aii). Such a mechanism can account for a wide range of styles of folding related to thrusts. The fold dihedral angles depend upon the amount of rotation before the thrust cuts the fold and this controls the angle between layers in the hanging wall and the thrust surface (Figure 9.24aii). Angular folds are shown, but the roundness depends upon the nature of the stratigraphy as well as the shape of the thrust surface. Once the fold is moving along the flat, further modifications are possible. If displacement is not easy then the fold might grow in a tank-tread-type process with the hinge line occupying different positions within the layering. In a more ductile regime simple shear may be concentrated on the short limb of the fold and change its shape.

Further variations on the thrust-fold theme are applicable to fold-belts where thrusts are seemingly very subordinate structures. Figure 9.25 shows the development of series of closely spaced thrusts with limited slip and related fault propagation folds. The restricted displacement on the thrusts means that the upper part of the stratigraphy is not breached by discontinuities and the dominant

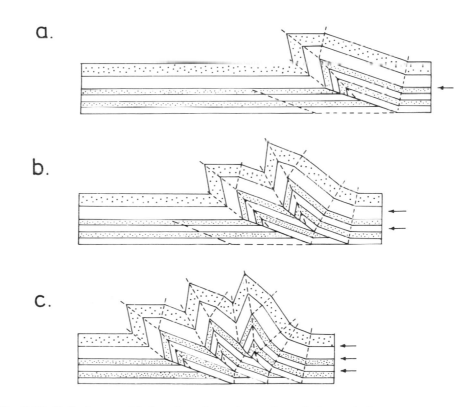

Fig. 9.25 If displacement is transferred to a lower thrust before a fault-propagation fold is completely faulted, the formation of the next thrust will steepen the hinge surface of the earlier fold. Repetition of this sequence of events progressively back-rotates (steepens) the earlier thrusts and folds

structures are folds. The same diagram shows a very important process in thrust tectonics. Progressive footwall collapse can lead to a steepening of earlier structures given slip amounts less than the length of the thrust blocks. This **back rotation** can occur on all scales and is the process responsible for the steeply dipping thrust slices found in many subduction complexes. Large rotations by this mechanism can convert reverse contractional faults into normal faults. If, in Figure 9.25(c), the next thrust to develop was some distance ahead of those shown, it would not add to the back rotation of earlier structures. Hence fold hinge surface attitudes could vary depending on thrust spacing.

To make the link between idealistic models of thrust structures and the real world, Figure 9.26 shows the type of map pattern created by a simple duplex and fault-bend folds. By using all the map information, a good subsurface prediction may be made; in the case shown, a basic assumption is that folds have constant stratigraphic thickness around the closures. On this basis the folds can be extended to depth and their geometry helps to locate buried ramp – flat transitions. Note also that the folds are gentle (dihedral angles around 130°) which would suggest the presence of fault-bend folds. Figure 9.27 is a large example of this style of folding in the classical Appalachian foreland fold-thrust belt. Key features include the relationship between thrusting and folding, both spatial and geometric. These regional examples are very similar to the stylized model (Figure 9.26). The parallel arrangement of the thrusts in

a. MAP

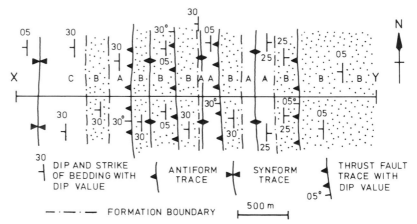

b. VERTICAL CROSS SECTION

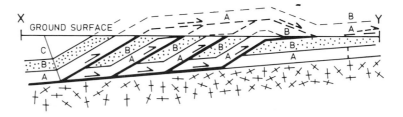

Fig. 9.26 A simplified map representation (a) of a fold – thrust belt. All the folding is fault-bend in type, hence surface data may be projected down to predict the structure at depth (b). Note the combination of symbol for thrust fault and its dip. The example has a regional dip of 5° to the west and shows a major decoupling between sedimentary cover rocks and crystalline basement – a very common situation in forelands to orogenic belts

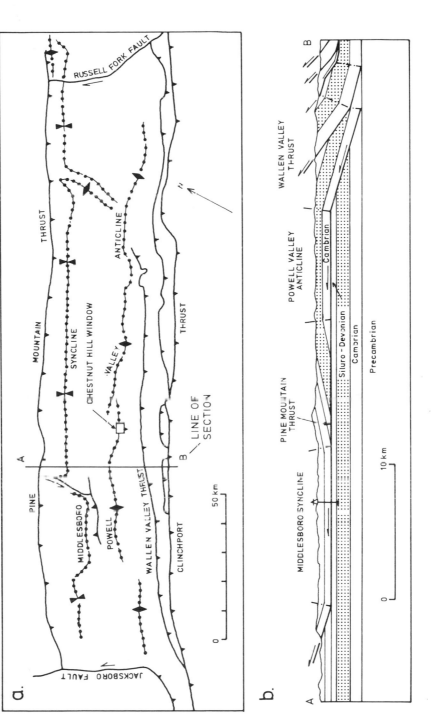

Fig. 9.27 Map (a) and cross-section (b) through a natural example of fault-bend folding in a thrust belt—the southern Appalachians. The Powell Valley Anticline and the Middlesboro Syncline can be readily understood in terms of the geometries shown in Figures 9.19 and 9.20. The Jacksboro and Russell Fork Faults are strike slip in nature subparallel to the overall transport direction

Figure 9.27 indicates that the overall displacement direction was towards the north-west. Subparallel to this transport direction are several strike-slip faults which support the interpretation of the dynamics.

9.8.2   EXTENSIONAL FAULTS

Extension by faulting in the brittle part of the crust is accommodated by one of three styles of behaviour (Figure 9.28):

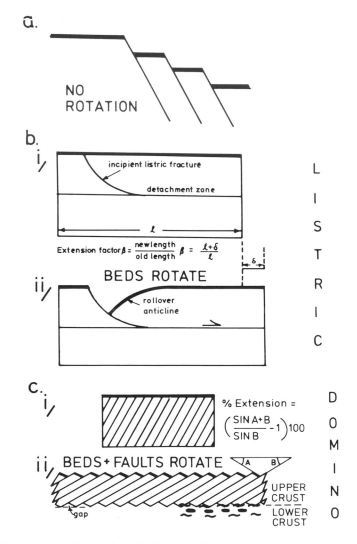

Fig. 9.28 Basic modes of extensional faulting. (a) Neither beds nor faults rotate and this is the style assumed in, for example, Figure 9.14. (b) Listric faulting where the fault maintains its geometry and the beds in the hanging wall rotate to keep in contact with the footwall thus creating the rollover anticline. The Beta factor quantifies the amount of extension from the initial state (i) to the final state (ii) as was done in Figure 9.16. (c) Domino faulting where both faults and beds rotate. The amount of extension can easily be calculated from the dip of the beds ($A°$) and the faults ($B°$). If the behaviour was totally brittle then the gaps at the base of the faulted zone would be a problem. However, this can be accommodated by a mixture of distributed fracturing and ductile strain at the brittle (upper) to ductile (lower) crust transition

1. **Non-rotational** planar faults.
2. Curved **listric** faults where **beds rotate**.
3. Planar **domino** faults where both **beds and faults rotate**.

Rotation, or the lack of it, is the basis of the classification. During extension, neither faults nor beds rotate in category (1) and the faults are essentially planar. The distinctive concave upwards-curved profile of listric faults allows these discontinuities to act as divides between stretched and unstretched regions. A further consequence is the folded layering (Figures 9.28bii, 9.29 and 9.30) in the displaced block — the hanging wall. Within an extended region, the domino style involving rotation of beds and faults, is probably the best means of allowing very high extensions.

The relative importance of the three styles is much debated but each one can dominate in areas of hundreds of square kilometres, and the analysis of a faulted extensional region should attempt to characterize the basic style (given that the map provides sufficient data). This debate is part of a wider discussion on lithospheric stretching, and at the scale of the whole lithosphere there are many competing

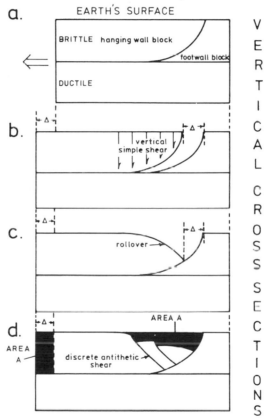

Fig. 9.29 (a) Major listric faults curve to become horizontal at the brittle/ductile transition level in the crust. Smaller examples sole out in weak layers (e.g. shale, evaporites). (b) Displacement of the hanging wall block as a rigid unit would create a large gap. (c) Rocks are too weak to support such an overhang and one response is internal deformation within the hanging wall to generate the rollover fold. The extension is Δ (delta). (d) Rather than a continuous response, the hanging wall deformation may be brittle on antithetic faults, perhaps with reduced amounts of folding. The black area A above the listric fault emphasizes the amount of sediment that would be needed to fill the basin created by the listric faulting. In pure extension the two areas labelled A should be the same — the system is balanced. The ductile lower section thins homogeneously during the extension. (Reproduced by permission of the Geological Society from R. A. Chadwick, Extension tectonics in the Wessex Basin, southern England, in *Journal of the Geological Society, London*, vol. 143, 1986.)

models to explain its extensional behaviour. The topic creates a link between structural geology, geophysics and sedimentology/stratigraphy, increasingly seen as natural bedfellows in modern studies of basin evolution. A stretched lithosphere is thinned and any continental crust component becomes involved in the extension. The process effectively replaces, at depth, light crust with denser mantle material and a consideration of isostasy says that subsidence should occur thus creating a sedimentary basin. The driving mechanism is the well-known horizontal displacements of the plate mosaic which may lead to complete continental rupture (continental drift) or which may be arrested to produce rift valley systems (Figure 9.31). The end-products of these processes are the tremendous accumulations of sedimentary rock on the continental margins (Atlantic or passive type) which eventually will be incorporated into the continents by orogenic activity. Mechanical stretching of the lithosphere has interesting thermal results in that hot material is taken nearer to the surface during extension. Cooling eventually restores the pre-stretching geotherms and this leads to a gradual (exponential decay) contraction of the lithosphere with more subsidence to match. Sedimentary infill to the basins reflects the overall tectonic history. Initially subsidence is rapid and fault controlled as the brittle top of the crust fractures (Figure 9.31). Extension stops and, as gradual cooling takes over, the later sediments truncate the earlier faults (Figure 9.31). Subsidence is greatest at the centre and decreases towards the margins leading to overlap at the edges and a profile referred to as a 'steer's head'. On the continental

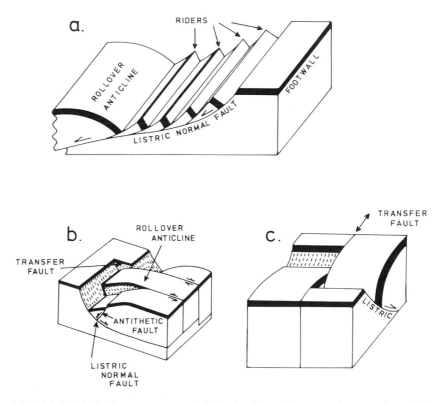

Fig. 9.30 (a) Listric faults can propagate into the footwall generating a series of riders. As each new listric fault is formed the earlier ones have to change shape so the basic principle of Figure 9.28(b) (beds only rotate) no longer holds. (b) Transfer faults strike parallel to the extension direction and allow separate blocks to behave somewhat independently. For the most part they are strike slip but, in the zone of rollover, folds have oblique slips. (c) Across transfer faults the sense of concavity of listric faults may change showing the decoupling these structures allow

margins 'break up' unconformities mark the change from active stretching to thermal contraction.

Listric faults create some interesting geometries. In scale they range from centimetric in soft-sedimentary deformation to kilometric during lithospheric stretching. Their curvature to a subhorizontal detachment surface (Figure 9.28b) is controlled in sedimentary basins by weak layers (shale, evaporites) or at the crustal scale by the brittle – ductile transition (Figure 9.29a) at 10 – 15 km depth. If the hanging wall block remained rigid during extension, a very large void would be created (Figure 9.29b). But, as rocks do not have sufficient strength to maintain such a gap, the blocks stay in contact and a **rollover anticline** is generated. Rather than simply folding, the hanging wall more commonly fractures at the same time as folding. Faults opposite to the main listric fault, antithetic faults (Figures 9.29 and 9.30), may dominate in the hanging wall. The rollover fold alternatively may be affected by a more symmetrical arrangement of faults in a **crestal collapse graben** (Figure 9.32a) which has a central downfaulted block. Additional variations occur in a form of footwall collapse where a series of listric faults propagate into unstretched regions (Figure 9.30a). The individual blocks bounded by these faults are referred to as **riders**. At large extensions they will have travelled considerable distances down their basal curved faults which requires the rider and its overlying fault to distort. We, therefore, can no longer claim that only bedding is rotated when dealing with groups of closely spaced listric faults.

At the time of extension, many fault surfaces intersect the earth's surface and each earthquake/displacement event is associated with creation of surface relief (fault scarps). Several diagrams show the results of faulting at the earth's surface (Figures 9.29, 9.30, 9.31 and 9.32). In particular, Figure 9.29(d) illustrates the cross-sectional area of a topographic trough associated with one style of listric faults. Syn-faulting sediments will be deposited in this depression and Figure 9.33 shows some of the consequences of growth faulting. Sediments deposited during a particular interval will be thicker in the hanging wall adjacent to the fault and will thin away from the fault (Figure 9.33). Younger beds/formations in the hanging wall will have progressively lower dips because they will have undergone less extension and, therefore, less rotation (Figure 9.33). The production of a fault-controlled topography may influence the sedimentation pattern depending upon the balance between rate of sedimentation (± erosion) and rate of fault displacement. Small fault scarps in areas of rapid sedimentation will not strongly influence sediment dispersal patterns. At the other extreme, slow sedimentation in zones of active faulting will generate

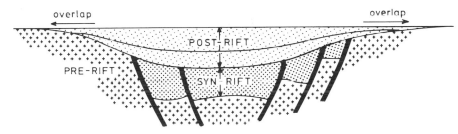

Fig. 9.31 In the formation of a major rift (tens to hundreds of kilometres wide), faulting related to the stretching happens in a brief period of time (<10 Ma). The thinned crust has to subside following isostatic considerations and syn-rift sediments infill fault-bounded troughs (grabens). The thermal pattern of the crust is disturbed by the stretching and, as cooling takes place, contraction allows more sediments (post-rift) to infill the basin. The post-rift sediments truncate the rift faults and cover a larger area than the fault-bounded basin. Progressive contraction expands the later basin giving rise to overlap (see Figure 11.6). (Reproduced by permission of the Geological Society from R. A. Chadwick, Extension tectonics in the Wessex Basin, southern England, in *Journal of the Geological Society, London*, vol. 143, 1986.)

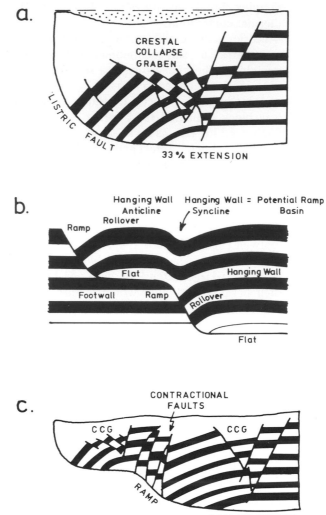

Fig. 9.32 (a) The folding that forms a rollover fold may not be completely continuous and extension of the layers on the fold outer arc generates conjugate normal faults. The central down-dropped part of this fault array is a graben (crestal collapse graben) and these effects are propagated to the surface as a depression which typically fills with sediment (stippled). (b) Listric faults often have ramp/flat alternations like contractional faults, particularly if they reuse earlier thrust faults. In a continuous response, each ramp has a rollover fold and the overlying layers are folded into a hanging-wall syncline. (c) A more likely fold + fault response to extension on a staircase trajectory. Two crestal collapse grabens form and a zone of contractional faults forms in this overall extensional environment. ((b) Reproduced by permission of the Geological Society from A. D. Gibbs, Structural evolution of extensional basin margins, in *Journal of the Geological Society, London*, vol. 141, 1984.)

pronounced topographic features like the rift valleys of Africa (Figure 9.34). In these continental terrains there are a large number of contemporaneous environments (fault-line scarps, ephemeral lakes, deltas, alluvial, flood plains, aeolian). Conglomerates from the fault-line scarps may only extend a few hundred metres from the fault and be adjacent to fine-grained lake sediments. Most of the sediment fill is influenced by the rollover anticline and many factors (e.g. climate, tectonic) combine to control features such as lake transgression and regression. Dramatic lateral facies variations are the order of the day in most fault-controlled basins.

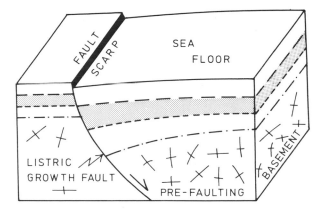

Fig. 9.33 Listric faults reaching the surface are active during sedimentation as growth faults. Equivalent formations are thin on the footwall block and thicker in the hanging wall but thin away from the rollover fold. The oldest syn-faulting sediments will have the steepest dip in rollover and dips decrease progressively as the sediments become younger; a fan of bedding is generated. Small fault movements, widely spaced in time, will not influence sediment dispersal, but very active faulting can produce large stratigraphic differences on either block

In any subsequent modification of a situation like Figure 9.31, either by further extension or by contraction, the effects could appear quite different depending on whether the pre-rift, syn-rift or post-rift geology is being considered (Figure 9.35). If the superimposed event were contraction (**inversion tectonics**) the amount of true contraction would only be shown by post-rift sediments (Figure 9.35). Shortening may restore the oldest of the syn-rift sediments back to their pre-extension position with no net change, thus masking the effects of the latest event. Net contraction on reactivated extension faults may totally obscure the important early extensional event. These difficult topics have to be faced in many orogenic belts which represent considerably foreshortened and collided continental margins which were once dominated by extensional structures.

Another aspect of reactivation is that many extensional faults rework earlier thrust faults. If the pre-existing thrust structure had a staircase trajectory, the hanging wall extensional structures will reflect this complexity (Figure 9.32b, c). Given that the hanging wall responds by folding (Figure 9.32b), one ramp at depth generates an extra rollover anticline and the development of a hanging wall syncline. If Figure 9.32(b) represents a near-surface situation then sedimentary basins will form in the hanging wall syncline (a **ramp basin**) and over the uppermost rollover anticline. Significant extension will displace the ramp basin away from the flat-ramp bend though the latter will continue to be the locus of subsidence; the result is a very asymmetric basin with youngest sediments towards the centre of the extended region. With a mixture of continuous and discontinuous deformation the ramp—flat configuration produces an extra crestal collapse graben for each ramp (Figure 9.32c). Also, paradoxically, the ramps generate contractional faults (Figure 9.32c) in a region of extensional tectonics.

Many regions will not have sufficient relief to show the true nature of listric faults particularly if the blocks are kilometres thick. Their recognition from maps alone depends upon identifying allied structures, especially the rollover folds. Careful studies of stratigraphic thickness variation may confirm the listric nature of a fault. Differences between stretched and unstretched regions should be apparent in a regional map, though groups of related listric faults may create difficulties in establishing which is the undeformed reference area. Analysis is greatly aided by subsurface data from drilling and seismic or reflection profiling.

176

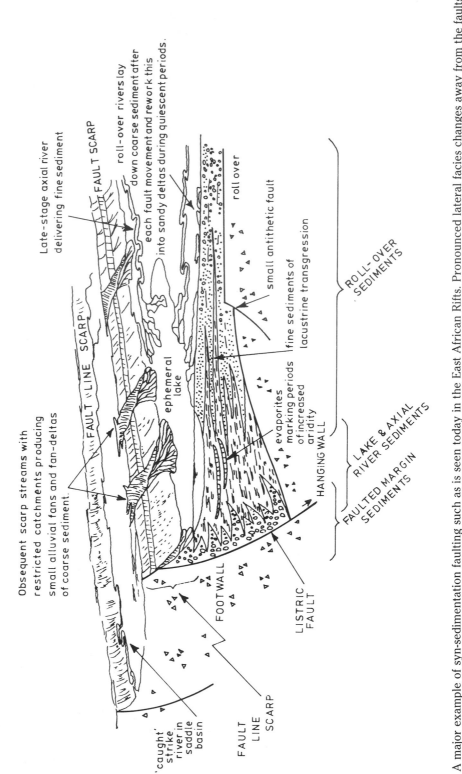

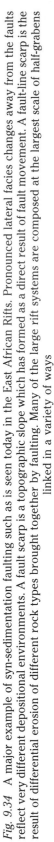

Obsequent scarp streams with
restricted catchments producing
small alluvial fans and fan-deltas
of coarse sediment.

Late-stage axial river
delivering fine sediment

FAULT SCARP

roll-over rivers lay
down coarse sediment after
each fault movement and rework this
into sandy deltas during quiescent periods.

FAULT 'LINE SCARP

ephemeral
lake

roll over

small antithetic fault

fine sediments of
lacustrine transgression

ROLL-OVER
SEDIMENTS

evaporites
marking periods
of increased
aridity

HANGING WALL

LAKE & AXIAL
RIVER SEDIMENTS

FAULTED MARGIN
SEDIMENTS

FOOTWALL

LISTRIC
FAULT

'caught' strike
river in
saddle basin

FAULT
LINE
SCARP

Fig. 9.34 A major example of syn-sedimentation faulting such as is seen today in the East African Rifts. Pronounced lateral facies changes away from the faults reflect very different depositional environments. A fault scarp is a topographic slope which has formed as a direct result of fault movement. A fault-line scarp is the result of differential erosion of different rock types brought together by faulting. Many of the large rift systems are composed at the largest scale of half-grabens linked in a variety of ways

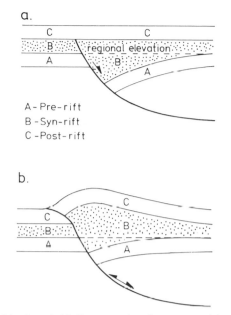

a.

A-Pre-rift
B-Syn-rift
C-Post-rift

b.

Fig. 9.35 Contraction (b) of an initially extensional structure (a), which mostly reuses the earlier fractures, is known as positive inversion tectonics. Only the post-rift sedimentary layering (C) can give a true measure of the superimposed shortening. Not all of the initial extension may be removed (lower beds B) giving net extension for this part of stratigraphy

Domino faults appear to be very common in some highly extended terrains. A simplistic view of the domino model (lefthand side of Figure 9.28cii) suggests a room problem because the tilting of the dominos leaves large gaps at their bases. However, if the faulting terminates at the brittle – ductile transition then the gap problem may be overcome by a mixture of more distributed faulting and ductile deformation (righthand side of Figure 9.28cii). Simple measurement of the dip of both the faults and bedding allows the amount of extension to be calculated (Figure 9.28c) because both features undergo rigid body rotation. A symmetrical response to lithospheric stretching would create a horizontal transition between a domino faulted upper crust and a more homogeneously deformed lower crust (Figure 9.28c). However, some regions appear to have major discontinuities dipping around 10° which has led to models of asymmetric lithospheric stretching. The role of transfer faults (Figure 9.30) may prove crucial in understanding these regions. Transfer faults allow adjacent districts to behave somewhat or totally independently and, therefore, extensional zones may vary in style considerably along strike. Across a transfer fault, rollover anticlines may have opposite senses of closure (Figure 9.30c) and even more fundamental changes are proposed particularly at the scale of continental margins. Perceptions of the nature of transfer faults also feed back into the interpretation of listric faults. Recent analysis of continental rifts in the early stages of development (Red Sea, East Africa) are suggesting that listric faults are curved in plan as well as section and have a 3-D spoon shape (Figure 9.36). Between faults with opposing senses of curvature are accommodation zones (basement highs) bounded by oblique slip fractures; these zones are the equivalent of transfer faults in Figures 9.30b, c and 9.36(a). Both models a and b (Figure 9.36) have been applied to rifts with small extensions, but at present the data base is too limited to say if both styles exist or whether there is a difference in interepretation of similar structures. The evidence from the East African Rifts strongly suggests major curvilinear faults as the dominant component and the alternations of zones with opposite concavity are well defined. The latter create sigmoidal sedimentary basins with an overall half-graben shape.

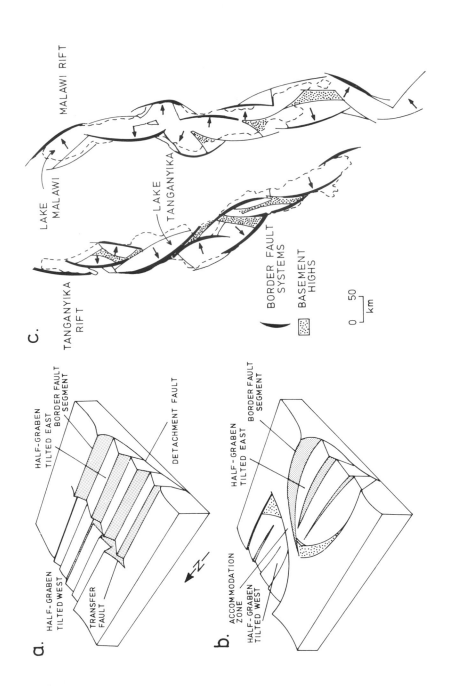

Fig. 9.36 At the scale of rifts (tens of kilometres wide) some regions have planar transfer faults that separate major zones of opposing listric fault concavity (a). Other regions are best interpreted in terms of curvilinear spoon-shaped listric faults that alternate concavity on accommodation zones (b). Several parts of the East African Rift system (c) are believed to be laterally linked generally alternating sets of curvilinear half-grabens.

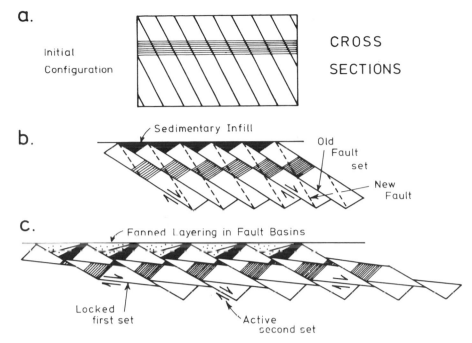

Fig. 9.37 Domino faults are initiated with dips around 60° (a). As they rotate the first fault set locks and a new set is initiated dipping at around 60° (b). During the first period of rotation triangular sedimentary basins form at the surface (b) and these progressively enlarge throughout the extension history (c). Again there is a fan of layering dips in these new basins with the greatest dips in the oldest units (c)

Because of their setting, domino faults will normally intersect the surface and create syn-faulting sedimentary basins (Figure 9.37b). Also at high extensions and low fault dips (30 – 40°), the earlier faults lock up and a new set of faults are initiated at dips of around 60° (Figure 9.37b, c). The new faults rotate with increasing extension, rotating all earlier features including the earlier syn-faulting sediments. The full sequence of syn-faulting sediments shows decreasing dip in the younging direction and thus creates a fan-like array (which will be faulted by new fractures, Figure 9.37c).

Figure 9.15(a) suggests that the map pattern of planar normal faults should be simple parallel arrays of fault sets with opposite dips perhaps broken up into compartments by strike-slip transfer faults. This treatment, however, considers that the influence of the intermediate principal stress (σ_2) is neutral. In contrast, if σ_2 is a compressional direction four fault sets are generated synchronously (Figure 9.38). Many interpretations of such patterns would invoke a polyphase history of two differently oriented stress fields operating sequentially to explain the observed geometry.

9.8.3   STRIKE-SLIP FAULTS

The basic notion of blocks sliding past one another (Figures 9.13 and 9.15c) has been known for a long time as has an awareness of associated sizeable displacements, but recent tectonic analysis now proposes thousands of kilometres of slip along some strike-slip zones. Also in the build-up to the goal of predicting earthquakes, this pattern of faulting includes most of the better-known structures on earth. Being one of the three types of plate boundary (transform) has considerably assisted the

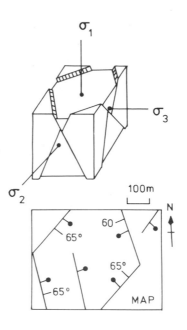

Fig. 9.38 Figure 9.15 assumed that the influence of the intermediate principal stress to be neutral, but, if it is compressive, four fault sets form in a single stress field instead of two. The map pattern created has normally been interpreted as two distinct phases of brittle failure with differently oriented stress fields

elevation to prominence of this fault category. An analysis of present-day plate boundaries shows that 60 per cent have significant strike-slip components and there is no reason to suspect that this was much different in the geological record. It is only now, in the late 1980s, that the importance of strike-slip displacements in ancient plate boundary zones (orogenic belts) is coming to light. Perhaps the general recognition of their importance has been hampered, paradoxically, by the diversity of specializations involved in the study of strike-slip systems. They make good sedimentation traps of interest to petroleum geologists, they generate very destructive earthquakes, they localize ore deposits, volcanic activity, hot springs and they are commonly geometrically complex. All this attracts sedimentologists, petroleum geologists, geophysicists, economic geologists, igneous petrologists and structural geologists, but it has taken time for coordination to develop.

At the simplest level we are dealing with nearly vertical faults, in single sets or conjugate arrays that have subhorizontal net slips (Figure 9.15c). When reading the literature on the topic you have to be familiar with the variety of synonyms that has been applied to this style of deformation, e.g. wrench, tear, transcurrent and lateral. In areas dominated by strike-slip tectonics one set of faults usually accounts for much of the differential displacement — principal displacement zones. Between two parallel faults the country rock is subjected to a rotational stress history (Figure 9.39) and resolving the stresses within the zone gives a maximum compression orientation at 45° to the bounding faults. The ellipse marked in Figure 9.39 was derived from a pre-deformation circle and is designed to emphasize the extension and compression directions. With large amounts of deformation within the fault block the ellipse would develop a high axial ratio and its long axis progressively becomes closer to the bounding fault. In the brittle part of the crust the main fault sets accommodate much of the displacement whereas deformation within the fault blocks may be limited. Under ductile conditions the strain will normally be more evenly distributed as a ductile shear zone (see Figures 7.3d and 9.13).

a. PLAN VIEWS

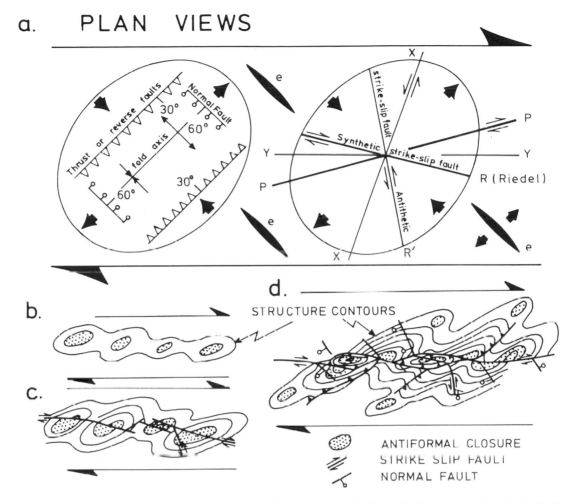

Fig. 9.39 (a) Strike-slip structures in a right-lateral (dextral) shear. An original circle is transformed into an ellipse which emphasizes the resolved extension and shortening directions. R, R', X, P, Y fractures are all strike slip; contractional plus extensional (dip-slip) faults can also be created at the same time. Extensional fractures (e) and folds are also likely to form R = synthetic Riedels; R' = antithetic Riedels. (b) The sequence of development of subsidiary structures varies, but commonly *en échelon* folds are the first expression which are then cut by synthetic Riedel strike-slip faults (c). As deformation increases a wider range of structures is created, mostly with *en échelon* patterns. In (b), (c) and (d) schematic structure contours on one bedding surface outline the general structure. Culminations (antiforms) are elongate domes that could be hydrocarbon traps

Sedimentary rocks at high crustal levels best display the complete range of responses to a shear couple of forces. Within a single fault block it is possible to find the full array of dip slip and strike slip illustrated in Figure 9.39(a) together with folds. The sequence of development will vary from place to place and not all sectors will produce all members of the array, but the scope for variety (complexity ?) is considerable! Of the strike-slip faults in the zone the synthetic (R) and antithetic (R') **Riedel shears** are easiest to understand. They form a conjugate pair about the resolved maximum compression direction and their senses of strike slip can be predicted following the reasoning applied to Figure 9.15. The synthetic Riedels have the same strike slip as the overall zone and the antithetic are opposite. The Y faults (sometimes referred to as D faults) are simply minor representatives of the principal displacement zone (PDZ, Figure 9.40) set within the major fault blocks. The X and P

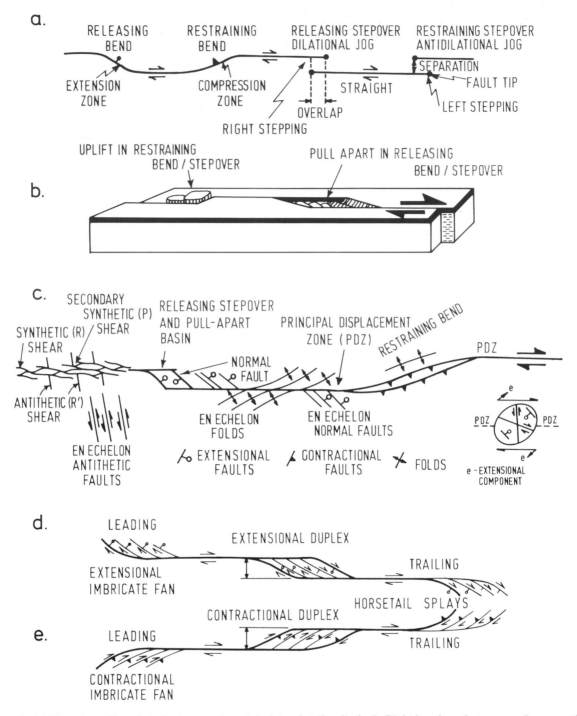

Fig. 9.40 (a) Plan view of bends and stepovers in a right-lateral strike-slip fault. Right bends and stepovers (in sympathy with the strike-slip sense) are releasing/dilational and create extension. Left bends and stepovers are restraining/antidilational and are the sites of compression. Overlap and separation of straights in a stepover control the detailed expression of extensional or contractional deformation. (b) A 3-D view of the consequences of non-planar strike-slip faults. (c) Different expressions of a strike-slip deformation zone in plan view. It may be an array of R plus P strike-slip faults with minor R' shears. It could be *en échelon* folds plus normal faults or simply a principal displacement zone (PDZ). Several other variations are possible. Also shown are the elements of structures in a releasing stepover (a pull-apart basin, e.g. Dead Sea) and a restraining bend (thrusts). The ellipse inset is to emphasize the extensional and contractional directions. (d) Plan view of right-lateral slip creating imbricate fans and right stepping to create an extensional duplex. The terms imbricate and duplex are taken from the thrust (contractional) fault literature where

Fig. 9.40 t

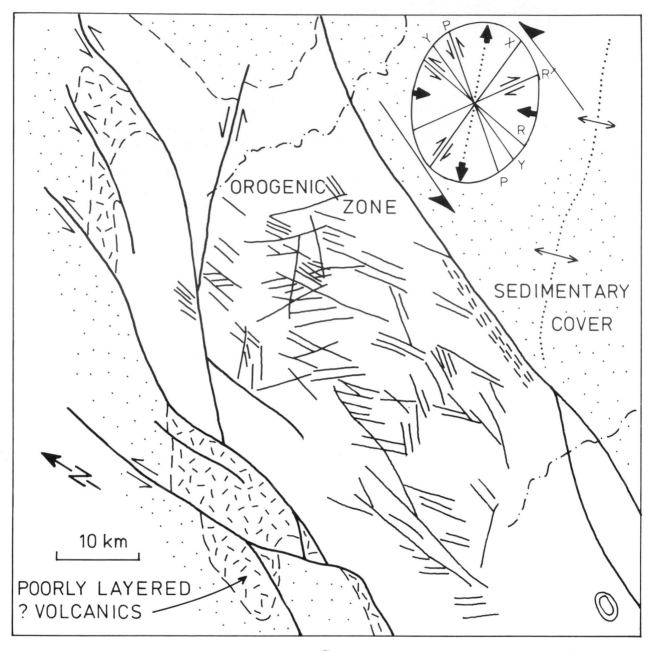

g

similar features are seen in cross-section. (e) Plan view of right-lateral slip with an imbricate fan (now contractional) and
left stepping so the duplex becomes contractional. (f) A Large Format Camera (LFC) image of part of an orogenic belt
(Halls Creek Mobile Zone, Western Australia) that has undergone late stage strike-slip deformation. The eastern edge (see
interpretation (g)) is a stereo view. The original negatives cover a strip of ground over 300 km long which gives
tremendous potential for gaining an overview of large-scale structures. Coupled with stero capacity, and good resolution,
LFC images are extremely valuable for regional interpretation. Note the varying expression of faults from thin
upstanding silicified ridges to negative linear troughs. (g) Structural interpretation of (f). Major faults are heavy lines and
the thinner lines within the orogenic zones (plutons and high-grade metamorphic rocks) are some of the fracture arrays.
The two ornamented units are cover sequences to the orogenic complex. The inset shows the ideal fracture patterns in
an ideal left-lateral strike-slip zone. The major faults on the west side, which offset the cover/basement boundary, are
best interpreted as Riedel faults as are many of the larger fractures within the orogenic zone. Along the eastern edge of
the orogenic zone, a single major fault separating basement and cover, is in the Y direction. Some of the complexity of
the fracture pattern within the orogenic zone could be smaller-scale shear zones within the regional shear zone

faults or shears are less well understood theoretically but have been observed in nature and in experimental deformation tests. The Ps are synthetic with respect to the regional shear sense and, if they occurred in contractional zones, would be thrust faults — a designation confusingly given to them in strike-slip terrains. The dip-slip faults (reverse and normal) and the folds are all easily related to the extension and compression directions. As shown in Figure 9.39, extensional fractures should trend at around 45° to the PDZ and in pure strike slip they would be vertical.

One of the most distinctive aspects of strike-slip zones is an **en échelon** arrangement of structures, that is, parallel arrays of folds or faults oblique to the general trend of the zones (Figures 9.39b and 9.40c). In areas of low strike slip and/or regions without well-developed principal displacement faults, the boundaries of the zones may be poorly defined. Gentle *en échelon* folds may be the only expression of incipient horizontal displacements (Figure 9.39b). With increasing deformation more and more of the associated structures develop (Figures 9.39c, and d, and Figure 9.40). Some strike-slip zones only contain arrays of synthetic Riedel shears, others just arrays of *en échelon* extensional faults, and others solely consist of *en échelon* antithetic Riedels; these monostructural arrays are not common but help to display the variety of strike-slip tectonics. Fault development within the zones most commonly starts with arrays of synthetic Riedels (Figure 9.40c) which are then linked by P faults (with perhaps some Y/D faults). Minor antithetic Riedel faults are often associated with this sequence of formation. If this complex array is not cut by a throughgoing principal displacement fault, the pattern is as seen on the left of Figure 9.40c though disruption by a major fault can put a lot of distance between the two halves of the array. With potential displacements of many kilometres, features that were initiated in one area at the same time on either side of the fault, may eventually become far separated.

Individual strike-slip faults rarely remain planar for long. Curvature is typically localized in pronounced bends but the faults also step sideways in a type of *en échelon* array (Figure 9.40a). If the sense of bending or stepping is the same sense as the net slip then extension occurs—see the pull-apart zone of Figure 9.40b—and these are releasing bends or releasing stepovers/dilational jogs (Figure 9.40a, b, c). Bending or stepping senses opposite to the slip sense leads to compressional zones— see the uplifted/thrusted zone of Figure 9.40b—in restraining bends or restraining stepovers/antidilational jogs. Figure 9.40a is a right slip PDZ, hence right-stepping or right-bending produces extension; left-stepping or left-bending creates compression. In a stepover, the nature of the structure generated between the straights depends upon the separation of the PDZs and the amount of overlap (if any).

Releasing bends and stepovers are ideal places to generate sedimentary basins and many of the younger (<100 Ma) major strike-slip faults have concentrated hydrocarbon resources in such features. The sometimes close association of restraining and releasing structures provides both provenance and depository for sediments. Also many strike-slip faults anastomose and the sense of convergence or divergence of strands together with slip sense dictates which lozenge-shaped blocks will be upthrusted to form topographic hights or downdropped to form basins. The surface expression of strike-slip faults is well known. In populated areas they are marked by horizontal offsets of rows of trees in orchards, footpaths, railway lines, fences, etc. Topographic indications include deflected drainage and offset canyons or similar features.

As an extension of geometric analysis in better-known thrust systems, comparisons with strike-slip faults have highlighted common features. Faults that diverge away from the PDZ generate **imbricate fans** with extensional or contractional components depending on the fan sense relative to slip sense (Figures

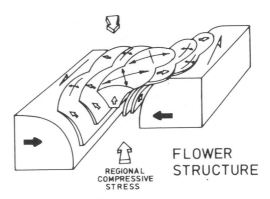

Fig. 9.41 Flower structures are curved faults decreasing their dip upwards. They are created by strike slip and either contraction (transpression — as shown) or extension (transtension). They are very common in bends and stepovers of pure strike-slip faults but can form in regional transpression or transtension

9.40d, e). These fans are understandably referred to by some authors as **horsetail splays.** Again following thrust-belt terminology, if bends or stepovers are linked by faults and these zones expand by the formation of more links, **a duplex** results. As with thrust duplexes the basic characteristics are of faults diverging and then converging to produce fault-bounded blocks. Extensional or contractional behaviour depends upon stepping/bending sense in relation to slip sense (Figure 9.40d, e). Fans are **leading** or **trailing** (Figure 9.40d, e) depending on the relation between the major displacement flanking strand and the imbricate fan.

Figure 9.40f, g is an example of an ancient strike-slip zone. The Large Format Camera image allows the width of the whole zone to be examined and the stereo capacity illustrates the variable topographic expression of faults; some are upstanding silicified zones though most are etched-out linear troughs. On the eastern edge of the orogenic zone is a major PDZ, whereas the western edge is an *en échelon* array of major Riedel faults. The resolution of these images, when enlarged, would provide a very detailed analysis of the fracture pattern over a wide region. On the interpretation, only the large subsidiary fractures, within the strike-slip zone, are presented. Some of the pattern might be explained by shear zones within shear zones where, for example, zones between major Riedels behave as discrete shear zones producing their own array of subsidiary faults.

Cross-sections through stepovers and bends are very deceptive. A vertical section through a pull-apart basin in a releasing overstep will show extensional geometry even though the amount of strike slip may be much greater than the dip-slip. The Dead Sea—Gulf of Aqaba (Elat) transform system is a classical example of extensional basins in a zone dominated by strike slip. In extensional and contractional duplexes,

Fig. 9.42 (on page 187) In strike-slip zones, large rotations of major fault blocks about vertical axes are now well documented. (a)(i) A conjugate set of faults was created by north-north-east—south-south-west compression. (ii) During strike-slip displacement on the San Andreas the whole region shown was subjected to a couple of forces. The earlier fractures were reused with opposing senses of strike slip. Even though the overall sense is right lateral, there are major left-lateral displacements within the zone. Between the blocks, complex adjustments have created extensional zones. (b) A small-scale (tens of centimetres) example where a dried-out crust over soil was affected by right-lateral slip. Note the dominant left-lateral displacement of the subsidiary structures and the complex thrust plus extension structures. The markers in this case are ploughed furrows in a field. (c) A model for the resolution of transtensional tectonics within a zone. Plate A moves relative to Plate B as shown by the large arrow on Plate A giving resolved strike slip along the deforming zone and extension across it. Slats, fixed by pivots, duplicate fault blocks. During displacement the blocks rotate in

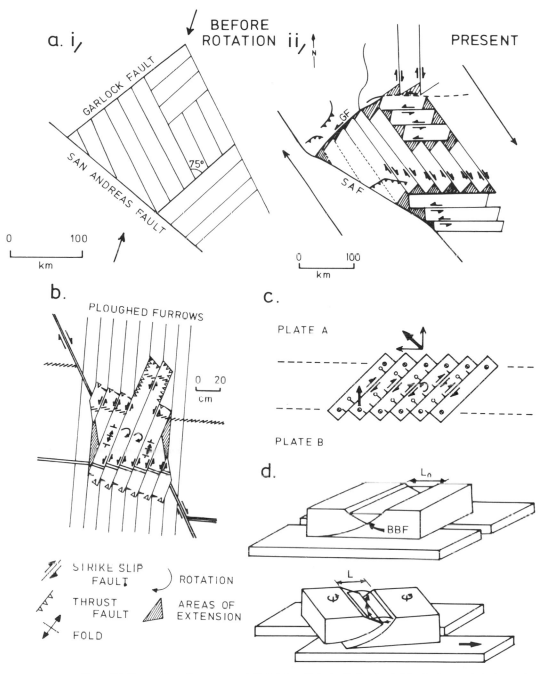

sympathy with the regional movement (left lateral), but all the fault blocks have right-lateral slip and would have extensional components (ball and bar symbol) in real examples. The net slip between adjacent blocks is shown by the large black arrow within the deforming zone. (d) An extension basin rotated by strike slip. The initial basin (width L_0) is defined by a listric normal fault BBF (basin bounding fault) and an antithetic fault in the hanging wall. During left-lateral slip of the region, the fault blocks either side of the basin rotate in sympathy. Slip on the BBF is much more complicated, first of all it continues to extend, then goes into pure strike slip (right lateral) and finally changes to hanging wall up, i.e contractional; the width of the basin changes accordingly. The slip history is traced by an arrowed line on the BBF surface. The final contraction stage is an example of inversion tectonics which will be expressed differently in different age sediments (cf. Figure 9.35). ((c) Reproduced by permission of the Geological Society from D. McKenzie and J. J. Jackson, A block model of distributed deformation by faulting, in *Journal of the Geological Society, London*, vol. 143, 1986.)

TRANSTENSION

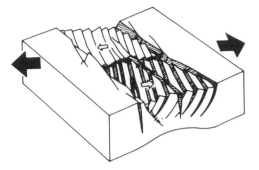

Fig. 9.43 Transtension will produce a complex array of strike-slip and dip displacements and fault styles and many faults will have oblique net slips. Transtension can either be regional or the localized response to a bend/stepover in a pure strike-slip system

the Riedel shears within the duplex have both strike-slip and dip-slip displacements. At depth these faults also link up with the main strand (PDZ) and in profile these faults have a convex upwards curvature (Figure 9.41) defining a **flower** structure. On parts of the San Andreas system, in restraining bends, considerable shortening has occurred and flower structures have thrust pre-Cretaceous granodiorite over Palaeocene sediments.

A characteristic of strike-slip zones is that rotation of large blocks about subvertical axes is common. Figure 9.42(a) gives a regional example where blocks 10 – 20 km wide have, on average, been rotated through 40° in the horizontal plane. In this example, at the junction of the San Andreas and Garlock Faults, the region was originally broken up into fault blocks by an earlier, near north—south oriented, compressive phase. Right lateral movement on the San Andreas Fault has been converted into a mixture of right and left lateral movements during the reworking of the earlier faults depending on their initial orientation. The extensional regions and the fault block contacts have been grossly simplified. Bulk rotation senses are commonly expressed as opposing strike-slip senses on subsidiary fractures within the deforming zone (Figure 9.42b, c, d). Figure 9.42(b) shows the effects of strike slip on a thin hard-pan layer in a ploughed field; all the rotations on the furrows are opposite to the overall rotation sense. This example also contains some interesting contractional and extensional structures as a consequence of the rotation. A similar style response is found in strike-slip zones which have extensional (transtension) components; the fractures in the deforming zone being antithetic to the regional displacement. The interaction of pre-existing structure and block rotations can generate some involved gelological histories. An extensional half-graben basin subjected to strike slip, given the correct relative orientations (Figure 9.42d), can initially continue to extend, then move into strike-slip displacement and eventually undergo contraction (inversion tectonics). As we saw earlier, the effects on syn-rift sediments during the block rotation event will be different to those seen in the pre- and post-rift units.

9.9 TRANSTENSION/TRANSPRESSION

Contractional structures in restraining bends and stepovers are localized zones of transpression (Figure 9.13) where typically the amount of strike slip considerably

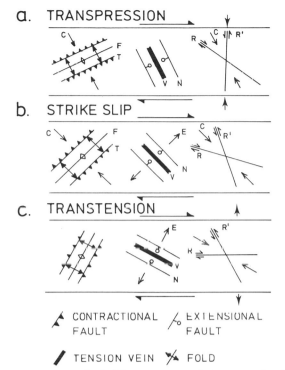

a. TRANSPRESSION

b. STRIKE SLIP

c. TRANSTENSION

CONTRACTIONAL FAULT

EXTENSIONAL FAULT

TENSION VEIN FOLD

Fig. 9.11 Variable arrangements of subsidiary structures in transpression (a) and transtension zones (c) relative to pure strike slip (b). With the additional compressive component across the transpression zone, the resolved shortening (C) is at high angles to the zone boundary. This skews the RR' conjugate pair and the extensional structures (V and N) to higher angles relative to the boundary. Contractional structures (thrusts and folds) lie closer to the zone boundary trend. Transtension has the opposite effect. C=compression; E=extension; V=vein; N=normal (extensional) fault; F=fold; T=thrust (contractional) fault; R=synthetic Riedel; R' =antithetic Riedel

exceeds the dip slip. Admittedly the localized areas may be several hundred kilometres long in a strike-slip zone thousands of kilometres in extent. This is the case in the 400 km long Transverse Ranges, a major topographic feature on a restraining bend on the San Andreas. This whole region is characterized by thrusting. We have also seen examples of **transtension** in releasing bends and stepovers along strike-slip faults. Transtension (Figure 9.43) will result in the oblique/*en échelon* arrangement of minor structures within the zone undergoing oblique extension.

At the plate tectonic scale, oblique convergence in a subduction zone is often resolved into thrust faults in the subduction complex and strike-slip faults in a zone nearer to magmatic arc. This partitioning of displacement styles accommodates the transpression but the spatial segregation may make the total effect difficult to decipher. Because 60 per cent of present-day plate boundaries are mixtures of strike slip with tension or compression, then we should expect analyses of the ancient record to reveal significant amounts of transpression and transtension.

In zones where transpressional and transtensional effects are more evenly distributed, the arrays of internal structures are differently oriented with respect to true strike slip (Figure 9.44). In transpression there is an extra component of shortening across the zone to add to the resolved compression within the zone generated by the strike-slip displacement. This means that the net shortening direction within the zone is at a higher angle to the zone boundaries (Figure 9.44a) with consequent effects on the orientation of the R and R' shears. Likewise the orientations of the dip-slip faults and the tension veins are modified. Extension across

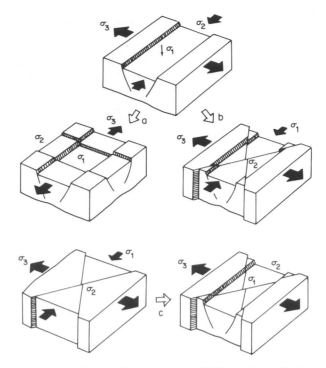

Fig. 9.45 Examples of polyphase brittle tectonics. (a) Extensional faults cut by orthogonal extension faults. Note that the intermediate and least principal stress orientations are interchanged from the first to the second extension. (b) Strike-slip faulting superimposed on an extensional fault geometry. Stress fields significantly rearranged though other orientations are possible. (c) Strike-slip faulting followed by extensional faulting. Ruling on fracture surfaces gives slip orientation

the strike-slip zone has the opposite effects on orientation, and the synthetic Riedel shears take up Y(D) attitudes; folds and contractional faults are at higher angles to the boundary, whereas extension veins and normal faults lie closer in orientation to the margins.

9.10   POLYPHASE BRITTLE TECTONICS

Most regions have undergone several distinct phases of brittle deformation with differently oriented stress fields. Remember, however, that apparently polyphase histories may be the result of progressive strike slip. Some of the possibilities are illustrated in Figure 9.45. An extended terrain may be reworked by later strike-slip deformation (Figure 9.45b) or extended in a very different orientation (Figure 9.45a). Naturally the later events need not be as orthogonal as shown here. A zone of strike slip may be later extended (Figure 9.45c) and if contraction effects are considered the range of permutations for just two events is increased. In the cases illustrated, reactivation is not considered. Extensional faults subsequently contracted may, in large part, be simply reused without the generation of new structures though some new fractures are likely. This topic has already been briefly mentioned under the heading of inversion tectonics.

10 IGNEOUS AND METAMORPHIC ROCKS: GEOMETRY AND MAP EXPRESSION

10.1 GENERAL CONSIDERATIONS

Geometrically igneous rocks provide quite a challenge. Formation boundaries in sedimentary sequences are fairly simple, typically being approximately planar at the 1 : 50 000 scale. Igneous masses are much less constrained in form and very irregular shapes are possible. Magmas, the products of partial melting at depth, rise buoyantly and eventually crystallize to produce igneous rocks. Those trapped during ascent (intrusive) and those that reach the surface (extrusive volcanics) display a vast range of morphologies. Chemically there is an equally vast diversity because melting occurs at different depths from different parent materials, melts rise at different rates (cm yr^{-1} to sonic velocities) and are held at different depths in magma chambers for varying lengths of time. During hiatuses in upward movement crystals settle out, changing the composition of the liquid that may eventually rise further. Consequently, to cope with the variety, classification of igneous rocks reaches nightmarish proportions. If you thought fault terminology was troublesome, wait for igneous petrology! It takes some considerable exposure to igneous petrology to know that a phonolitic tephrite is a lava.

Igneous rocks on many map keys are seemingly relegated to 'also rans' and typically scant information is given. However, for someone with the basics of igneous classification, the clues are powerful. Granites, gabbros, diorites and granodiorites (Table 10.1) are likely to form large intrusive masses, whereas dolerites, aplites and lamprophyres, commonly occur as thin intrusive sheets, and basalts, andesites and rhyolites are lavas. Most map-interpretation courses are not synchronized with other branches of the course so the bare necessities of igneous classification are given here. Problems are compounded in several ways: there are three fundamentally different ways of classification and, within each, international agreement on terminology and subdivision is slow in coming. A broad-brush crude scheme based on chemistry uses the SiO_2 content of the rock (Table 10.1), but there are many more variations using for example alkali content (Na, K) or the ratio of alumina to alkalis. Each additional approach mushrooms the number of names. Classification by colour index (light = felsic, dark = mafic, Table 10.1) is as broad as the SiO_2 system. Colour is generally dictated by the proportions of light to dark minerals, but some acid rocks are black despite being 95 per cent felsic minerals (quartz and alkali feldspar). Mineralogical classification is the most reliable but sophistication brings with it complexity. Coarse-grained high-silica igneous rocks are distinguished by the ratio of quartz to alkali feldspar to plagioclase feldspar; mafic minerals are largely ignored. The mineralogical basis has to be different for other parts of the chemical spectrum and even then there are quite a few oddball cases, e.g. magnetite lavas, carbonate lavas, etc.

Grain size is the additional parameter in Table 10.1. Glassy and fine-grained rocks are almost always extrusive though rare intrusives are in this category. Grain size of

Table 10.1 Classification of common igneous rocks

	Grain-size composition			
	Acid, high silica >63% SiO$_2$ (felsic)	Intermediate <63% >52%	Basic, low silica <52% >45% (mafic)	Ultrabasic <45% (ultramafic)
Glassy	Obsidian			
Fine (<1 mm)	Rhyolite (dacite)	Andesite	Basalt	Komatiite
Medium (1—5 mm)	Microgranite	Microdiorite	Dolerite	
Coarse (>5 mm)	Granite (granodiorite)	Diorite	Gabbro	Peridotite
Extrusion temperatures	(700 – 800 °C)		(1000 – 1200 °C)	(to 1700 °C)

intrusives largely depends upon the size of the body and coarse grain mostly means large bodies, medium grain size equals a small intrusion. The buoyant ascent of silicate liquids takes them to colder surrounds. A very small cubic kilometre intrusive mass would halve the temperature difference between itself and the country rock in 100 years; a very fast cooling rate which does not allow large grains to grow. Factors other than cooling rate may become important. In pegmatites, where crystals can be metres long, volatiles play a major role in crystallization. Thick basalt flows may have medium-grain cores tens of metres thick, but if the extrusive nature is clear the term dolerite is not used. Medium and coarse grain size were once equated with depth of emplacement, but coarse granites emplaced as large bodies under a couple of kilometres cover can be higher level intrusions than many dolerites.

From the map analysis point of view, several generalizations may be related to chemistry and allied characteristics. Viscosity strongly influences the form of intrusive bodies and extrusive accumulations. Viscosity depends on pressure, (*P*) and temperature *(T)*, the nature of volatiles, magma chemistry and the crystal plus bubble volume; all these parameters change with crystallization and so does viscosity. Basic to ultrabasic magmas are very fluid and can occupy narrow fractures, whereas acid magmas are very viscous tending to form large intrusions. Viscosity also influences rate of ascent so there is more opportunity for the slow viscous granitic magmas to crystallize before reaching the surface. Hence in acid-dominated terrains intrusives are much more common than extrusives. Once magmas are extruded, basalts can flow great distances (300 km) as thin sheets but sticky rhyolites pile up and rarely flow more than 3 km. Because generation sites are different, basic magmas have a much lower volatiles content than acid magmas. Eruption style is strongly controlled by the amount of volatiles; basic eruptions are mainly lavas, acid magmas mainly erupt explosively generating fragmental pyroclastics resulting from the rapid expansion by decompression of volatiles in the ascending magma. Lava flow morphology is also a function of effusion rate and environmental factors such as slope and terrain at the time of eruption.

10.2 INTRUSIVE SHEETS

Igneous rocks occupying approximately planar fractures have a sheet geometry (Figures 10.1 and 10.2) which is easy to analyse with our earlier construction

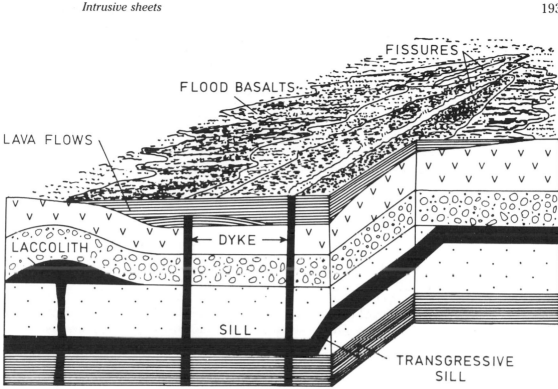

Fig. 10.1 A schematic representation of flood basalts, which have swamped a pre-existing topography, and feeder dykes (transgressive sheets). The generally concordant sills locally cut across bedding. The view of the laccolith is somewhat romanticized as many are flat-topped sheets of limited area surrounded by a monocline or circular to elliptical fracture

methods. Structure contours can be drawn for the boundaries the intrusions make with the country rocks (wall rocks) and attitudes (dips and strikes) determined. Some Geological Surveys use a special symbol to indicate the attitude of such boundaries but this is rarely used. Usually an intrusive contact is simply shown by the standard line for a lithological boundary, hence its igneous nature has to be determined from reading the lithology key and recognizing the igneous material. Many intrusive sheets are either parallel (concordant) or perpendicular (discordant) to bedding which simplifies the analysis of form. These relations, however, are not mandatory and intermediate intrusion/layering angles are not uncommon.

Concordant sheets are **sills** and are mostly but not exclusively basic in composition. Famous examples are the Tasmanian Jurassic dolerites and the Palisade sill, New Jersey. Individual sills reach 400 m in thickness and, in complex branches, 800 m vertical thicknesses of dolerite are recorded. Many dolerites are injected at over 1000 °C into county rocks at less than 200 °C. This temperature contrast chills the outer 10−20 m of dolerite which becomes fine grained and a similar amount of country rock is baked (**contact metamorphism**). These zones may be capable of representation on 1 : 10 000 maps but will not appear at smaller scales. Differentiation during crystallization produces an ultrabasic layer, a few metres thick, above the chilled base, a core of dolerite with gradual changes of mineralogy to a more acidic upper 20 per cent or so of granophyre. Some 1 : 50 000 to 1 : 250 000 maps will show the granophyre (Figure 10.3) and where the sill is intact this will give younging. However, even in areas that are virtually free of regional tectonism during intrusion, the granophyre appears to be very mobile and often moves from its site of generation (Figure 10.3). More ultrabasic liquids will have variations on this differentiation theme giving igneous layering of dunite/pyroxenite/dolerite from old

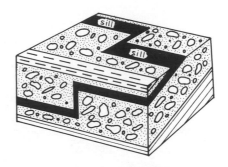

Fig. 10.2 A block diagram to indicate the potential influence on the outcrop pattern of a locally transgressive sill

to young. Sills usually form under less than 3 km of overburden and inject along bedding because the magma pressure (largely a buoyancy effect) exceeds the weight of the overlying rocks and is, hence, capable of lifting its lid. This is a positive dilation and the complex form of some regionally concordant intrusives can be understood if this is kept in mind. From A to B in Figure 10.3, dolerite has lifted the overlying rocks bounded by north – south-trending fractures probably generated at the time of intrusion. Note the opposite sense of strike separation on the subparallel faults at A and B. Also the sense of strike separation further south on the fault from A is opposite, thus emphasizing that emplacing varying thicknesses of dolerite was the main factor controlling the deformation of the country rock. Many of the isolated blocks of country rock seen in map view are probably isolated in the third dimension forming **xenoliths** or rafts (see also Figure 10.7). Sills locally transgress layering, often abruptly (Figures 10.1 – 10.3), but a transgression may be gradual, cutting bedding at a few degrees. Because most joints are bedding perpendicular the transgressive sill connections should typically follow this pattern, but if pre-existing joints are of a reverse fault pattern then low-angle links can form. Near the western margin of Figure 10.3, the regionally subconcordant Hart Dolerite cuts bedding at right angles over an outcrop distance of about 7 km. Another change from sill to discordance is located north of B. Similar patterns are seen on the stereopair of vertical air photographs (Figure 10.4) with the most abrupt transgression being between points A and B. Note how the blocks of country rock would fit back together if the dolerite were removed. The restoration would not be perfect and there are indications that pure dilation alone cannot explain the map pattern and perhaps some displacement parallel to the intrusion margins contributed.

Subplanar sheets discordant to bedding are **dykes** (Figure 10.1). Most are metres to tens of metres wide but the Great Dyke of Zimbabwe averages 5.8 km width over its 480 km length. All planar sheets are intruded subperpendicular to the least principal stress (σ_3) and imply that the magma pressure was greater than the sum of σ_3 and the tensile strength of the rock. Regional extension is the usual driving force for the majority of dykes which may form swarms of subparallel arrays particularly noticeable in gneissic basement terrains. These may be transcontinental in scale affecting areas of hundreds of thousands of square kilometres. Interaction between local and regional stress fields produces more complex patterns. For example, a vertical cylindrical volcanic neck will generate radiating vertical dykes in a regionally non-stressed environment (Figure 10.5). A far field tectonic stress pattern will cause a preferred orientation to develop around the neck, losing the truly radial arrangement. Again dilation is the most likely emplacement process though replacement (see Figure 7.5) may be important if the country rocks are high temperature (>500 °C).

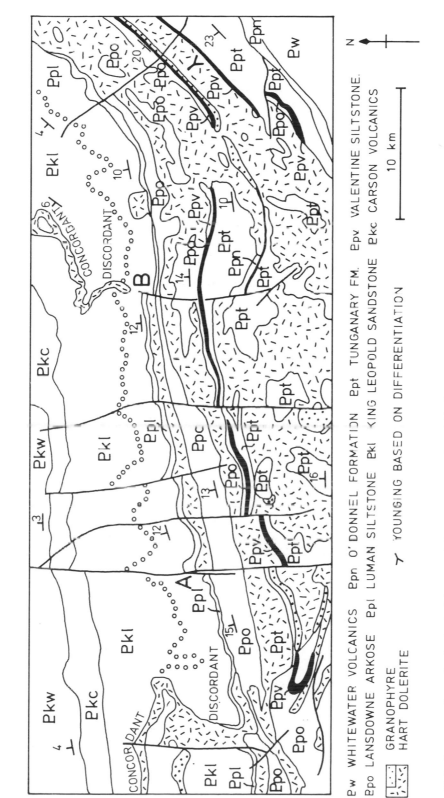

Pw WHITEWATER VOLCANICS   Ppn O'DONNEL FORMATION   Ppt TUNGANARY FM.   Ppv VALENTINE SILTSTONE.

Ppo LANSDOWNE ARKOSE   Ppl LUMAN SILTSTONE   Pkl KING LEOPOLD SANDSTONE   Pkc CARSON VOLCANICS

GRANOPHYRE
HART DOLERITE

⊥ YOUNGING BASED ON DIFFERENTIATION

Fig. 10.3 An extract from the Geological Survey of Western Australia's 1 : 250 000 Lansdowne map. A sequence of sediments and volcanics has been injected by regionally concordant dolerite sheets but there are many local transgressive relations. In the main intrusive mass, the isolated areas of country rock are probably isolated in 3-D forming xenoliths. Igneous differentiation generated a more felsic liquid in the late stages of crystallization, and where this granophyre remained *in situ* it gives information on younging. However, some irregular distribution of granophyre shows that this late stage liquid was very mobile. From A to B, dolerite injected by lifting the overlying sediments which shows the general dilational style of emplacement

Fig. 10.4 a

b.

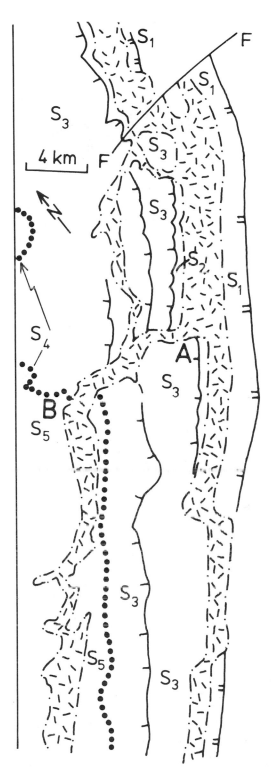

Fig. 10.4 A vertical stereopair of air photographs showing a sill which is locally discordant. In this climate dolerite is intensely chemically weathered and forms a negative feature, thus the transgressive portion is highlighted. Sedimentary layer S_2 is also easily eroded hence separating this from the dolerite is difficult. Recognition of layer S_2 shows that the generally concordant section of dolerite does cut layering at a few degrees. If the dolerite were removed the two sides of the intrusion would fit together fairly well but not perfectly, suggesting some displacement parallel to the walls, i.e. not pure dilation

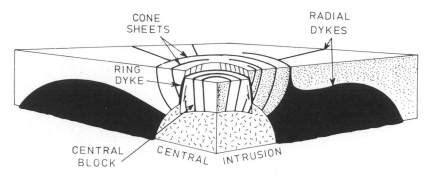

Fig. 10.5 A 3-D view of cone sheets, ring dykes and radial dykes above a stock-like intrusion. Cone sheets need upward displacement to create the dilation necessary for their emplacement and ring dykes are the opposite. The radial dykes result from the stresses built up around a pressurized plug

Reference to dykes and sills becomes difficult when the host rocks are not layered (e.g. granitic plutons or homogeneous gneisses). Under these circumstances many authors use dykes for steeply dipping planar bodies and sills for gently dipping sheets and there is a marked reluctance to refer to such bodies as veins. This has been extended by some workers to apply to intrusions into layered rocks irrespective of their relations to layering. To these people a horizontal sheet perpendicular to bedding would be a sill despite its highly discordant nature — watch out for how the terms are used!

Curved sheets in the form of partial cones are common above shallow intrusions (Figure 10.5). They are known as **cone sheets** if the cones point down but those closing upwards are **ring dykes**. With cone sheets, dilation on the bounding fractures occurs if the inner blocks are pushed up relative to their outer neighbours. Magma pressure from the central intrusion can drive this uplift process, but the generation of cone sheets themselves relieves the pressure by allowing magma to escape and naturally leads to ring dykes. These cause dilation on the conical fractures if the inner blocks fall and seem a logical consequence of the behaviour of a large shallow intrusion. Basic magmas vented at the surface by cone sheets are not violently eruptive (the Hawaiian fire fountains seen on the TV news are mild) and the integrity of the roof to the central intrusion is essentially maintained. A similar but much more catastrophic history occurs over an acid intrusion (see later discussion of caldera collapse in Section 10.7 on pyroclastics).

Ring dykes may be up to 2 km in thickness, 20 km in diameter and can create dramatic outcrop patterns. Good relief is necessary to distinguish between the vertical to outward dip of ring dykes and inward dip of cone sheets. Ring dykes are generally near vertical, whereas cone sheets often converge at moderate angles. Some ring dykes are crowded with small xenoliths suggesting enlargement by magma erosion of the walls. This mechanism overcomes the problem of vertical ring dykes where neither uplift or downdropping will create a space by dilation for the magma to occupy. It is possible that examples without xenoliths formed this way, followed by sinking of the inclusions through the magma. Cone sheets are typically metres in thickness but may occur in swarms occupying more than 50 per cent of the outcrop. Single centres, with multiple ring dykes and cone sheets, do provide complex outcrop patterns, but this is compounded (Figure 10.6) where several centres overlap and overprint (e.g. Ardnamurchan, Scotland, and the Younger Granite province of Nigeria).

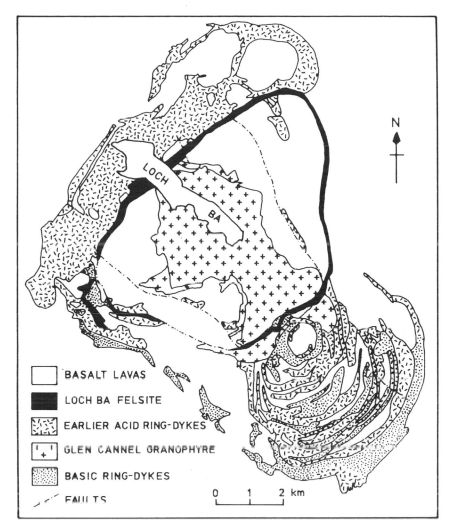

Fig. 10.6 An example of a map of a ring-dyke complex (British Geological Survey, Tertiary igneous complex, Mull, Scotland). The near-closed loop of the Loch Ba Felsite encloses parts of the basalt lava stratigraphy at lower levels than outside the dyke. The involved outcrop pattern of the earlier acid ring dykes shows the typical nature of these complexes

10.3 LACCOLITHS

These are dome-shaped concordant intrusions with a planar base which may be slightly discordant to the host layering (Figure 10.1). Normally they are limited in area to between 1 and 5 km^2 with dimensions up to 5 × 2 km and they are rarely more than 500 m thick. In plan laccoliths are circular to subcircular. The majority appear to have intruded nearly horizontal sediments under shallow cover (<3 km). Figure 10.1 is the somewhat romanticized view with a gentle continuous curvature to the top surface. Most in fact are flat topped with sharp deflections of layering between 30 and 90° at their margins so they are bounded by a circular monocline. Rather than a kink the change from roof to side may be a fault which dies out quickly (tens to hundreds of metres) upwards. Magmas involved are normally viscous acid alkaline compositions accounting for the limited extent. It is proposed that they evolve from a thin sill which at a critical area, because of the viscous magma, finds it easier to increase volume by pushing the roof up rather than continuing to flow outwards. The intrusion shape may, however, be partially controlled by a weakness in the cover or propagate at near full thickness. Only detailed field data can resolve the mechanism for each locality. Presumably in most cases the magma injects via a single feeder pipe.

10.4 LOPOLITHS

These are considered to be generally concordant saucer-shaped basic intrusions, the approximate inverse of laccoliths having flat tops and continuously curved bases. They are essentially broad, upward-opening, funnel-shaped intrusions, but the existence of this intrusion style is now being questioned. Many major igneous bodies originally interpreted this way are now known to be sills, plutons, etc. One such reinterpretation is represented by the Bushveld. It is the world's largest igneous body with an area of $66\,000\,km^2$ and thicknesses up to 9 km. It is also the largest repository of magmatic ore deposits containing Pt, Cr, V, Ni and Ti ores generated by magmatic differentiation.

10.5 LARGE IRREGULAR INTRUSIVES

A **pluton** is simply defined as a large intrusive body (Figure 10.7). The majority are acid to intermediate in composition. Basic to ultrabasic bodies are more likely, though not exclusively, to be sheet-like in form. A more comprehensive pluton definition is:

> 'a large intrusive body of one composition or several related compositions intruded approximately synchronously and contained within a single contact'.

The definition had to be broadened because composite bodies are the norm (Figure 10.8). However, several aspects of the definition are hard to determine from map evidence done. Map keys will rarely give sufficient information to define compositional relationships between different phases or enough chronometric resolution to show assembly of elements within a short time-span. Related compositions may evolve in one place by the separation of crystals and liquid during crystallization, but the liquid may inject at much higher levels. Tracing links may be difficult without detailed geochemical and petrological evidence. Fractionation *in situ* during crystallization will produce a zoned multicomponent body and though boundaries are gradational they are not usually shown as such on maps. In the Loch Doon Granite (Figure 10.9) this is indicated by a zone of 'transitional granite', but is normally suggested by a regular concentric arrangement of igneous units in contrast to the complex patterns of multiphase plutons (cf. Figure 10.8). Generally plutons are steep sided with flatish roofs, being box-like in cross-section with a rapid change from roof to side (Figure 10.10). Typically they are circular to slightly elongate in plan with diameters around 10–20 km, though rare examples up to 50 km are known. If the body is a near-perfect cylinder it is referred to as a **stock**. On a map, isolated

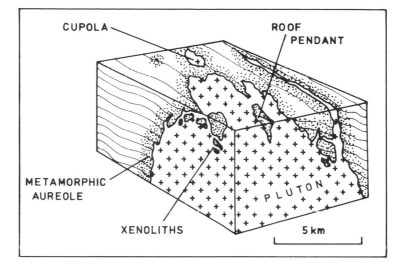

Fig. 10.7 Block diagram of a discordant pluton cutting the layering in the country rock and forming a jagged boundary partly controlled by fractures in the host rock. Xenoliths are isolated fragments of country rock whereas roof pendants are not detached. Cupolas are dome-shaped culminations of the intrusive body. Stippling in the country rock indicates the limits of contact metamorphism

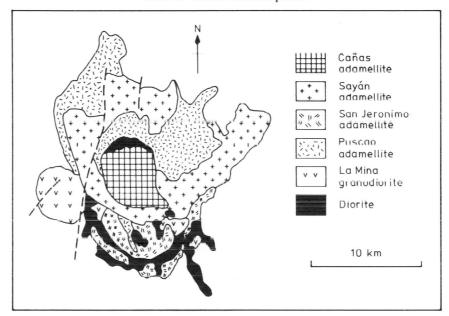

Fig. 10.8 An example of a composite pluton from the coastal batholith of central Peru (Cobbing, E. S. and Pitcher, W. S., 1972 *Jour. geol. Soc. Lond.*, **128**, 440). Geochemistry, radiometric data and field relations show that the different members of this body are essentially one intrusive event. The country rocks are volcanics and other intrusives; delineation of a pluton commonly requires more than map evidence. (Reproduced by permission of the Geological Society from E. J. Cobbing and W. S. Pitcher, The coastal batholith of central Peru, in *Journal of the Geological Society, London*, vol. 128, 1972.)

fragments of country rock within a pluton could be totally isolated xenoliths or **roof pendants**, downward projections of the roof into the body (Figure 10.7). The latter should have structural trends similar to adjacent country rock but some xenoliths founder without rotation so discrimination may not be possible. A small satellite body, if it is connected to the main pluton, is a **cupola**.

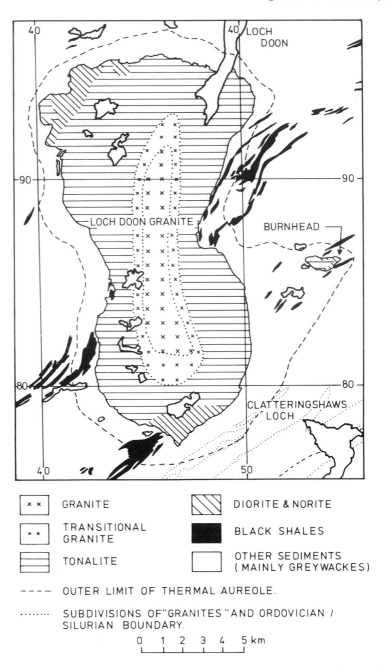

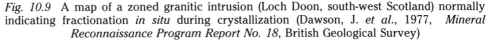

Fig. 10.9 A map of a zoned granitic intrusion (Loch Doon, south-west Scotland) normally indicating fractionation *in situ* during crystallization (Dawson, J. *et al.*, 1977, *Mineral Reconnaissance Program Report No. 18*, British Geological Survey)

Many large acid to intermediate plutons are intruded at 700 – 900 °C into rocks at temperatures below 400 °C. If the map key only mentions sedimentary or igneous terms (shales, tuff, granite, andesite) then the country rock was most likely less than 200°C when the pluton arrived. Heat is transferred to the wall rock and contact metamorphism may be developed up to 3 km from the intrusive margins (Figure 10.7). The rocks in the metamorphic aureole are hornfelsed (baked) and new mineralogies form in proportion to the maximum temperature reached. This forms the basis for zonation of the aureole but many maps only represent the contact

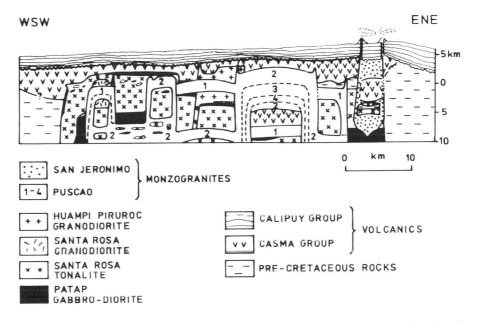

WSW   ENE

SAN JERONIMO ⎫
PUSCAO 1-4 ⎭ MONZOGRANITES

HUAMPI PIRUROC GRANODIORITE
SANTA ROSA GRANODIORITE
SANTA ROSA TONALITE
PATAP GABBRO-DIORITE

CALIPUY GROUP ⎫
CASMA GROUP ⎬ VOLCANICS
PRE-CRETACEOUS ROCKS ⎭

Fig. 10.10 Schematic cross-section of the Coastal Batholith of central Peru showing the piston-like shapes of successive multiple intrusions. Volcanic and intrusive activity were synchronous with the plutons injecting the cover created by the magmatic activity (Myers, J. S. *Bull. geol. Soc. Amer.*, **86**, 1209–20)

effects in very general terms. In mudrocks there may be a spotted outer zone, an intermediate andalusite (chiastolite) zone and an inner cordierite – biotite – andalusite zone which is well hornfelsed. Mapping contact metamorphic zones is complicated by the different mineralogical expression of the same temperature in varying rock types. Sizeable change in outcrop width of the aureole or zones might be related to varying attitude of the contact (see the area around Burnhead, Figure 10.9). These patterns may also help to suggest if satellite intrusives are connected to the main body. **Metasomatism** — loss and/or gain of chemical components in the country rock from either magmatic fluids or circulating ground water — is shown by exotic lithologies on the map key, e.g. tourmaline, fluorite and topaz rocks, and cassiterite-bearing carbonates which are susceptible to replacement. **Skarn** is the general term given to these materials. If the granitic intrusive is severely altered by metasomatism it become a **greisen**.

Batholiths are clusters of plutons that are usually arranged in linear arrays. The Coastal Batholith of Peru which was assembled over a period of 70 Ma is 1600 km long, 100 km wide and is about 60 per cent occupied by intrusions. In some cases there is a close association between batholiths and intermediate to acid volcanism. In the Andes, up to 7 km of Mesozoic andesites to rhyolites, and directly derived clastics, are associated with batholith emplacement which in large part intrudes the cover that it created (Figure 10.10). Other batholiths are found in more deeply eroded parts of orogenic belts (mountain belts) where it is not uncommon for 30 per cent or more of the outcrop to be granitic intrusives. Granitic in this sense is used to cover the spectrum of intrusives from intermediate to acid and not just granites as defined mineralogically.

The very large volumes of granitic intrusions in some terrains clearly begs the question, 'How was space made available?' The answer is simple in strike-slip regions where we have seen how large openings are created. But a more general emplacement mechanism is needed. Plutons may either be concordant or discordant with their country rock and these relations are important in analysing the emplacement process. Mechanisms fall into two categories: (1) **permissive**; (2) **forceful**. Forceful intrusions shoulder aside the country rock and play a very active role in creating space for themselves. Permissive intrusion involves the passive displacement of country rocks which creates discordant relations (Figure 10.11). One permissive style is **piecemeal stoping** which occurs where magma enters fractures and isolates blocks of country rock which generally sink thus allowing magma to advance upwards. This is a high-level process in the brittle part of the crust resulting in jagged or blocky margins. Major subsidence of a central block of a ring complex will allow permissive intrusion of magma into a piston-shaped area above the central block. This is known as cauldron subsidence which if repeated many times can create composite plutons (Figure 10.10). The original central block of country rock is rarely seen because it sinks to great depths, but when exposed is considerably broken up by faulting. Multiple cauldron subsidence on one or several overlapping sites produces complex piston shapes in cross-section with younger pistons breaking up earlier ones (Figure 10.10). The outcrop patterns in areas of high relief are equally complex. Though these are dominantly permissive mechanisms they were initiated by forceful doming driven by the magma pressure which fractured the roof triggering a reduction of pressure, in turn leading to the subsidence and permissive process.

Forceful intrusions pierce the country rock. Their simplest expression is in brittle rock where a cylinder of rock is pushed up ahead of the magma which is, therefore, a discordant body. Given expected ascent velocities, erosion rates during the rise of the magma would not completely remove the surface blister and along a batholith a high topography is to be expected. Piercing (Figure 10.12) is the essence of **diapirism** yet, paradoxically, conformable relations between country rock structures and the intrusive are the most diagnostic map criteria for this process (Figure 10.13). Diapirs are bodies that have risen buoyantly under gravitational influence to pierce and displace overlying rocks (Figure 10.12). Some authors would restrict the term diapir to bodies that were emplaced in the solid state (e.g. salt domes), but the similarity with many granitic intrusions shows that magmas should be included (they may have been very sticky crystal mushes at the time of intrusion). Density contrast is the primary driving force as is seen in shale and mud diapirs in rapidly deposited sediments as well as salt and igneous diapirs. Salt is less dense than the lithified and compacted overlying sediments and flows upwards as finger-shaped bodies though along faults it may form sheets. Salt diapirs probably show more discordance (piercement) than comparable igneous structures because they do not soften up the country rock by metamorphism. Most salt diapirs are sourced in thick evaporites, and whole sections of stratigraphy may become mobilized including minor non-evaporite components. In the Pre-Betic diapirs of southern Spain all of the Triassic units (shale, gypsiferous mudstone, limestone and dolerite) became upwardly mobile. A map shows Triassic formations in intrusive contact with a range of Lower Cretaceous to Mid-Miocene age rocks.

For granite, terms such as mushroom-shaped, inverted tear drop or Montgolfiere balloon, describe how most workers envisage the morphology of diapirs. A tadpole-shaped body is considered to separate from a zone of partial melting (migmatite) in the lower crust (Figure 10.14) and rise by forcibly compacting and shouldering aside the country rock. The latter near the site of generation is ductile and capable of high strains. Diapiric ascent involves the tadpole maintaining its dimensions as it rises. However, there is growing evidence that many concordant granitic intrusives rise through the lower and most of the middle crust as narrow pipes and sheets, and, if they freeze in relatively ductile crust, expand *in situ* like a balloon by continuous additions of magma. Fluids rising ahead of the intrusion, together with its contact thermal effects, aid the process by softening up the country rock ready for strain. At

Fig. 10.11 Discordant granitic plutons on a regional scale, Grampian Highlands, Scotland. The country rocks are regionally metamorphosed but lithostratigraphy has been well established despite the tectonic overprint. Particularly obvious is the truncation of the Tayvallich Subgroup by intrusions but other parts of the stratigraphy are equally affected

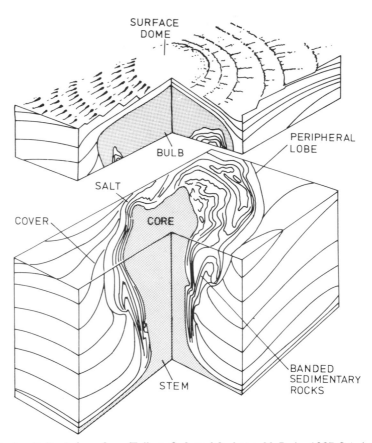

Fig. 10.12 A salt diapir from Iran (Talbot, C. J. and Jackson, M. P. A., 1987 *Sci. American*, pp. 58 – 67) shows the piercement nature of these structures and the degree of concordance between intrusions and country rock structure that can be generated adjacent to the intrusion. Very complex structures are found in this contact zone which can be very highly strained

these intermediate crustal levels there can be interesting interactions between regional thermal and deformation effects, and contact strain and temperature effects adjacent to the diapir. Both ballooning *in situ* and diapiric ascent produce smooth pluton outlines and nearly concordant relations with country rock structures because of the high strains next to the intrusions. Internal foliations (preferred orientation of minerals) will also be approximately concordant with the shape of the pluton (Figures 10.13 and 10.14). The exact significance of the foliations is often equivocal, even from field evidence. Mineral alignment may be generated by rotation of phenocrysts during intrusion-related flow of a largely liquid host or by strain in the solid state. For a strongly discordant body the former process is likely and, where there is evidence of high strain in the envelope, foliations probably represent solid-state deformation. Strains may be so high that granitic gneisses are produced. At this stage, if the intrusive nature is obscured by deformation, it is commonly very difficult to eliminate any of the emplacement explanations. Domal-shaped bodies can be produced by the interaction of two generations of upright folding, single fold events with very curved hinge lines or by ductile extensional features (boudins). To discriminate between these possibilities requires much information that is typically in short supply on Geological Survey-style maps. In particular, information is needed about strain intensity variations and the orientation of the maximum stretching direction. On diapir margins, stretching is subvertical and radially disposed with horizontal stretching on the crest. This may be indicated on maps by the pattern of mineral

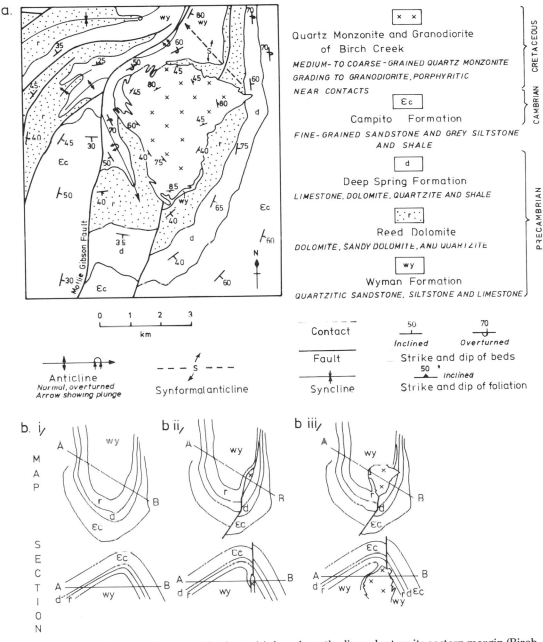

Fig. 10.13 A generally concordant granitic pluton (a) though partly discordant on its eastern margin (Birch Creek Pluton, California; Nelson, C. A. and Sylvester, A G , 1971, *Bull. geol. Soc. Amer.*, **82**, 2891 – 904). Part (b) shows the map and cross-section views of the intrusion history. The current interpretation of this body is that it was emplaced as a sheet-like body (b)(ii) that ballooned (expanded) *in situ* (b)(iii)

lineations which are considered to be equal or close to the maximum stretch orientation in the rock.

Generally-concordant plutons may be the product of permissive intrusion or wholesale punching out of an overlying cylinder followed by a regional strain. At moderate to high crustal levels, solid granite will normally be more viscous than the host rocks which will be moulded by the strain around the more rigid intrusion. The final pattern mimics that seen around diapirs or ballooned plutons. Information on most published maps will not allow these possibilities to be discriminated and it may

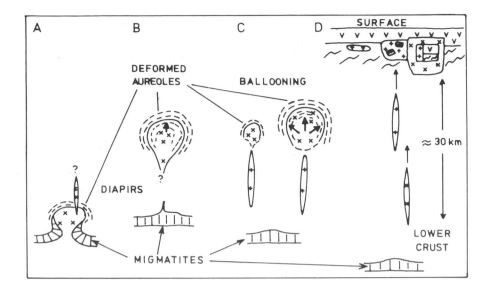

Fig. 10.14 A schematic representation of the various modes of emplacement of granitic bodies. Diapiric bodies rise through the crust at approximately their final dimensions whereas a ballooning process (expansion at the limit of ascent) is more likely. High-level intrusion in the brittle crust are also likely to be fed by tabular sheets. The 30 km crustal thickness may be extended to 70 km in collisional orogenic belts and Andean-type continental margins

be impossible to determine the relative timing of intrusion and regional deformation. In most metamorphic terrains the problem is compounded by several intense deformation events and a particular pluton may post-date the first regional deformation but be strained by the second.

10.6   EXTRUSIVE PROCESSES

Volcanic processes generate variety that seems to have few bounds. Published maps at scales less than 1:50 000 rarely do justice to the expected complexity of these terrains and sensible portrayals may not even be possible on 1:10 000 maps. Most map scales simplify life by not being able to reveal the full tortuous geometry of volcanic regions. Brief examination of most modern volcanic districts will emphasize the coeval generation of several very different rock types. Single short-lived eruption cycles can produce lavas, several styles of pyroclastics (fragmented lavas), both of which in turn can be eroded and resedimented under marine and/or non-marine conditions. Some of the lavas and pyroclastics could enter or be erupted under the sea or large lakes to add yet another dimension to the range of permutations. Volcanic products are the result of many factors:

1.   Lava composition — viscosity is broadly proportional to SiO_2 content.
2.   Shape of the volcanic vent — linear, point source, multiple points or ring dyke control.
3.   Lava temperature, composition and amount of volatiles.
4.   Supply of lava — rate of supply in eruption and length of dormant periods.

5. Environmental factors — control erosion style and rate, and also access of water to magma.
6. Tectonic setting — active faulting may destabilize volcanic edifices, strike slip may disrupt volcanoes or preferentially localize them.

Some examples of the interaction of these factors are shown in Figures 10.15 and 10.16. Small-volume lava flows on an actively eroding volcanic cone (Figure 10.15a) will be confined to the pre-existing topography — **valley accommodated**. Larger volumes of lavas eventually lead to **valley overspill**, creating flows with an irregular base but a smooth top if the lava had low viscosity. Valley accommodation of pyroclastics and lava flows can create very complex stratigraphic relations. Viscous silicic lavas (dacites and rhyolites) have steep flow fronts (Figure 10.16) and any later deposits (lavas, pyroclastics epiclastic sediments) will have to fill in this topography before covering the flows (Figure 10.15b). Figures 10.15c, d show other stratigraphic complexities of volcanic regions indicating the challenge posed in mapping these areas. On a regional scale, volcanoes are either point sources or from linear arrays; in both cases the igneous influence is laterally limited and all these terrains must pass outwards into zone of normal sedimentation (Figure 10.17). Major lateral facies variations of this style can be complex areas of inter-tonguing volcanic and non-volcanic materials.

10.7 FRAGMENTAL VOLCANIC ROCKS

It is only in the last few years that a more consistent approach has emerged for the analysis of the fragmental products of volcanic complexes. Hence old maps may be confusing, incomplete or misleading in the information they provide on this topic. A **volcaniclastic** rock is an aggregate of fragments of volcanic parentage no matter what fragmentation process was responsible. **Pyroclastic** aggregates are formed by explosive volcanic activity and deposited after being transported by processes directly related to the eruption mechanism. Once a pyroclastic sediment is deposited it may then be reworked by normal surface weathering and erosion during a volcanic dormant period. If this history can be proved, the final product is **epiclastic** though the general term volcaniclastic still applies. Pyroclastic rocks are sediments because deposition of the volcanic particles is controlled by hydrodynamics where shape, size and density are the important parameters. Their boundary position between igneous petrology and sedimentology left the study of pyroclastics in the wilderness for a long time, but now they are deservedly attracting considerable attention.

Many pyroclastics (Table 10.2) start to form within the volcanic conduit as expansion of volatiles accelerates a mixture of pyroclasts and gas to high velocities which may be supersonic. The gas thrust, which is a function of the pressure gradient between vesiculating magma and the atmosphere, may project particles ballistically to heights of a few kilometres but hundreds of metres is more typical. However, eruption columns of ash above silicic magmas do reach 55 km so other mechanisms must take over. Air from around the column (Figure 10.18b) is entrained and heated until the whole mixture becomes buoyant and rises convectively. At the top of the columns, upper atmosphere winds laterally disperse the low-density mixture as a

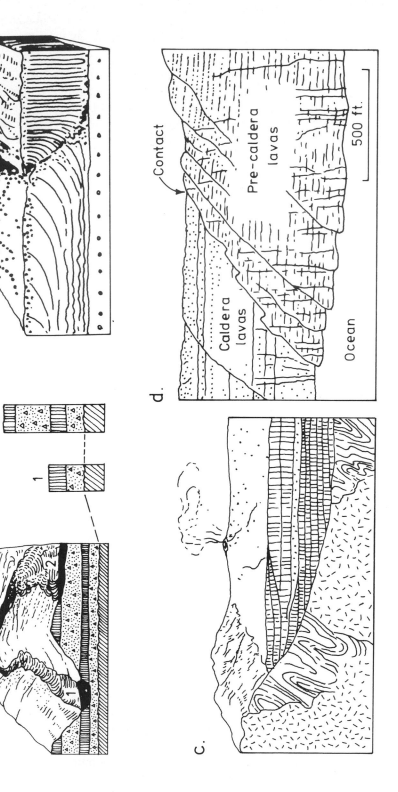

Fig. 10.15 (a) Valley accommodated lava flows can create major problems in correlations based on lithology alone. (b) Rhyolite flows are very sticky and have steep sides. Later flows will be contained by the earlier flow topography and possibly very different lava types of different ages will occur on the same horizon. (c) Flood basalts typically swamp earlier topography because of the rate of extrusion and amount of lava produced. (d) Calderas may be depressions kilometres across which become filled with lavas, pyroclastics or sediments, very different in nature to the lateral equivalents outside the caldera

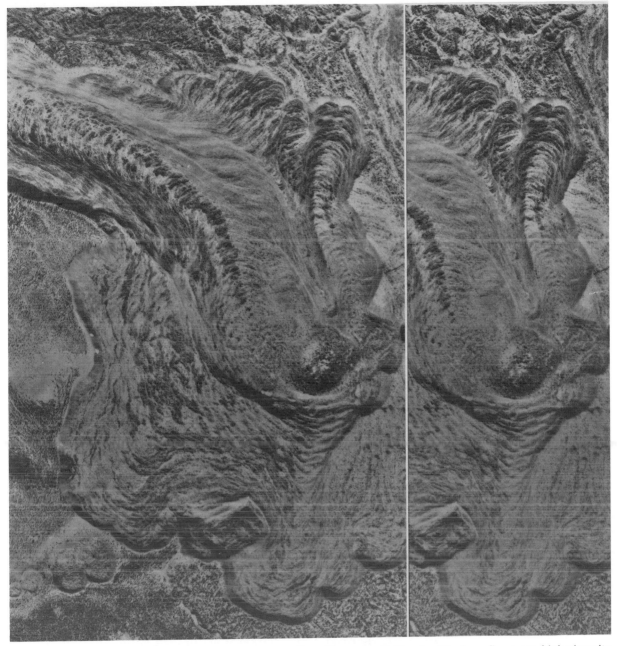

Fig. 10.16 Vertical stereopair of air photographs, Big Glass Mountain, California. The lava flows are high-viscosity obsidian with steep flow fronts and irregular tops characterized by concentric ridges called ogives. The folds in the flow tops formed while the lava was in motion as a result of a very viscous cooled outer layer overlying a less viscous hotter core to the flow

plume and the pyroclasts eventually drop to the ground as **fall deposits**. These may be 10 cm thick 2000 km from the source and typically thin systematically away from the volcano. This regular pattern may be interrupted if rain flushes the ash out prematurely. Any material ballistically ejected from the vent with a sideways component also contributes to fall deposits. Under some eruption conditions the majority of material in a column does not go fully into the convective phase and starts to fall out around the margins of the column (Figure 10.18c). This creates a gas – solid dispersion with quite a high particle concentration which, on reaching the

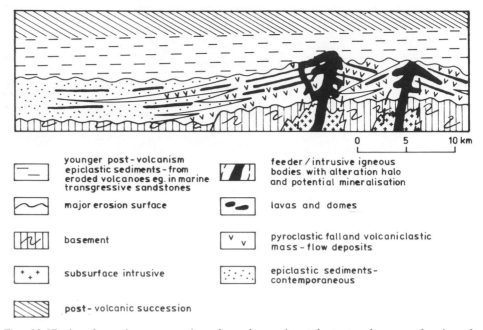

younger post-volcanism epiclastic sediments - from eroded volcanoes eg. in marine transgressive sandstones

feeder / intrusive igneous bodies with alteration halo and potential mineralisation

major erosion surface

lavas and domes

basement

pyroclastic fall and volcaniclastic mass - flow deposits

subsurface intrusive

epiclastic sediments - contemporaneous

post - volcanic succession

Fig. 10.17 A schematic cross-section through continental stratovolcanoes showing the variety of facies associated with the volcanic event and the consequent potential for complex lateral facies relations. Preservation of the stratacones themselves has probably been overemphasized as these structures are gravitationally unstable and prone to collapse on a massive scale

Table 10.2 Grain-size classification of volcanic fragments and fragmental volcanic rocks

Grain size (mm)	Pyroclasts	Pyroclastic rocks	Non-genetic rock term
64	Blocks and bombs	Pyroclastic breccia and agglomerate	Volcanic breccia
2	Lapilli	Lapillistone	Volcanic conglomerate
0.063	Coarse ash	Coarse tuff	Volcanic sandstone
0.004	Fine ash	Fine tuff	Volcanic siltstone
			Volcanic claystone

ground, spreads out as a density flow at speeds up to 160 km s^{-1}. Flows of this type have travelled 225 km and have overridden 600 m high obstacles 60 km from the source — quite an event! **Flow deposits** are formed which are sometimes referred to as ignimbrites. Ash flow is another common term, but because large pumice fragments are involved not all particles are ash sized and this nomenclature is inappropriate. On deposition, the pyroclasts are still hot and the middle to lower parts of the flows compact, plastically flattening pumice clasts into streaks; welded tuff is used to describe these materials. A typical violent eruption has elements of simple ballistic fall deposits, air fall deposits from dispersed columns and flow deposits. Volcanologists also recognize **surge** deposits which are similar to flow deposits but come from dispersions with low particle concentrations. Needless to say a very detailed key would be needed to allow this level of analysis from a map alone.

Explosions purely controlled by the properties of the magma are referred to as magmatic processes. If water from the sea, a lake, or groundwater makes a

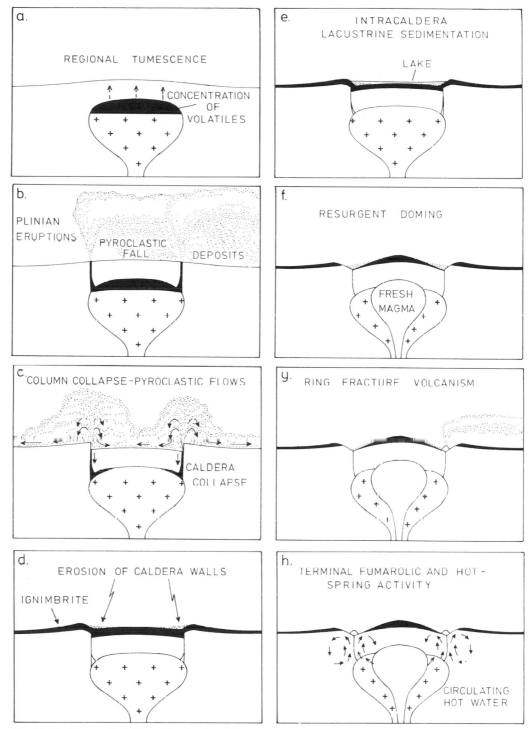

Fig. 10.18 The typical life-cycle of a rhyolitic volcano. Doming over the intrusion fractures the roof and releases pyroclastic eruptions. The drained magma chamber can no longer support the roof which collapses on a ring dyke to form a caldera which traps great thicknesses of pyroclastic deposits relative to beyond the caldera. A lake normally forms in the caldera (or sea-water breaks through) and the walls become eroded. If resurgence occurs the caldera sediments are deformed and eroded. In late stages, once the magma has lost most of its volatiles, rhyolitic lavas are extruded around the ring fracture but these rarely extend more than 3 km. A final stage is hot spring activity which can generate very rich epithermal gold and silver deposits

significant contribution to the fragmentation of the lava, it is a phreatomagmatic process. Explosions driven by steam alone are phreatic. Very intense fragmentation is characteristic of phreatomagmatic eruptions which produce high proportions of fine ash. Other fragmentation styles are autoclastic where lava enters a body of water and shatters on quenching to create a hyaloclastic deposit. During the subaerial flow of a lava, the chilled outer portion may autobrecciate to form a flow breccia. The fragmentation process may be difficult to decipher in the limited information seen on most map keys.

10.8 VOLCANOES — GEOMETRY AND HISTORY

There are ten basic styles of volcanic activity:

1. Basaltic shield volcanoes;
2. Flood basalts;
3. Scoria cones;
4. Maar craters;
5. Stratovolcanoes;
6. Multivent intermediate to silicic centres;
7. Rhyolitic volcanoes;
8. Submarine spreading centres;
9. Seamounts;
10. Subglacial events.

The latter three are somewhat specialized or remote so they will not be considered further even though the products of (8) and (9) cover two-thirds of the earth's surface.

10.8.1 BASALTIC SHIELD VOLCANOES

These mostly have central vents and are close to circular in plan reaching a maximum diameter of 100 km. They create enormous structures, such as those in Hawaii, standing 9 km above the level of the surrounding abyssal plains. Frequent voluminous eruptions only take about a million years to build 95 per cent of a shield volcano, though subsequent minor activity may extend the full active period by several million years. Almost all of the structure is basaltic lava which is responsible for the characteristic 10° slopes though the oceanic examples have steeper submarine slopes. Small calderas up to 5 km across occur at the summits and caldera collapse events may be related to phreatomagmatic eruptions that disperse ash over the entire volcano. In an otherwise uniform mass of shield basalt these form very useful stratigraphic markers (Figure 10.19a).Though commonly placed in a central vent category, most shields have large rift fractures (Figures 10.19a, c) that extrude much lava. However, the centre dominates as it overlies the principal magma chamber. Gravitational instability is the likely explanation for the rifts which are probably initiated when a volcano is as yet unbuttressed by a neighbour, e.g. Kilauea, Hawaii (Figure 10.19).

Hawaii is a good illustration of map patterns on young volcanoes (Figure 10.19a). Radial outcrop patterns dominate, though the reasons for this are quite varied. On Mauna Kea Volcano which has been dormant for 3600 years, the

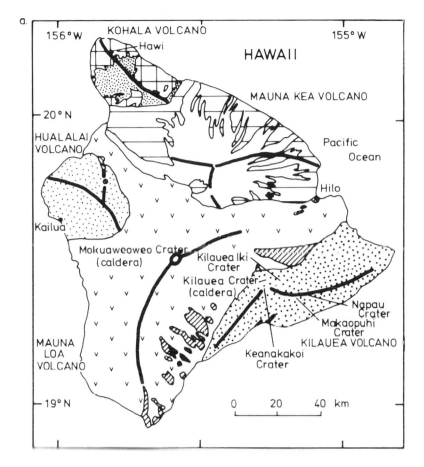

Fig. 10.19 a (Caption on following page.)

strong radial pattern is a function of erosion (see also Figure 10.22). The oldest formation (the Hamakua Volcanics) were covered by a thin veneer of post-shield alkalic basalts (Laupahoehoe Volcanics). Subsequent erosion has exposed the older unit in valley bottoms in drainage controlled by the cone-shaped topography. On Mauna Loa Volcano, a 20 m ash formation (Pahala Ash) separates the stratigraphy into two thick basalt formations. The internal complexity of each of these lava formations is indicated by the record of historic eruptions which form part of the Kau Basalt (cf. Figure 10.19a, c). Mapped at the formation level, the two basalt units look deceptively simple. Stratigraphic procedures can hide a lot of irregularity and perhaps falsely convey a picture of simplicity of limited use to volcanologists. Maps of individual lava flows would be expected to show a dominantly radial pattern.

10.8.2   FLOOD BASALTS

Formerly known as plateau basalts, these can create vast provinces nearly a million square kilometres in area at around a kilometre in thickness. Most lava is fed from vertical fracture zones (Figure 10.1) where individuals may be 30 km long and 20 m wide. Single flows may travel 300 km and be up to 35 m thick. The record volume for a single flow is 1500 km³ and at this rate it is easy to see how a large province could be formed in a couple of million years. Most topographic slopes on the tops of lava

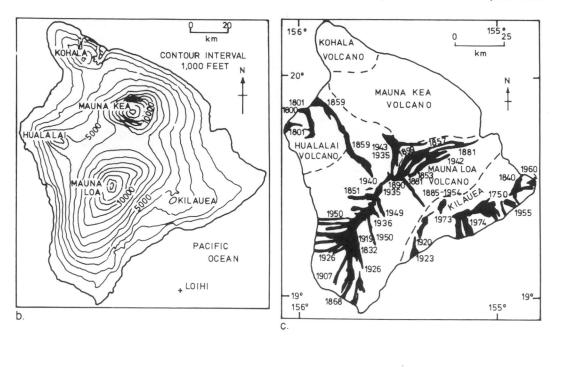

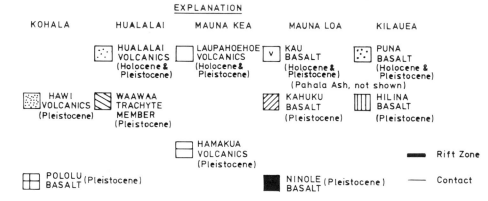

Fig. 10.19 (a) Geological map of Hawaii illustrating the radial outcrop pattern typical of young volcanoes. There is a tendency to interpret all of these patterns in terms of valley accommodated lava flows and pyroclastic deposits. However, on Mauna Kea the radial style is caused by valleys cutting down through the youngest part of the stratigraphy into the older Hamakua Volcanics. (b) Contour maps of Hawaii showing the subcircular topography of basaltic shield volcanoes. Loihi is the currently active submarine volcano emphasizing the overall direction of younging in the Hawaiian chain. (c) Historic lava flows on three of Hawaii's volcanoes. On Mauna Loa the complexity of these flows is hidden by the stratigraphy which places them in one lithostratigraphic formation (Decker, R. W. *et al.*, 1987, *U.S. Geol. Surv. Prof. Paper*, No. 1350)

flows are less than 2° though some zones tend towards central vent behaviour creating local shields with slopes up to 5° (Figure 10.20). Fire fountains along feeder fractures produce spatter cone deposits and some more widespread pyroclastic activity erupts sporadically. On erosion, these thick piles of lava are very resistant forming positive topography indicating the derivation of their earlier name even though they rarely erupt on plateaux. Flat-lying flood basalts, when eroded, produce stepped topographic profiles (**trap topography**) because the vesicular flow tops are

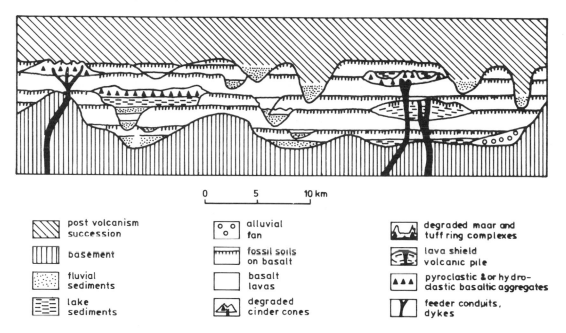

Fig. 10.20 Schematic cross-section through a pile of flood basalts. Local zones behave like shield volcanoes and build steeper slopes. The overall morphology is strongly controlled by the length of dormant periods, but these tend to be short and activity may be concentrated in just a few millions of years. Note that, to show the lateral and vertical relationships, a considerable vertical stretch has been employed, the typical pile being about a kilometre thick

less resistant than the massive flow cores. A contributing factor is weathering between flows, any resulting soil is also easily eroded.

10.8.3 SCORIA CONES

These are perhaps the most common form of volcano but their small size (around 200 m high, <1 km diameter) reduces their importance. Scoria is a general name given to vesiculated basic to intermediate pyroclasts and in a scoria cone most are lapilli to bomb sized. The cones are easily eroded and rarely survive more than 4 million years. They start with >30° slopes but these are reduced quickly. Lavas and spatter deposits are minor components of the cones.

10.8.4 MAARS, TUFF RINGS AND TUFF CONES

These are volcanic craters which in active and relatively recent regions form the second most common volcanic element. However, they have very low preservation potential and, when they survive, are difficult to recognize in the ancient record particularly from map data alone. Maars (Figure 10.21a) are cut into pre-eruption surfaces and thus into pre-eruption rocks which are exposed in the crater walls. Around the crater is a 10 – 40 m thick low rim of ejectamenta, a mixture of pyroclasts and up to 80 per cent country rock fragments. Crater diameters range from 100 m to 3 km and reach depths of 200 m. The vesicle free style of the pyroclasts shows a phreatomagmatic origin and explains many maar features. Maars overlie **diatremes** (tuff pipes) which link at depth to dykes, relationships revealed by deep mining of

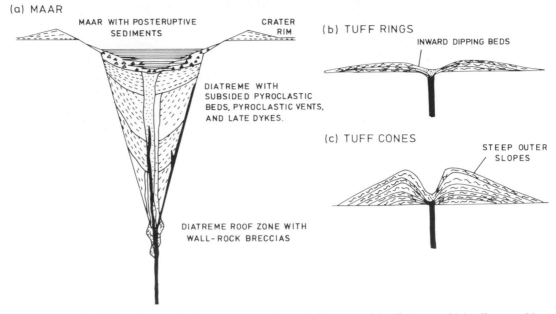

Fig. 10.21 Generalized cross-sections through (a) maars, (b) tuff rings and (c) tuff cones. Maars are cut into the pre-eruption surface and the apron of ejected material can be nearly all non-volcanic material

diamondiferous pipes. Tuff rings and cones (Figures 10.21b, c) have craters which sit on or above the pre-eruption surface and the differences are mainly in the slope angle and, therefore, height. Because they do not penetrate the country rock, the fragmental apron is 95 per cent igneous in composition. Maars, tuff rings and tuff cones, are associated with basic to acidic activity.

10.8.5 STRATOVOLCANOES

These are the classic near-conical volcanic landform though the symmetry is often disturbed when parasitic vents compete with the central vent's production rate. Most stratocones are andesite to basaltic andesite but the magma range can be from rhyolite to basalt. Fuji, probably one of the most perfect examples, is basic. Cone slopes vary from 15 to 33° and the shape is controlled by an alternation of short lava flows and pyroclastics. Base diameters range from just over 1 km to nearly 90 km and heights above surrounding plains range from 600 m to 5 km (Figure 10.22). The largest structures are very substantial, reaching volumes of around 250 km^3. The volcanic edifices are very unstable and reworking (debris flows, mud flows, stream erosion, etc.) is a major process. In island arcs, 80 per cent of the rocks are volcaniclastic and lavas are very much in the minority. Current thinking indicates that the inherent instability of the cones considerably reduces their preservation potential and perhaps the highest 5 km examples are least likely to survive. Collapse of the cone is to be expected, perhaps driven by injection of new magma to bulge out a sector or by phreatomagmatic explosion or by gravitationally controlled avalanches. In one 10-minute event, a 70° segment of Socompa Volcano, Chile, collapsed and debris was carried over 40 km reaching velocities of 100 km h^{-1}. Blocks 2 km long and 500 m thick, were displaced several kilometres and the overall deposit is a very poorly sorted chaos with blocks of all sizes and many compositions

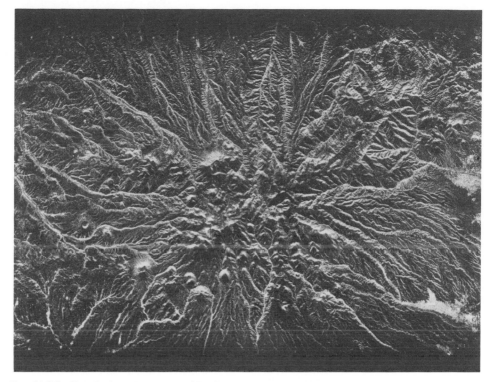

Fig. 10.22 Shuttle imaging radar (SIR-A) image of an andesitic stratovolcano in north-west Iran. The image covers 50×65 km and clearly displays the radial drainage associated with young volcanoes. Kilometric-scale parasitic cones are found on the flanks of the major volcano; one of these is nearly perfect in form and is so young that erosion has had little opportunity to modify its shape. In contrast the central caldera of the main cone is not recognizable because of severe erosion

including much pre-eruption material. The facies relations for active stratovolcanoes (Figure 10.17) will be very significantly modified by collapse events. Hence their expression in the ancient record will be more cryptic. Individuals have histories spanning a million or so years and groups often occupy 20 Ma. Their hallmark is very variable style, duration and frequency of eruption. The distribution of fine-grain pyroclastic fall deposits is dictated by the prevailing wind.

10.8.6 INTERMEDIATE-SILICIC MULTIVENT CENTRES

These are very large roughly conical volcanoes without a central vent. Instead they have a complex core zone perhaps 10 km in diameter with a cluster of ten or more vents. Radial patterns to valley accommodated lava flows, pyroclastic flows and reworked clastic deposits, are dictated by the overall cone shape. They suffer the same instability problems that typify stratovolcanoes.

10.8.7 RHYOLITIC VOLCANOES

These are the big league of volcanic features though to the casual observer they do not look like volcanoes. Modern examples are vast low shields up to 200 km in diameter with a central caldera depression up to 60 km across containing scattered low rhyolitic hills. Magma injection (Figure 10.18) is believed to fracture the roof (ring fractures) and the tapped magma erupts explosively leading to massive outpourings (up to $3000\,km^3$) of pyroclastic flow deposits (ignimbrites) which cover most of the shield. Caldera collapse over the drained magma chamber synchronous with the pyroclastic flows can create trapped intracaldera ignimbrites 1200 m thick, even though the same deposits are only a few tens of metres thick on the flanks (Figure 10. 23). Several caldera collapse events, separated by hundreds of thousands to a million years, are possible. With such lengthy dormant periods the flow and fall deposits can be very significantly modified by normal surface erosion. After caldera collapse a regular sequence of events is commonly observed (Figure 10.18). Initially the near-vertical caldera walls erode to form wedges of sedimentary breccias and often a lake forms in the caldera and thick lacustrine sediments are deposited. Minor resurgence of magma can deform this complex leading to more erosion within the caldera. Even without resurgence, small rhyolitic lava flows are extruded from many points on the ring fracture and these may coalesce to form a complete circle of lava. The change from pyroclastic eruptions to lavas is a function of volatile loss from the magma chamber during the earlier destructive eruptive events. As the igneous activity wanes hot springs develop along the fractures and this has localized many very rich silver and gold deposits (e.g. Andes).

10.9 METAMORPHIC REGIONS

Many large areas are characterized by rocks transformed texturally and mineralogically in the solid state from their initial sedimentary or igneous condition as a result of being caught up in tectonic processes. Metamorphism means a change of form and was originally defined as isochemical but is now applied to situations where compositions do not change appreciably. Some changes of chemistry are almost inevitable as fluids are driven out of sedimentary rocks in progressive metamorphism. Retrogressive metamorphism likewise involves a fluid that will move chemical components around the system. Metamorphic changes reflect the superposition of new P, T conditions subsequent to the formation of the parent sedimentary or igneous rock. In continent/continent collisional zones, contraction generates thrusts that cut through the crust causing large-scale duplications that rapidly increase the load on the underlying segments. Rapid in this context is in relation to the poor thermal conductivity rocks and thrusting, therefore, buries cold rock which eventually heats up; new mineralogies reflect this history. In extensional terrains isotherms are quickly (geologically) brought nearer the surface and, again, the thermal disturbance brings about mineralogical changes. For both extension and contraction (+ potential strike-slip contributions), strain occurs in the ductile and transitional parts of the crust producing new fabrics.

Local (contact metamorphic) changes around intrusions have been briefly mentioned and we now turn our attention to effects which extend up to hundreds of thousands of square kilometres. Shallow burial diagenetic conditions are normally excluded from discussions of regional metamorphism though boundaries are blurred. The application of a stress difference to a mudrock can create a slate at 150 °C, whereas many sedimentary lithologies would have to be heated to over 300 °C before their mineralogy would clearly indicate a metamorphic event. The progression slate, phyllite, schist, to gneiss, is one of grain size and increasing spacing

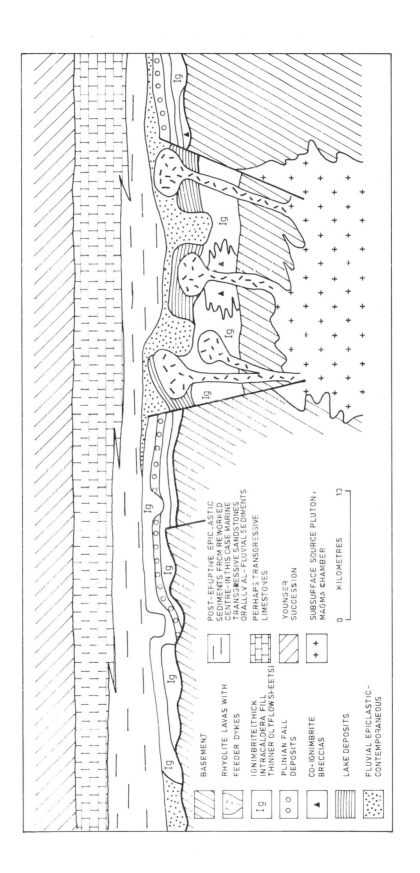

BASEMENT

RHYOLITE LAVAS WITH
FEEDER DYKES

Ig   IGNIMBRITE (THICK
INTRACALDERA FILL,
THINNER OUTFLOW SHEETS)

PLINIAN FALL
DEPOSITS

CO-IGNIMBRITE
BRECCIAS

LAKE DEPOSITS

FLUVIAL EPICLASTIC-
CONTEMPORANEOUS

POST-ERUPTIVE EPICLASTIC
SEDIMENTS FROM REWORKED
CENTRE-IN THIS CASE MARINE
TRANSGRESSIVE SANDSTONES
OR ALLUVIAL-FLUVIAL SEDIMENTS

PERHAPS TRANSGRESSIVE
LIMESTONES

YOUNGER
SUCCESSION

SUBSURFACE SOURCE PLUTON,
MAGMA CHAMBER

0   KILOMETRES   10

Fig. 10.23 Schematic cross-section through part of a rhyolitic volcano; one flank is shown complete and the other would be fairly symmetrical about the central caldera. As with most volcanic regions, lateral facies changes are dramatic particularly between inside and outside the caldera. Ignimbrites, the products of pyroclastic flows, may be many times thicker within the caldera in comparison to coeval deposits on the volcano flanks. Regional lateral facies changes mean that eventually the volcanic region will give way to normal sedimentation.

Table 10.3 Classification of metamorphic rocks

Bulk composition	Common minerals	Foliated	No mineral alignment
Pelitic (mudrock)	Quartz Muscovite Biotite Feldspars Aluminium silicates Garnet	Slate Phyllite Schist Gneiss	Pelitic hornfels
Basic (basalt, etc.)	Chlorite Amphibole Plagioclase Pyroxene	Greenschist Amphibolite	Basic hornfels Basic granulite[a]
Quartzo-feldspathic (granite)	Quartz Feldspars Mica	Schist Gneiss	Hornfels Acid granulite[a]
Quartzose-psammitic (quartz sandstones)	Quartz (mica)	Quartzite	Quartzite
Calcareous limestone ± (impurities)	Calcite Ca-garnet Ca-pyroxene Wollastonite	Marble Calc-silicate rock	Marble Calc-silicate hornfels (originally impure limestones)

[a] These are somewhat unfortunate terms but well established. Do not confuse them with granulite facies — used in this content they refer to the lack of fabric.

of cleavage/foliation planes (Table 10.3). The main influence on this series is increasing T, but other variables can be important including the fluids in the system. Intensity of metamorphism (grade) varies from place to place. At the very lowest grades many of the mineralogical changes are subtle and are rarely recorded on Geological Survey-style maps.

Metamorphic studies are critical in the analysis of tectonic history. At their simplest they can, for example, distinguish the high P low T conditions of subduction zones from high T low P effects in island arcs. More detailed study can give $P-T-t$ information on how P and T have changed with time thus getting to explanations of the metamorphism (collision, extension, etc.). However, these studies are sophisticated and only the basics are presented on map keys. Old maps will normally only indicate the metamorphic state by referring to schists or gneisses, and many give no further lithological information. Indeed many venerable British maps of regional metamorphosed areas refer only to pelites (mudrocks) and psammites (quartz arenites) and say nothing more about the metamorphic products. Reconnaissance maps may only refer to meta-basalts, meta-greywackes, etc., which says very little about the metamorphic state. One step in the right direction is the listing of mineralogies, e.g. garnet-staurolite-quartz-biotite schist. A general knowledge of the stability fields (in P and T) of metamorphic minerals — see, for example, the areas occupied by andalusite, kyanite and sillimanite in Figure 10.24 — will then give an idea of the metamorphic grade and allow any variations on the map sheet to be crudely monitored. More information is provided if **metamorphic zones** are delineated on the map by boundaries known as **isograds** (lines of equal grade). The zones are based on the first appearance of an **index mineral** which gives its name to the zone. In a region of intermediate pressure, original mudrocks show the following progression of zones with increasing temperature: chlorite, biotite, garnet, staurolite, kyanite and sillimanite. Each zone indicates the maximum P/T conditions

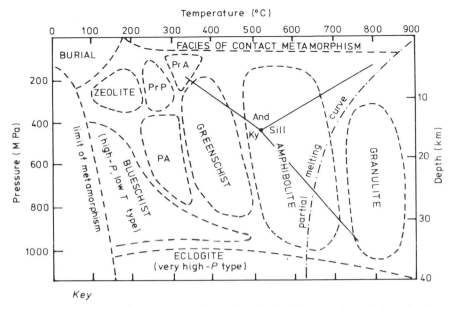

Fig. 10.24 Representation of the main metamorphic facies in terms of P/T space. Boundaries are gradational. A vast amount of data on mineral stability fields were used in constructing this diagram; the information on the aluminium silicates (andalusite, kyanite, sillimanite) are used to indicate the style of relationships used. PrA = prehnite – actinolite facies; PrP = prehnite – pumpellyite facies; PA = pumpellyite – actinolite facies

for that area. Given a reasonable distribution of appropriate rock compositions, metamorphic zones are easy to map but the concept is rather qualitative. Isograds in many cases are isotherms because the majority of important metamorphic reactions involve devolatilization and these are temperature sensitive. Some isograds, however, are more influenced by confining pressure and even the general notion of grade has to be handled carefully. In their original form many isograds are subhorizontal but, if a major deformation event post-dates peak metamorphism, then complex isograd geometries can be generated.

A much more exact specification of metamorphic conditions is that of facies. A metamorphic facies is a collection of metamorphic mineral assemblages repeatedly associated in space and time where a constant relationship exists between mineral assemblage and bulk composition of the rock. The facies scheme subdivides P,T space into a limited number of fields (Figure 10.24) each of which occupies a T range of $100 - 200°C$ and pressure ranges up to 800 MPa (1 kilobar = 100 MPa). All metamorphic rocks formed within one of these P,T fields belongs to that metamorphic facies and a vast data bank has accumulated to record the mineralogies of all the disparate bulk rock compositions for all the facies. Rocks of the same chemical composition have the same mineral assemblage if they were generated in the same metamorphic facies. For maps at scales of 1 : 50 000 and better, it is only rare examples that would contain representatives of more than one facies. On regional scale maps (1 : 250 000 or less), facies distribution, if given, is normally shown on very small inset or margin maps because the patterns are often fairly simple. Even with facies information it is difficult to make interpretations of the tectonic setting without detailed comments on the map key. In fact little of the metamorphic history of a moderate to high-grade terrain is presented on maps. Most of these regions have several major deformation events, one of which will normally occur before the peak of metamorphism and one after. Intermediate deformations in

Table 10.4 Chronological sequence of events in the Ameralik Region (see Figure 10.26)

Age	Event
57 Ma	Intrusion of 'red dykes' around the mouth of Ameralik
c. 1150 Ma	Mild thermal event shown by fission track ages
1500 – 1600 Ma	Metamorphism and metasomatism that caused partial recrystallization and is recorded in Rb/Sr mineral ages
2000 – 1800 Ma	Intrusion of basic dykes (MD dykes), faulting with retrogression under greenschist facies conditions
c. 2550 Ma	Intrusion of post-tectonic granites and pegmatite sheets (Qôrqut Granite Complex)
2700 – 2600 Ma	Ductile deformation under amphibolite facies conditions. Formation of linear belts of intense deformation and elongate basins and domes. Intrusion of granitoid sheets (including the Qârusuk dykes)
2800 Ma	High-grade metamorphism that reached granulite facies conditions
2880 – 3100 Ma	Intrusion of diorites, tonalites, granodiorites and minor trondhjemites, the protoliths of the Nûk gneisses, polyphase deformation
2980 Ma	High-grade metamorphism which culminated in granulite facies conditions
	Intrusion of anorthosite – leucogabbro – gabbro complexes into Malene supracrustal rocks.
	Extrusion of subaqueous basic volcanic rocks with related subvolcanic intrusions, deposition of sediments (Malene supracrustal rocks)
	Intrusion of basic dyke swarms (Ameralik dykes)
c. 3600 Ma	Metamorphism that reached granulite facies. Intrusion of big-feldspar granodiorites and subordinate ferrodiorites (Amitsôq iron-rich suite)
c. 3750 Ma	Intrusion (syntectonic?) of voluminous tonalites, the protoliths of the Amitsôq layered grey gneisses. Intrusion of granite and pegmatite sheets, deformation and metamorphism
c. 3800 Ma	Extrusion of basic and subordinate felsic volcanic rocks, deposition of chemical sediments and subordinate felsic and pelitic sediments, intrusion of gabbroic sheets (Akilia association). Correlated with the Isua supracrustal rocks

time may be synchronous with peak *P,T*, or slightly before, or slightly after. Detailed laboratory work can unravel these sequences but often none of this appears on the map key or legend. High-grade gneiss terrains, that may have been buried to depths of 50 – 60 km, take a long time to reach the surface. During this time they may be subjected to one or more orogenic overprints and become very complex in their thermal and structural histories. Quite a number of such terrains generated in the Early Archaean have been through orogenic reworking in either or all of the Archaean, Proterozoic and Phanerozoic (e.g. Table 10.4). Major events in such old terrains can be placed in order by studying overprinting relationships but absolute ages rely on radiometric methods (geochronometry).

Migmatites are important components of very-high-grade regions. Migmatite literally means mixed rock, in this case a mixture of igneous and metamorphic components. Typically migmatites are viewed as the product of ultrametamorphism where partial melts have not moved far from the point of generation. This places them at the boundary of igneous and metamorphic petrology. Isolated igneous veins in a gneiss do not constitute a migmatite. Before the name is applied a significant igneous component should be present though hard limits are not prescribed. The newer igneous part is the **neosome** and the host (meta-sedimentary or meta-igneous) is the **palaeosome**. Neosomes are normally quartz and feldspar-rich (**leucosome**) but can be mafic (**melanosome**). The geometry of the neosome varies enormously, reflecting injection/partial melting processes but much is controlled by strain modifications, partially molten rocks being very ductile. An extensive terminology has developed around this topic, but on most maps migmatite complexes are dealt with on the macroscopic scale and are represented as single units; the complexity being on too small a scale to be resolved. Not all migmatites are the product of ultrametamorphism. Some plutonic complexes, especially if intruded at depth during orogenesis, can be transformed into migmatites. Acid to intermediate intrusives, if criss-crossed by many pegmatite, granitic and/or aplitic veins, and then highly strained, produce large areas of classic migmatite. Regional map relations may give a clue to the genesis of a particular set of migmatites. Those seen as the pinnacle of metamorphism should have progression of increasing metamorphic grade leading up to them. The deformed plutons may show the vestiges of compositional zonation and/or multiple intrusion, though the significant modification of shape by regional strain may make recognition a difficult task.

10.10 GEOMETRY OF REGIONAL METAMORPHIC ROCKS

Geometrically, medium to high-grade metamorphosed regions present major challenges in 3-D thinking which are on the outer limits of this introductory text. Intricate outcrop patterns in these areas may be the result of three or four separate deformation events (D_1, D_2, etc.) each generating structures on the kilometric scale. Most examples are further complicated by additional minor events which, locally, may produce large structures. Heterogeneity on all scales is the name of the game. Low-grade regions and contact aureoles of forcefully intruded plutons can be equally challenging and your guard should never be dropped. Chapter 8 gives most of the analytical methods relevant to low-grade regions and here we mainly consider superposed or polyphase deformation together with some of the effects of high strain. Superposition of fold generations can easily be understood where two orogenic belts cross, but within single orogens the geometry is also controlled by several deformation events.

The basic approach is to concentrate on two deformation events ($D_1 = F_1$; $D_2 = F_2$) (Figure 10.25) and worry about variations later. Two sets of upright folds, with hinge lines trending at high angles to one another, give the simplest pattern. Obviously the D_2 folds in Figure 10.25ai do not exist as shown; they are the folds that would have developed in the absence of D_1 effects or where D_1 left the layering horizontal. In such **cross-folding**, where antiforms of the two generations are superimposed, the maximum culmination of the hinge lines creates a **dome** with dips radiating out in all directions (Figure 10.25aiii, aiv). Similarly, two synforms crossing give rise to depressions which are structural **basins**. The intermediate cases of synforms crossing antiforms and vice versa bring out **saddles** or **cols** in the system. A synformal saddle where two fold trends of different intensity cross each other at fairly high angles is illustrated in Figure 5.3. Basin and dome interference structures have been likened to egg cartons in their general form. Not shown in Figure 10.25

are the effects of significant strain syn-D_2 which would generate very elongate domes and basins, complicating the recognition of the interference nature of the structure. Also only one F_1/F_2 angle is shown. With angles between the two sets of hinge lines less than 90°, the resultant domes and basins become markedly skew in outcrop. The simple models of Figure 10.25a show persistent folds of both generations yet we expect folds to die out laterally; such additional factors account for the natural product as illustrated in Figure 5.14. Despite the variations, the characteristics of dome and basin structures allow recognition of a class of interference patterns referred to as **Type 1**. Rather than being the product of the two distinct phases of deformation, a single event of shortening of all directions within near-horizontal layers will also generate domes and basins. It should be noted that Type 1 patterns are also found in very-low-grade, essentially non-metamorphic, cover sequences. For our figured example (Figure 10.25a), the D_1 folds face upwards and in the resulting interference product all beds are right-way-up and all folds face upwards.

Type 2 interference patterns are examples of another cross-folding style where hinge lines of the two events are at high angles. Rather than two upright fold events we now have an upright event superimposed on near-recumbent tight folds (Figure 10.25b). Similar relationships of the two generations but in different orientations relative to the horizontal would have the same 3-D effect; the attitudes used here give the simplest expression of the style. During the second deformation, F_1 hinge lines are folded as are the limbs and hinge surfaces of D_1 folds. Because of the tightness of the first folds both limbs are folded into an antiform if crossed by a D_2 antiform and vice versa. It is this feature that creates the most distinctive Type 2 characteristic of arrowhead outcrop patterns; Figure 10.25biii is sliced horizontally to give an impression of the map pattern (Figure 10.25biv). In the D_2 culmination on the right of Figure 10.25biii, the D_1 antiform in the upper limb of the D_1 fold is plunging at a much lower angle than the D_2 antiform in the lower limb of the D_1 fold. These plunge differences explain the variation in curvature of the inner and outer parts of the arrowhead. This also highlights how early fold geometry exerts an influence on the later structures. Both Types 1 and 2 outcrop patterns have closed loops defined by formation boundaries, marker beds and/or structural trends. The map of the Type 2 pattern (Figure 10.25biv) is based on F_2 being superimposed on upwards-facing D_1 folds such that pre-D_2 fold limbs alternated in younging from gently dipping right-way-up layers to more steeply dipping upside-down layers. Bending the F_1 hinge lines disperses them and does the same to minor hinge lines and intersection lineations (see map Figure 10.25biv). Arrowhead or mushroom outcrop patterns may appear freakish but they are common in orogenic belts because early folds are often near recumbent and later folds are typically upright. Highly strained examples, when stretched out, become difficult to recognize because shapes are severely modified. Again a non-orthogonal F_1/F_2 hinge line relationship gives rise to skew outcrop patterns with lop-sided arrowheads.

Type 3 interference (Figure 10.25c) is **coaxial** because F_1 and F_2 hinge lines share the same bearing. Early hinge lines, in contrast to Type 1 and 2 styles, are not refolded though F_1 limbs and hinge surfaces are folded. Outcrop patterns of Type 3 interference are not closed but converge and diverge, closed loops only occur as a result of topographic interactions. The potential for varied topography cutting an interference pattern to create nightmarish outcrops is staggering. Careful analysis must be employed using sections and/or profiles. Again Type 3 patterns are common because of the attitude variation from early to late folds in many orogenic belts. A cautionary note must be added at this point. A full 3-D understanding of just a two-event interference requires a lot of information. It is possible to find some 2-D slices through a Type 2 pattern that look like Type 3, so only having a map view could be

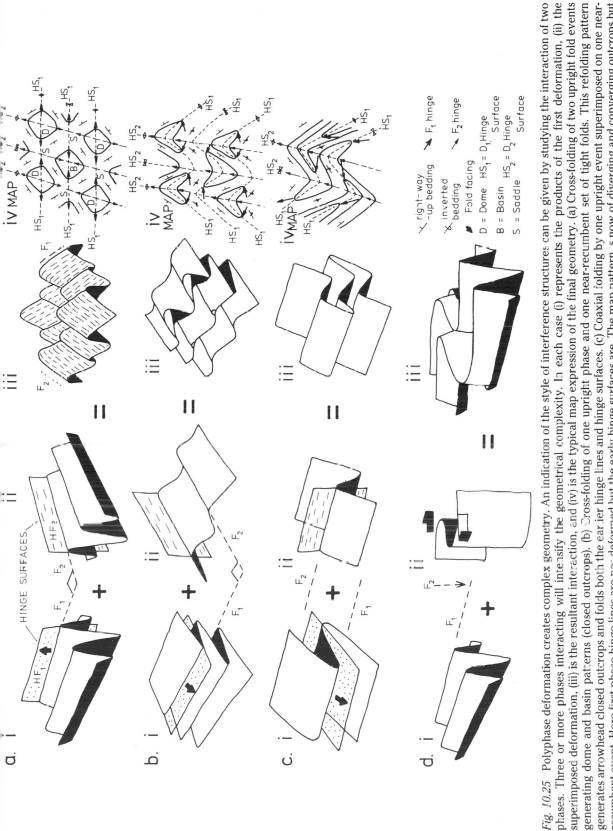

Fig. 10.25 Polyphase deformation creates complex structures can be given by studying the interaction of two phases. An indication of the style of interference structures can be given by studying the interaction of two phases. Three or more phases interacting will intensify the geometrical complexity. In each case (i) represents the products of the first deformation, (ii) the superimposed deformation, (iii) is the resultant interaction, and (iv) is the typical map expression of the final geometry. (a) Cross-folding of two upright fold events generating dome and basin patterns (closed outcrops). (b) Cross-folding of one upright phase and one near-recumbent set of tight folds. This refolding pattern generates arrowhead closed outcrops and folds both the earlier hinge lines and hinge surfaces. (c) Coaxial folding by one upright event superimposed on one near-recumbent event. Here first-phase hinge lines are not deformed but the early hinge surfaces are. The map pattern is now of diverging and converging outcrops but not closed loops unless topography adds to the complexity. (d) Illustrates a fairly common situation in high-level (low-metamorphic grade) fold belts where upright folds are re-deformed in shear zones with strike-slip displacement.

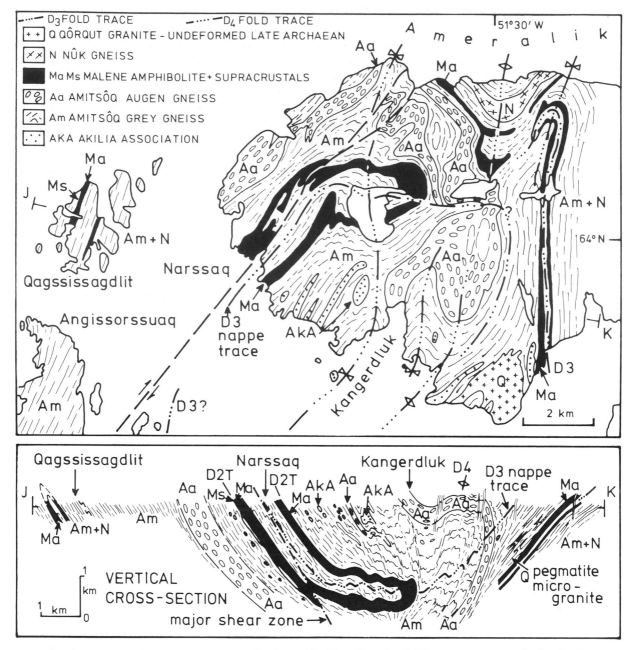

Fig. 10.26 A map of a high-grade gneiss terrain, Ameralik, West Greenland. The outcrop pattern is simpler than you would expect from such a polyorogenic region (see Table 10.4); this is partly a function of the scale of representation, but it is also a function of high strain in several of the events flattening out discordant relationships

very misleading. Figure 10.25(d) shows upright folds superimposed by strike slip movements; a fairly common sequence of events in high level metamorphic belts, and cover sequences above reactivated orogens.

Figure 10.26 is a fairly representative example of a high grade gneiss terrain (**crystalline basement**) metamorphosed to granulite and amphibolite facies. The map shows several characteristics of these terrains. (1) relatively simple outcrop pattern geometry, (2) a predominance of gneisses derived from granitic precursors (**protoliths**), and (3) little in the way of definitive chronologic evidence. Structurally the bulk pattern is a **Type 3** fold interference best appreciated on the cross-section but also apparent on the map. If a more regional overview had been taken the north/ south upright folds (D_4) would have been seen to form domes and basins here interpreted to be the result of one phase of deformation (cf Figure 8.32). However, two high strain events (D_1 and D_2) are not represented by folds on the map and their recognition is dependent on less direct evidence. Foliation traces in the Amitsôq Gneiss are folded around the closures of the isoclines (D_3 folds) which shows that these folds do not belong to the first phase of deformation. Again we are heavily dependent on the key which has labelled fold traces as D_3 and D_4 drawing our attention to the presence of two earlier events.

Major events in the region's history span 1.2 billion years (twice the length of the Phanerozoic) but little of the events can be deduced from map evidence alone. We see the oldest rocks (Akilia association, Table 10.4) as isolated blebs and slithers surrounded by voluminous Amitsôq Gneisses. Because the Amitsôq precursors were intruded as batholiths the areas of Akilia rocks are probably rafts but high strain has certainly played a part in creating their morphology and may have caused the fragmentation (see boudinage, Figure 10.31). Field observations suggest that the Malene supracrustals (rocks formed at the earth's surface) were probably formed on an eroded basement of Amitsôq Gneiss and the Akilia association which had already undergone granulite facies metamorphism. Therefore one complete orogenic cycle (deep burial, high strain, uplift) predates the Malene rocks; geochronometry gives ages of 3600 Ma for the orogenesis and 3050 Ma for generation of the Malene supracrustals. Tectonic interleaving of the Amitsôq and Malene rocks by isoclinal folding and thrusting (D_2) generated the parallel layering seen on the map. Another plutonic event and syntectonic (D_3) emplacement generated the Nûk Gneisses; the isoclines on the map belong to this second orogenic cycle. A less intense deformation (D_4) was followed by essentially post-tectonic and discordant Qôrqut Granitic intrusions. This sequence of events (Table 10.4) was largely derived from field observations followed by radiometric work which added the chronometry and confirmed the field relations. Multiple orogenois is typical of gneissic terrains and results in very complicated geology. The example chosen in fact starts with the oldest rocks yet located on earth so there had been plenty of time for tectonic reworking. Research in the mid-1980s has added to the story by showing that this gneissic region represents the amalgamation of three crustal fragments that evolved in geographically separate areas. Not all of the elements of Table 10.4 were generated in any one fragment; such revelations are destined to become common place as plate tectonics impacts on continental geology. Figure 10.27 gives a stereo triplet illustrating a set of refolded folds in a high-grade gneiss terrain in another part of Greenland. As with the Ameralik example (Figure 10.26) the large folds probably post-date one or more high strain events. The photographs clearly show the attenuation/thinning of fold limbs that can occur in these high metamorphic grade areas. At high temperatures there is a convergence of physical properties and this probably accounts for the lack of tonal contrast on the image.

More than two large-scale deformation events creates considerable complexity in the form surfaces, e.g. Figure 10.28. Here major subhorizontal folds (Little Moose Syncline and Canada Lake Nappe) are refolded by three different generations of steeply inclined folds though the D_3 effects are quite restricted in areal extent. Analysis of such situations on Survey maps is hampered by a lack of information. Hinge surface traces for the several deformation events are rarely labelled separately though there is a move to display these data on tectonics sketch-maps in the margins of the main map.

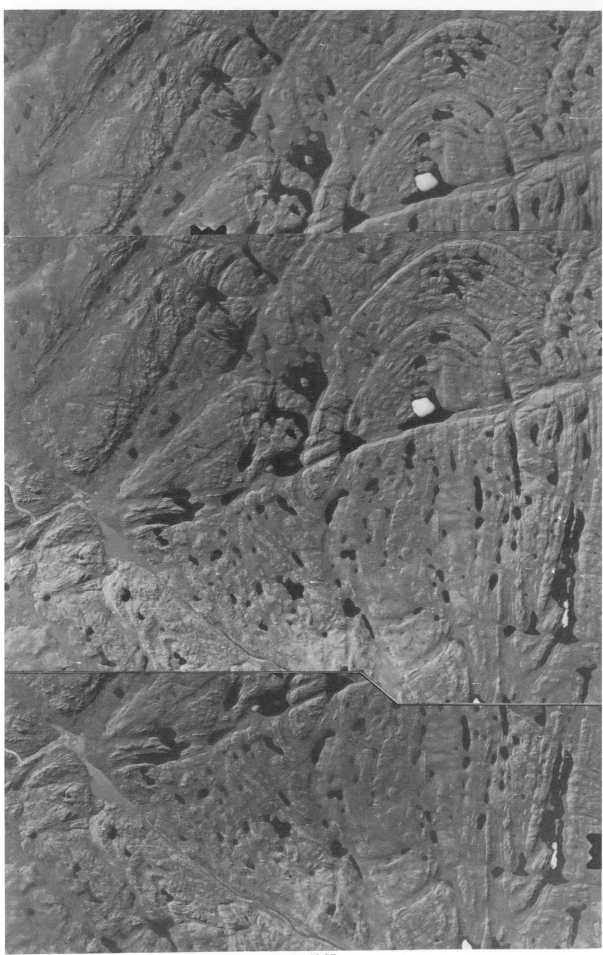

Fig. 10.27

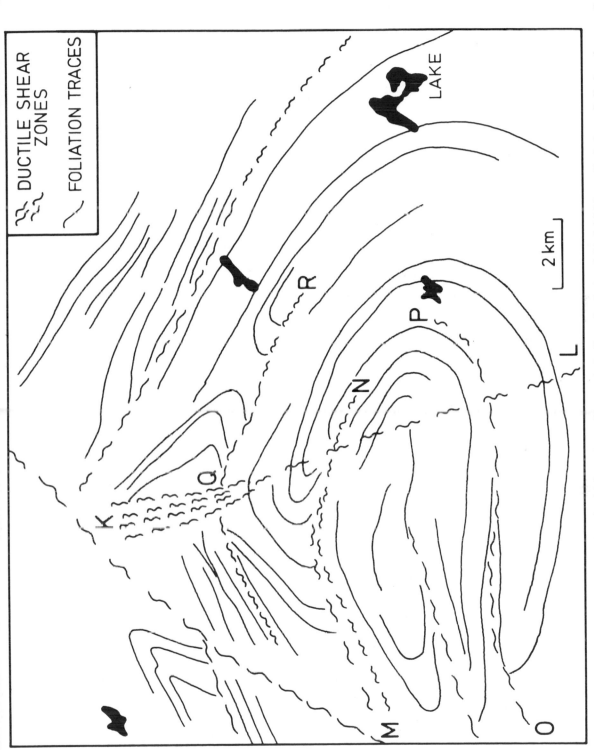

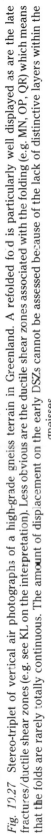

Fig. 12.27 Stereo-triplet of vertical air photographs of a high-grade gneiss terrain in Greenland. A refolded fold is particularly well displayed as are the late fractures/ductile shear zones (e.g. see KL on the interpretation). Less obvious are the ductile shear zones associated with the folding (e.g. MN, OP, QR) which means that the folds are rarely totally continuous. The amount of displacement on the early DSZs cannot be assessed because of the lack of distinctive layers within the gneisses

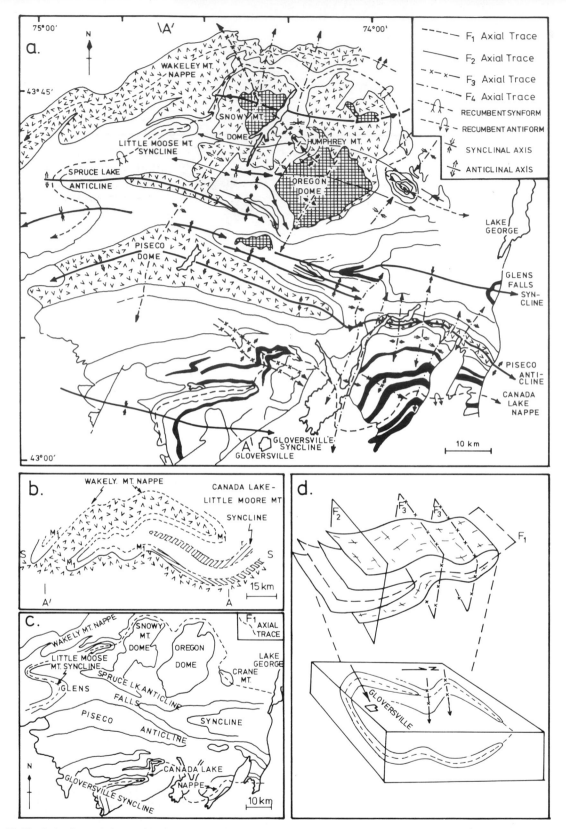

Fig. 10.28 Polyphase deformation in the high-grade metamorphics of the southern Adirondacks, USA (McLelland, J. and Isachsen, Y 1980, *Bull. geol. Soc. Amer.*, **91**, 208–92). (a) An outline map of the lithostratigraphy of the regions showing the traces of the major folds. (b) A cross-section along the line of AA′ on map (a) with the omission of some minor map units. The D_1 isoclinal folds are clearly displayed as are the later open and upright folds. (c) The major fold structures. (d) A block diagram of the D_1 Canada Lake Nappe and its refolding by folds of D_2 and D_3 age; there is not much difference in orientation between these trends but the authors have distinguished them as separate events

Complicated outcrop patterns that are not simply a function of irregular topography should alert you to the presence of polyphase deformation. Combinations of variable bedding, younging, hinge line and intersection lineation, data will help to confirm the basic geometry. Each deformation event may produce its own planar fabric (cleavage, schistosity, foliation, gneissosity) where D_1 produces $S_1, D_2 = S_2 \ldots D_n = S_n$. Unfortunately not many Survey maps differentiate separate generations of foliations, but where this is done it immediately draws attention to the polyphase history and greatly facilitates the analysis. On the maps of Figures 10.25aiv, biv, and civ, for the simple case of two interfering events, S_1 readings would follow the traces of D_1 fold hinge surfaces and S_2 would parallel D_2 hinge surfaces. Clearly the later fabric will be fairly constantly oriented, whereas the earlier foliations can be very dispersed particularly in Type 2 and 3 patterns. The practicalities of mapping such regions are enormous and very detailed work is required coupled with a sophisticated feel for 3-D geometry. The pinnacle of data transfer in this field would be fold and cleavage vergence given for each deformation event, but this may be some time in coming on published maps.

We have already seen that ductile shear zones (DSZs) are localized zones of high strain (see Figure 7.3d) which may become intense and produce mylonites and ultramylonites. They are characteristic features of high-grade gneiss terrains particularly when these regions have been partially reworked in later orogenic events. In most cases the DSZs are belts of lower metamorphic grade than the host rocks, and metamorphic retrogression is a characteristic of these structures normally implying that they are associated with the introduction of fluids into otherwise dehydrated regions. Individual DSZs display a gradually intensifying strain (reflected in foliation development) towards the centre of the zone (see Figures 7.3d and 10.29a). Idealized models show progressive curvature of the foliation traces into the DSZ (see Figures 7.3d and 10.29a, b, c) though abrupt deflections are commonly observed. The sense of deflection gives the shear sense (comparable to slip sense in fault analysis) and the direction of displacement between adjacent blocks is indicated by the lineation data. On maps this should be shown as a mineral lineation or perhaps as the linear alignment of object (xenoliths, cobbles, grains, etc.) long axes.Lineations down the dip of the DSZ will either be associated with normal or reverse senses of movement, and lineations along the zone strike are examples of strike parallel displacement with either right or left senses. Foliation parallel to the DSZ boundaries is a C-surface and oblique foliations are S-surfaces (Figure 10.29b). Within a DSZ both surfaces (S and C) are commonly developed together (Figure 10.29b) which gives another means of determining the sense of shear (rotation) on the DSZ. Three-dimensional patterns of DSZ arrays are variable. DSZs with the same sense of shear may splay off one another at low angles and then rejoin (Figures 10.29d and 10.30c) though a very similar pattern is generated by intersecting conjugate DSZs where the two trends have opposing shear senses. Detailed analysis of the overall 3-D pattern and displacements is usually impossible from Geological Survey-style maps though the amount of data present is increasing. Old maps treat DSZs in very different ways. 'Lines of movement' is one description of DSZs in a gneiss terrain, a not very informative comment. Sometimes DSZs are referred to as 'straightening zones' which reflects the degree of organization imposed by the high strain. Pre-existing structures (folds, faults, plutons, etc.) of any orientation will be skewed round into the DSZ trend and this is the major means of identifying a DSZ from outcrop patterns (Figure 10.30). DSZs may be kilometres wide and one example, 25 km wide and 1300 km long, in the Canadian Precambrian has displacements across it of as much as 700 km. This major structure requires tectonic-scale maps (1:1 million) and the convergence of many techniques for its identification. Its best definition is on an image-processed map of

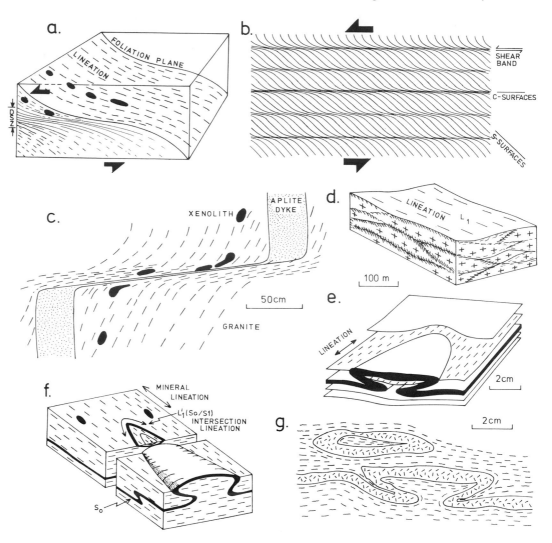

Fig. 10.29 (a) A 3-D view of a ductile shear zone where a continuous foliation plane has been revealed to show its contained lineation which gives the displacement direction. The curvature of the foliation into the DSZ gives the sense of movement (shear criterion). (b) In the high-strain part of a DSZ, C-surfaces develop parallel to the boundaries of the DSZ. The relationship between S- and C-surfaces also gives shear sense. (c) Pre-DSZ features may be highly attenuated in DSZs and in areas of poor exposure might appear to be disrupted. (d) In 3-D DSZs anastomose and enclose lozenges of rock little strained by the DSZ deformation event. The branching DSZs may be splays with the same shear sense (regional simple shear) or conjugate arrays with opposing shear (regional pure shear). (e), (f) Minor folds with greatly curved hinge lines are very common in DSZs. High strain greatly exaggerates any small curvature of hinge lines and creates pointed closures (sheath folds) that are elongate in the direction of the mineral lineation. (g) Cross-sections near the closure of a shear fold give closed outcrop patterns and away from the closure give folds of opposing vergence along single planar enveloping surfaces

aeromagnetic data. Structures of these dimensions are the basement expression of features like the present-day San Andreas Fault which at depth must now be forming a wide zone of foliated rock, a DSZ.

A relatively recent discovery in DSZs is the pronounced curvature achieved by some fold hinge lines (Figure 10.29e, f). Near 180° U-turns are not uncommon and are a function of very high strain. Small departures from rectilinear hinges are

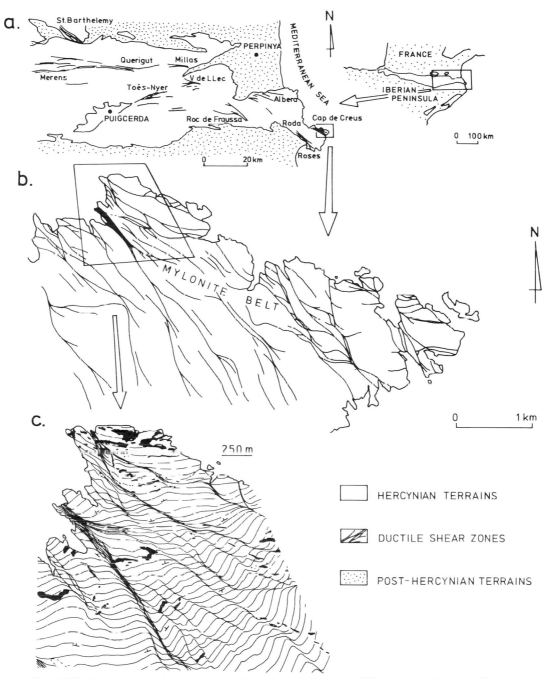

Fig. 10.30 An example of a map pattern in an area of abundant DSZs, eastern Pyrenees (Carreras, J. and Casa, J. M. 1987, *Tectonophysics*, **135**, 88). (a) Regional tectonic setting and location map. (b) Cap de Creus mylonite belt of branching DSZs showing how these structures dubdivide the region into a number of blocks or lozenges. (c) A detailed map of the regional foliation and its deflection into the DSZs, all of which are dextral in the strike component of shear sense

exaggerated by the intense strain and interesting geometries result. The terminations of folds are commonly highly pointed domes (Figure 10.29f) giving rise to the name **sheath folds**. Cross-sections near the point show circular or elliptical outcrop patterns like Type 1 interference (Figure 10.29g) but this is in a single progressive deformation event. Away from the closure tip, cross-sections show asymmetric fold

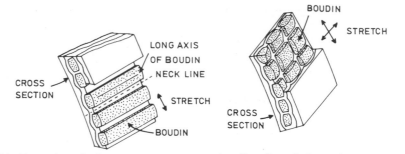

Fig. 10.31 Extension of a layer creates segmentation (*boudinage*). A much greater extension in one direction than the other generates finger (sausage)-shaped boudins, whereas approximately equal extension gives rise to tablet-shaped boudins (chocolate tablet)

pairs of opposing vergence along planar layers or within planar enveloping surfaces (Figure 10.29g); a feature that is hard to explain with simple models of superposed folding. Kilometric-scale sheath folds are known from the Alps and the early Proterozoic of Canada. Some parts of Figure 10.29 (d, e, g) are based on natural examples and scale is indicated; the same style, however, repeats from millimetric to kilometric scales. Photographs or sketches of field situations, exposures, hand specimens or thin sections need a scale because most structures are scale invariant, the same shapes are repeated whatever the scale. Sometimes I do not know whether I am looking at a satellite image or a photomicrograph of a thin section until the scale bar is read.

Extension of layers during strain can create segmentation of the more viscous units (Figure 10.31), thus providing an extra structural dimension to metamorphic regions. The individual segments are **boudins** which may be finger (sausage)-shaped, tablets or anything between. A viscous layer embedded in a ductile medium, stretched beyond its original length, will pull apart. Equal extension in all directions within a layer produces tablet shapes whereas strong extension in one direction within a layer, with no dimension change in the other, gives the linear boudin form. Some boudins are angular in cross-section, some are markedly tapered, shapes that depend, amongst other parameters, on viscosity contrast between boudin and matrix. A very ductile host will flow between the boudins and produce a series of folds (scar-folds or bending-folds). Again scale ranges widely from millimetric to kilometric; large boudins many kilometres in length having been mapped and similar structures have been observed in thin sections.

11 UNCONFORMITIES—HELPFUL GAPS IN THE GEOLOGICAL RECORD

Unconformities are gaps in information but, perhaps surprisingly, in terms of getting a chronology from a geological map they are invaluable reference points. Their usefulness was alluded to in Chapter 6 on chronology but their full impact could not have been appreciated before the discussions on folding, faulting and igneous intrusion. Unconformities, recognizable at the map scale, act as something of a watershed in the history of a region; folds, faults, intrusions, etc. before the unconformity can be separated in time from similar events (and products) post-dating the unconformity. A reasonably comprehensive definition of an **unconformity** describes it as:

> 'a surface of erosion and/or nondeposition between rock bodies, representing a significant hiatus or gap in the rock record'.

Many definitions emphasize interruption of deposition for a considerable span of time, and gaps, in the stratigraphic succession. These place undue emphasis on breaks in sedimentation when intrusive and metamorphic activity may occupy significant proportions of any gap in deposition. The definition presented is not watertight and particularly open to interpretation is the length of a 'significant' hiatus or gap. To the splitter of hairs every bedding plane is a hiatus of some significance, but thankfully we can apply the criterion of map recognition; if it cannot be represented on a geological map we are not interested. In between the hiccups denoted by bedding planes, and significant gaps (unconformities), there are **diastems**, still brief breaks in sedimentation but longer than bedding plane hiatuses.

11.1 UNCONFORMITY TYPES

The approach adopted here is to use unconformity as the general term for a significant break in the rock record and then to specify the type of unconformity using a series of qualifiers. This style of nomenclature was recommended by an international body in the 1960s but has yet to catch on so the more general terminology is also given (in brackets).

1. **Angular unconformity**: an angular difference between the attitude of bedding on either side of the surface of unconformity (Figure 11.1a).
2. **Heterolithic unconformity** (Nonconformity): younger sedimentary rocks resting on plutonic igneous rocks or a high-grade metamorphic complex (gneisses). A marked contrast of lithology and conditions of formation are the primary requirements (Figure 11.1b).
3. **Non-depositional unconformity** (paraconformity): bedding on either side of the surface of unconformity is parallel but there is no evidence of erosional activity at the interface (Figure 11.1c).

4. **Parallel unconformity** (disconformity): bedding on either side of the surface of unconformity is parallel but erosion at the interface is evident (Figure 11.1d).

In any discussion about unconformities, the category being described must be clearly specified. It is not sufficient to use the word unconformity on its own. Also, for example, if an angular unconformity is present then further qualification, such as low-angle or high-angle, may be necessary to give any reader a good impression of the feature.

11.2 PROCESSES

A brief look at the mechanisms that create unconformities will hopefully give a better understanding of their role in map interpretation. Examples are taken from the marine environment but many processes discussed are equally applicable to continental erosion and tectonics. The simplest case is the accumulation of several formations in a marine basin (Figure 11.2aA). Tectonic activity interrupts the depositional process by tilting the layers or by a combination of tilting and uplift. Whatever mechanism allowed the original basin to subside and act as a sediment trap, the tilting may completely rework the tectonic regime thus keeping the region positive and being eroded for a long time. Alternatively the tilt may be a minor event only temporarily halting the basin subsidence. In both cases, the gap at the surface of unconformity is the time taken for the tilting and the uplift, that is, the length of time the area was a positive feature. Renewed deposition to create an angular unconformity may be brought about by a rise in the sea-level or a return to tectonic subsidence; both will allow a marine transgression. World-wide changes of sea-level (**eustatic**) have been viewed with new interest since the development of plate tectonic theory. It was long recognized that glaciation could influence sea-level by storing water as ice at the poles and then releasing it. Major ice-ages are known from the Quaternary, the Permian, the Ordovician and at either end of the Proterozoic. Tectonic patterns can, however, induce much more frequent variation. If mid-ocean ridges form a longer network in one time interval than another, then the extra volume of submerged mountain chains in the oceans displaces sea-water on to the continents causing sea-level to rise. Varying the sea-floor spreading rate with a fixed length of mid-ocean ridge also causes fluctuations; a fast rate gives hot buoyant ridges which stand higher above the average ocean floor and again displace water on to the continental shelves to cause widespread transgressions. Only careful analysis, usually on a regional basis, will reveal the cause of renewed deposition and from small areas it is usually impossible to distinguish between a eustatic rise in sea-level or a tectonic subsidence; for this reason we normally talk about relative changes in sea-level without specifying if the land is falling or the sea is rising. Layers 10 – 12 in Figure 11.2aD are shown with a horizontal attitude as a function of uniform subsidence or eustatic sea-level rise. A subsequent tilt (rotation) of the whole area would give the younger succession a dip and result in a geometry similar to Figure 11.1a. In some cases it is possible for the later rotation to cancel out the first tilt to leave the oldest beds horizontal.

Figure 11.2b shows a classical sequence of a sedimentary basin affected by folding (compression). Subsequent or concurrent uplift takes the folded rocks above wave

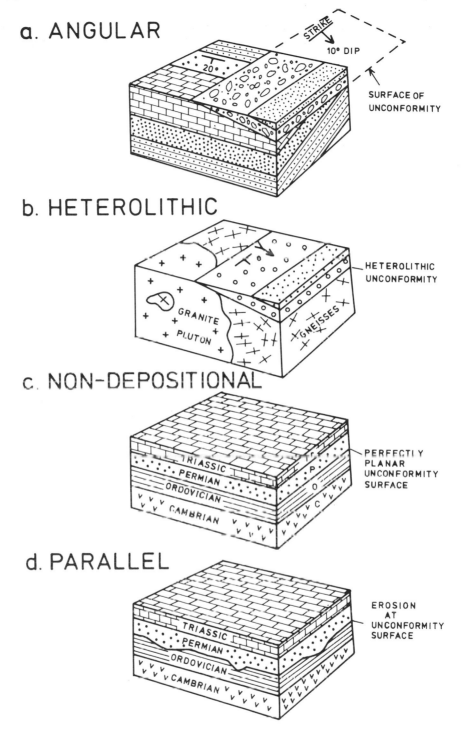

a. ANGULAR

STRIKE

10° DIP

20°

SURFACE OF
UNCONFORMITY

b. HETEROLITHIC

HETEROLITHIC
UNCONFORMITY

GRANITE
PLUTON

GNEISSES

c. NON-DEPOSITIONAL

TRIASSIC

PERMIAN

ORDOVICIAN

CAMBRIAN

PERFECTLY
PLANAR
UNCONFORMITY
SURFACE

d. PARALLEL

TRIASSIC

PERMIAN

ORDOVICIAN

CAMBRIAN

EROSION
AT
UNCONFORMITY
SURFACE

Fig. 11.1 (a) Different attitudes of younger and older bedded sequences creates an angular unconformity where formation/formation contacts in the older rocks are truncated at the unconformity surface. (b) Sediments resting on either high-grade metamorphic rocks or intrusive igneous rocks form a heterolithic unconformity. (c) A perfectly conformable sequence in terms of layering attitude with a significant time gap but no signs of erosion defines a non-depositional unconformity. (d) Here also, layering attitudes above and below the surface of unconformity are parallel but there is evidence of erosion. This is a parallel unconformity

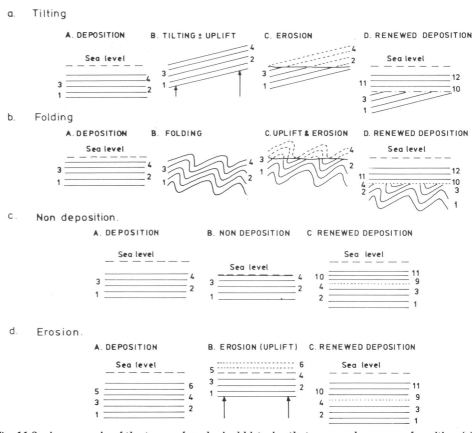

Fig. 11.2 An example of the types of geological histories that can produce unconformities. (a) Deposition of sediments, tilting (with perhaps regional uplift), erosion, followed by renewed deposition as a consequence of sea-level rise or regional subsidence. (b) History as (a) but the deformation was folding instead of simple tilting. (c) Sediment deposition is interrupted by a period of non-deposition and then continues such that there is no angular discordance. If the surface of unconformity is smooth, a parallel unconformity is generated, but if erosion is evident at the hiatus it is a non-depositional unconformity. (d) A more involved history than (c) in that some of the earliest sequence is removed by erosion. This could, however, leave a smooth unconformity surface (parallel unconformity) but is more likely to be non-depositional in character with some signs of erosion at the break

base to undergo erosion which truncates the folded layers. A eustatic rise in sea-level or reactiviated tectonic subsidence will form a sedimentary sequence which has an angular unconformity at its base. The gap in this geological history is from the folding until the deposition of formation 10. If the folded rocks have been metamorphosed, the mineralogy they contain may provide evidence for the near-maximum depth of burial which will give a very approximate minimum estimate for the duration of uplift. In many orogenic belts, collisional processes of differing intensities are responsible for thickening the crust by thrust stacking. This load depresses the base of the crust and takes cool rocks down to higher-temperature environments where metamorphic transformations take place. Gradually isostatic equilibrium is approximately restored by uplift of the orogenic zone which brings the metamorphic rocks to the surface. Remember that if this process exposes high-grade gneisses then we are dealing with a heterolithic unconformity when sediments are deposited on top.

In Figure 11.2c, d are two scenarios that will generate either a non-depositional or a parallel unconformity. These two styles have conformable attitudes for both the

Fig. 11.3 A once planar unconformity surface deformed by later tectonism. The geometry could be analysed by structure contours on the unconformity. The earlier folds (pre-unconformity) would have been tightened in the later tectonism. Folded angular unconformities are illustrated in Figures 5.3, 11.10 and 11.11

younger and older successions. The sequence of events shown in Figure 11.2c is more likely to produce a smooth contact between the two groups of rocks which is a non-depositional unconformity. However, very shallow water at the intermediate stage might, perhaps under storm conditions, scour and channel the top of layer 4 which will give a parallel unconformity. Figure 11.2d involves significant erosion and is more likely to form a parallel unconformity but still could result in a smooth surface of unconformity of a non-depositional type.

11.3 SHAPES

Unconformities are surfaces that can vary in shape. The simplest case is a perfect plane (Figure 11.1a, b, c) which if inclined will have straight, parallel and evenly spaced structure contours. Even the most regular natural example will show some undulation to give slight variations in the trend of the structure contours and spacing. Departures in planarity are either (1) a function of the mode of formation of the unconformity (Figure 11.5), or (2) the result of superimposed deformation (Figure 11.3 and 11.4). A compressional event after the deposition of the younger succession will fold all earlier features and an intense overprint can isoclinally fold the surface of unconformity. If the rocks older than the unconformity were already folded, then the final geometry either side of the erosional surface will be very different. If the later deformation is associated with ductile processes, then any angular discordance could be severely modified. Angles between layering above and below the unconformity could be reduced (Figure 11.4) or increased depending upon the initial geometry and orientation with respect to principal strain axes. Unconformities within the metamorphic interiors of orogenic belts are commonly severely modified.

Inherent irregularity in a surface of unconformity is best shown in a buried landscape situation (Figure 11.5) where an irregular topography is inundated by detritus from its own hills or mountains. Many of the classic examples of this process formed in desert environments where wadis become infilled to preserve some of the dramatically steep slopes typical of these conditions. More commonly the surface of unconformity has slight irregularities filled in by pockets of conglomerate (Figure 11.6). This figure also shows several other aspects of unconformable (not uncomfortable as it sometimes comes out in a typescript!) relationships. **Overstep** describes the inevitable result of angular unconformity where the units in the younger of the two successions lie on different formations of the older. **Overlap (onlap)** is the result of an expanding basin of deposition such that each successive formation in the younger rocks extends beyond the formation immediately below. Onlap is the term

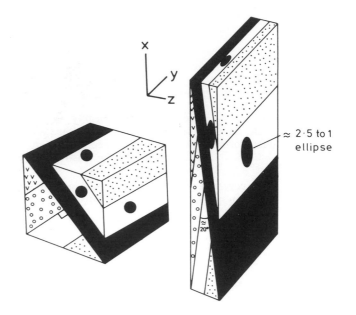

Fig. 11.4 The cube represents the pre-strain situation of an angular unconformity where the youngest sequence had been rotated through 45°. Homogeneous strain (constant volume) was then applied involving 60 per cent shortening in the Z direction and 150 per cent extension in X. The 90° angular discordance has been reduced to 20°, significantly altering the geometry. Strains within orogenic belts can be much larger than shown here

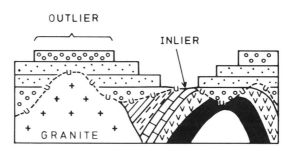

Fig. 11.5 A non-planar unconformity surface representing a buried landscape. Subsequent erosion has expressed the old succession in the valley, creating an inlier (old rocks surrounded by young). The converse, an outlier, is formed by the youngest bed in the overlying sequence being isolated on a mesa. Note that the unconformity changes from angular to heterolithic over the granite

used in the United States whereas overlap is a British term; onlap is probably winning the popularity stakes. American usage considers both overstep and onlap to involve overlap in its more general, non-scientific meaning! Overlap (onlap) is a relationship between different formations above an unconformity whereas overstep is between formations above *and* below an unconformity. The terms are not mutually exclusive as one formation may easily overlap (onlap) lower parts of the same succession and at the same time overstep units below the unconformity (Figure 11.6). Overlap is the hallmark of a transgression.

Unconformities are particularly good at creating **inliers** and **outliers** (Figure 11.5 and 11.7); inliers are outcrops of older rocks completely surrounded by younger rocks and outliers are outcrops of younger rocks isolated by a surround of old.

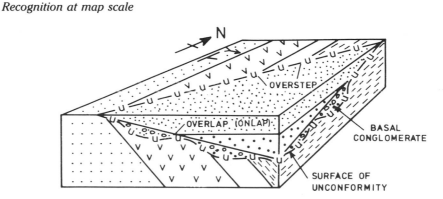

Fig. 11.6 Surfaces of unconformity are commonly slightly uneven with pockets of conglomerate underlying more continuous stratigraphic layers. Overstep is where one layer at the base of the young sequence rest on two or more formations of the old sequence at an angular unconformity. Overlap/onlap is the consequence of an expanding sedimentary basin within the young succession; a younger formation occupies a larger area, thus it overlaps the lower formation

However, surfaces of unconformity are not necessary for the formation of these features (see the outlier in Figure 11.5) but the terms are very useful labels for what may be large areas of outcrop delineated by unconformities.

11.4 RECOGNITION AT MAP SCALE

The most easily recognized type of unconformity is angular. Analysis of the dip and strike reading of bedding on the map may show the discordance in attitudes (Figure 11.1a) but in practice the truncation of formations by overstep will be the primary evidence (Figures 11.6, 11.7 and 11.8). For an angular unconformity the following combinations are possible for the younger and older successions:

1. Same strike but different dip amounts;
2. Different strikes and same dip amounts;
3. Dip and strikes both different.

For a simple tilting event, each formation boundary in the older rocks will be truncated along a single line of intersection between the formation and unconformity surface. If, however, the older rocks have been folded pre-erosion then it is possible that the older formation boundaries may meet the unconformity at several places (see the back bed in Figure 11.3). The relationship looks simple on the diagrams, but finding them in the morass of data of a Geological Survey map is a very different matter requiring practice/experience. Careful searching of the map may be necessary to distinguish these two different styles (tilt or folding) of pre-unconformity structure. Some angular unconformities form rapidly — a million years or so — and, despite the brief gap in the geological record the discordance allows ready recognition. Also note that a dipping younger succession means post-unconformity tilting or folding and is another event in the geological history of the region (see Cambrian dip in Figure 11.8). Parasitic folding could further complicate the picture if folding post-dates the youngest sequence; because of the many different bedding dips

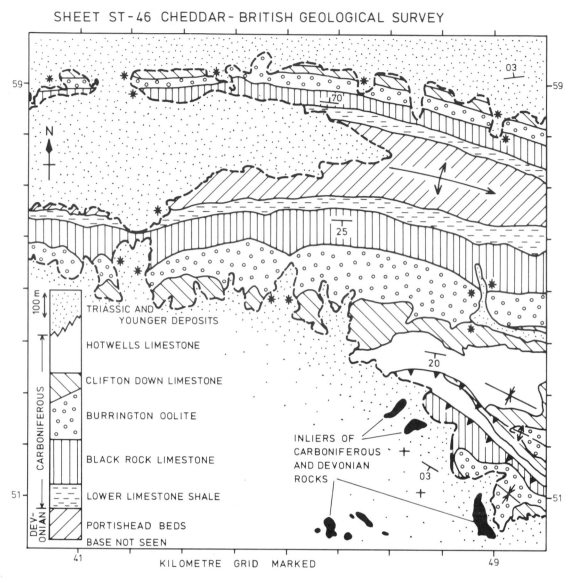

SHEET ST-46 CHEDDAR- BRITISH GEOLOGICAL SURVEY

Fig. 11.7 An abstract from the British Geological Survey map of Cheddar. The Carboniferous and Devonian sequences were folded then uplifted and eroded under arid continental conditions. The angular unconformity between the folded rocks and the nearly horizontal Triassic is very uneven in shape. Topographic information would be needed to prove this, but the trace of the unconformity between the folded rocks and the nearly horizontal Triassic is so irregular it strongly suggests a buried landscape style. Asterisks give examples of overstep. Both the folds and the thrusts are truncated by the unconformity. Small inliers to the south of the main basement outcrop are useful to give some information on the structure of the older rocks below the cover. Note how the map view of the outcrop widths of formations in the Carboniferous varies according to the limb dip of the major anticline

and strikes possible around plunging folds, an analysis of the orientation data on the map may not be strightforward.

Heterolithic unconformities by definition should be quite obvious. A sediment sitting on a plutonic rock (left-hand sides of Figures 11.1b and 11.5) would normally imply that several kilometres of country rock cover to the pluton have been eroded. A greater hiatus is indicated by sediments resting on gneisses (Figures 11.1b and 11.8) which are representatives of lower crustal rocks exhumed from depths normally greater than 20 km. The Assynt District (Figure 11.8) of the North-west Highlands of

ASSYNT DISTRICT

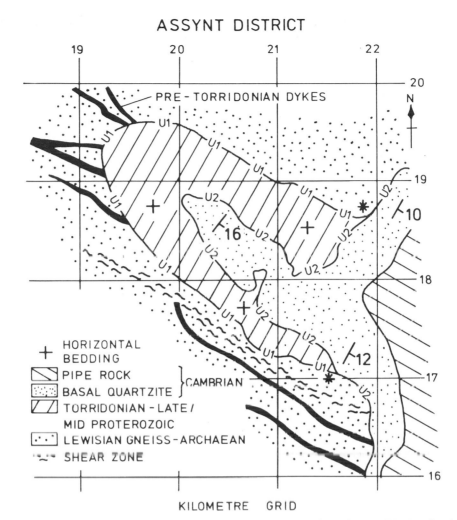

Fig. 11.8 An abstract of the British Geological Survey map of the Assynt District, Scotland. Lewisian gneisses (Archaean) were injected by igneous dykes and cut by ductile shear zones before the deposition of the Torridonian sediments (unconformity U1). A regional rotation tilted both units 10–15° to the north-west. Following uplift and erosion, renewed deposition formed the Cambrian succession as near-horizontal sheets. The base of the Cambrian (unconformity U2) cuts across the Lewisian/Torridonian boundary (asterisks) as well as truncating the pre-Torridonian dykes and shear zones in the Lewisian. A final regional rotation tilted the Cambrian and restored the Torridonian to near horizontal

Scotland is a classic district for the study of thrusts and unconformities. The Torridonian in Figure 11.8 is a near-horizontal red-bed succession of Mid to Late Proterozoic age lying on Archaean Lewisian gneisses. The Torridonian has well-preserved sedimentary structures whereas gneisses were metamorphosed under lower crustal conditions at temperatures between 800 and 1000 °C; unconformity U1 is heterolithic. The Cambrian Basal Quartzite rests on both the Torridonian sediments (angular unconformity) and the Lewisian gneiss (heterolithic), so surface U2 is composite in type changing its spots from place to place. Besides the pronounced difference in conditions of formation between the Lewisian and Torridonian, some igneous intrusions were emplaced into the gneisses before the red-bed sediments were formed; the truncation of the dykes (Figure 11.8) is a graphic pointer to the trace

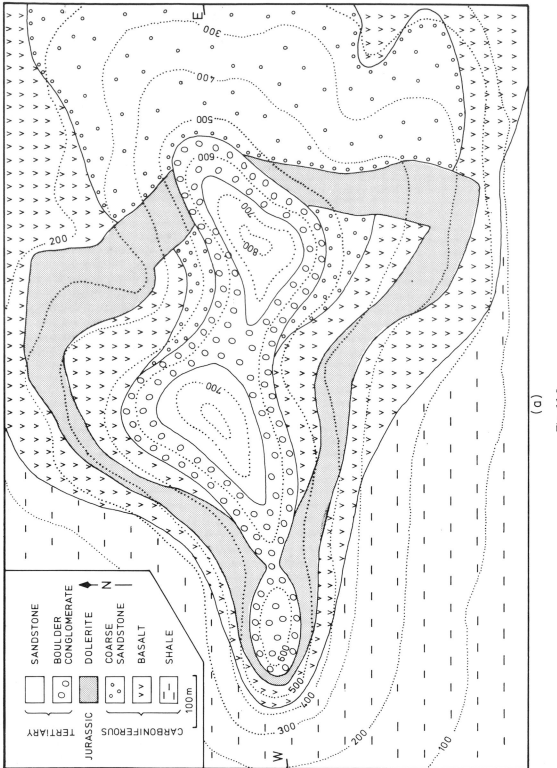

Fig. 11.9 a

SANDSTONE

BOULDER
CONGLOMERATE

DOLERITE

COARSE
SANDSTONE

BASALT

SHALE

TERTIARY

JURASSIC

CARBONIFEROUS

N

100 m

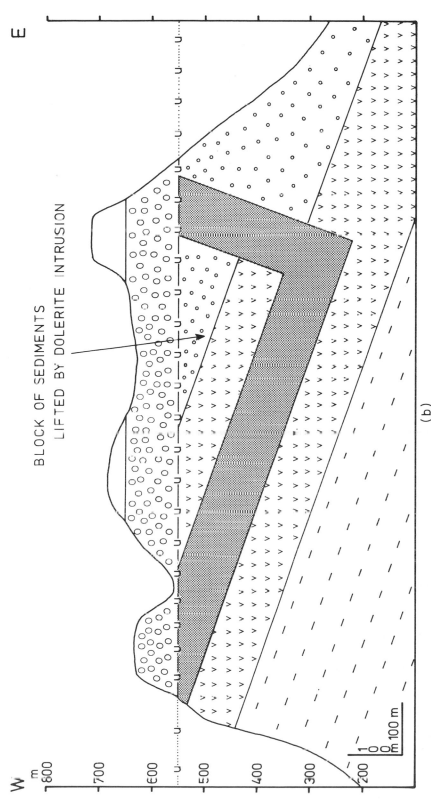
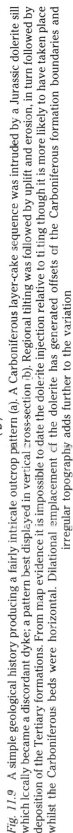

BLOCK OF SEDIMENTS
LIFTED BY DOLERITE INTRUSION

E

W
m
800
700
600
500
400
300
200

1
0
0 m   100 m

(b)

Fig. 11.9 A simple geological history producing a fairly intricate outcrop pattern (a). A Carboniferous layer-cake sequence was intruded by a Jurassic dolerite sill which locally became a discordant dyke; a pattern best displayed in vertical cross-section (b). Regional tilting was followed by uplift and erosion, in turn followed by deposition of the Tertiary formations. From map evidence it is impossible to date the dolerite injection relative to tilting though it is more likely to have taken place whilst the Carboniferous beds were horizontal. Dilational emplacement of the dolerite has generated offsets of the Carboniferous formation boundaries and irregular topography adds further to the variation

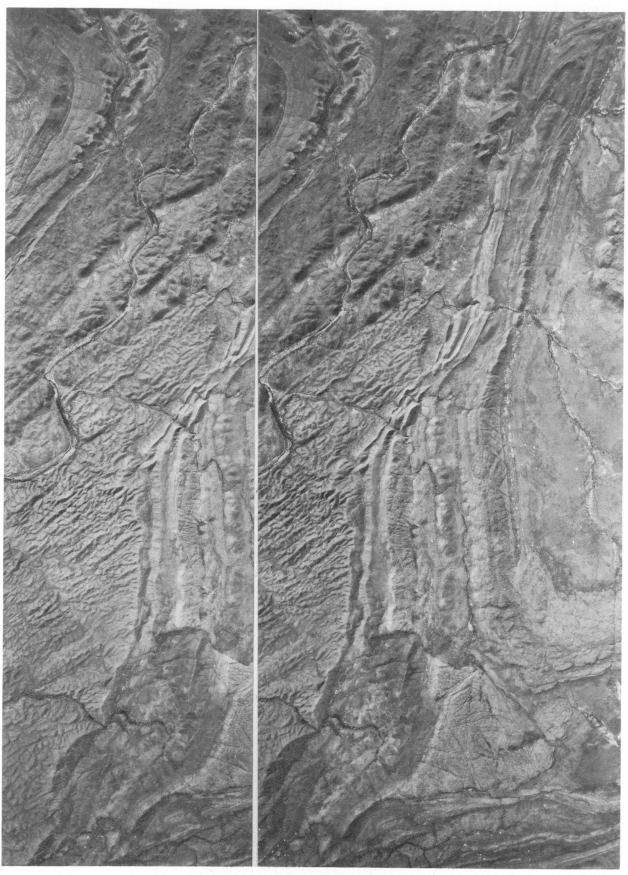

a.

Fig. 11.10 Stereopair of vertical air photographs (a) and geological interpretation (b) which covers a slightly larger region. A tightly folded sequence of thinly bedded resistant lithologies (sandstones) alternating with weak rocks (slates) is truncated by a sequence with pronounced stratigraphic continuity. The structural grain in the older rocks is at about 45° to the strike

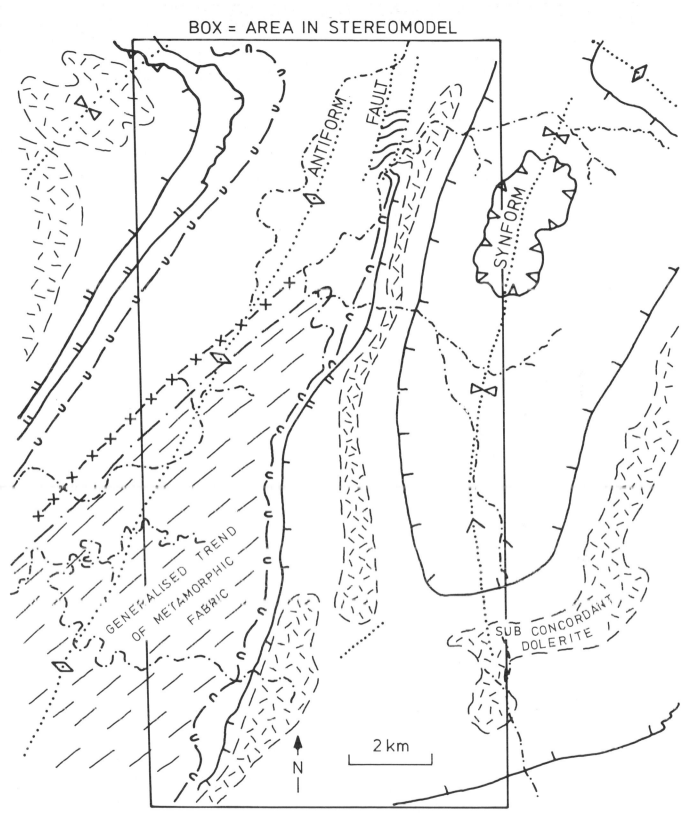

BOX = AREA IN STEREOMODEL

ANTIFORM

FAULT

SYNFORM

GENERALISED TREND OF METAMORPHIC FABRIC

SUB CONCORDANT DOLERITE

2 km

N

b.

of the younger succession which itself is folded together with the surfaces of unconformity.
Two major units can be recognized below the angular unconformity. The northern unit has
very thick internal layering, is darker in tone and has eroded very differently to the sandstone/
slate unit. Within the younger succession, offsets of some layering, but not of adjacent units,
suggests igneous intrusion (cf. Figure 10.3)

of the unconformity. The basal Cambrian unconformity is a near-planar surface which can be mapped for tens of kilometres with remarkable continuity. In contrast the basal Triassic surface on the Cheddar sheet (Figure 11.7) is very irregular. The tongue-like extensions of the Triassic into the folded succession are old wadi-like valleys; this is a buried landscape.

In the Assynt District, truncation of the pre-Torridonian dykes clearly draws attention to the position of both unconformities (U1 and U2). Figure 11.9 again shows hyperbyssal intrusions, that, because of their pre-unconformity injection, help to define the nature and position of the hiatus in the rock record. The intrusion style illustrated shows the common situation of a concordant body locally transgressing the layering of the host succession. Note the offset of the basalt/coarse sandstone boundary generated by the increase in volume associated with the arrival of the dolerite sill by dilation. Considering with interaction between geological and topographic surfaces, it is worth pointing out that the offset is more obvious on the shallower slopes to the south of the highest hill than to the north on steeper ground.

Non-depositional and parallel unconformities can only be recognized on a map with the aid of the legend. Layering either side of the unconformity surface has the same attitude (both have the same dip and strike if later rotations have taken place) and hence no truncations occur. The resolving power you have at your disposal to identify such an unconformity is entirely dictated by the precision of the information provided on the map's stratigraphic column. Even for the Phanerozoic, published maps rarely give biostratigraphic zones and not all go as far as giving ages/stages. It is possible that breaks of 5 – 10 Ma may go unrecognized and in the Precambrian this may be stretched to many tens of millions of years or more.

Quite a number of angular and heterolithic unconformities are notoriously hard to find intact in the field. Such surfaces are major heterogeneities within the rock body and commonly become the focus of later events. In particular sills will preferentially intrude along unconformities, also faults will make use of the anisotropy they induce. Contractional faults seem to take great delight in removing evidence for angular or heterolithic unconformities thus creating a tectonic contact in place of a stratigraphic 'unmoved' boundary. Over strike lengths of tens or even hundreds of kilometres, unconformities may be obscured by overprints and evidence for their original nature may only be preserved in small areas.

On remotely sensed images, in the absence of ground data, non-depositional and parallel unconformities cannot be recognized. The presence of angular and heterolithic varieties should be detectable employing the same criteria as noted for map analysis (Figures 11.10 and 11.11). A sediment resting on an eroded granite may not be as obvious on a remotely sensed image as it is on a map which gives age relations on a legend. This situation on an air photograph might require, for safe recognition, the younger sediment to be seen to cover some of the intrusion's country rock at an angular unconformity. Even slight angular discordances, under favourable conditions, might be clearly displayed by remote sensing (Figure 11.11). One note of caution; care must be taken to distinguish between faulted contacts and unconformities. Erosion etches out some unconformities in a fashion typical of many faults and their photo expressions can be very similar. Again the synoptic view of remote sensing is of some help. Faults may well be associated with 'drag' effects, rotation of blocks between anastomosing splays, wide zones of brecciation, veining and characteristic patterns such as *en échelon* arrays. The absence of such features should point to an unconformable relationship.

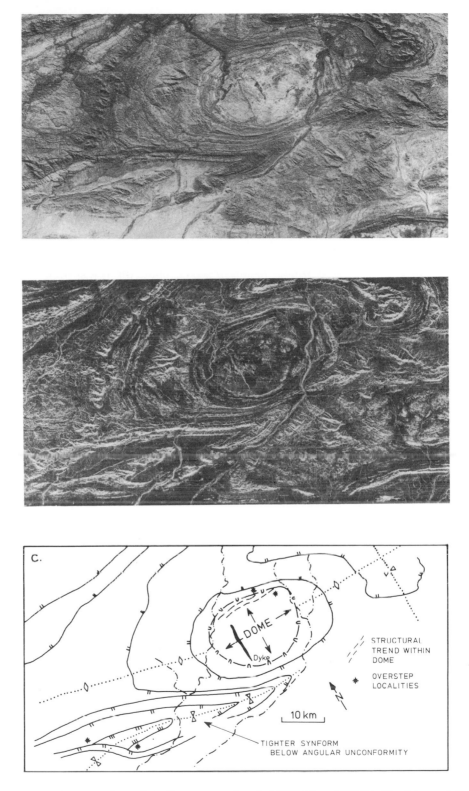

Fig. 11.11 A comparison of similar areas on Landsat MSS (a) and SIR (b). The interpretation for the SIR image (c) shows two pronounced angular unconformities. The unconformity within the Archaean dome can be seen in the MSS image but not clearly. This emphasizes the usefulness of radar for structural studies

11.5 STRATIGRAPHIC NOMENCLATURE

Many of the systems in the chronostratigraphic scale (see Figure 6.4) were first
defined by the presence of unconformities marking their start and end. Since then
the need for virtually continuous sedimentation across these boundaries, has led to
the abandonment of most original type areas. However, debate has continued about
the usefulness of unconformity – bounded stratigraphic units. Eventually (1987) the
International Subcommission on Stratigraphic Classification recommended a
nomenclature for general adoption. A **synthem** is a unit bounded above and below
by unconformities of regional extent (Figure 11.12). Synthems may be composed of
any type of rock (igneous, metamorphic or sedimentary) and their only diagnostic
criterion is the presence of bounding unconformities. Recognition of synthems is not
dependent on lithostratigraphic, biostratigraphic, chronostratigraphic or any other
stratigraphic means of classification; this is a new classification scheme distinct from
any of those discussed in Chapter 6. The boundaries of unconformity-bounded units
can be parallel to the boundaries of the stratigraphic units (lithostratigraphic,
biostratigraphic, etc.) they include, but most commonly there are angular
differences. In 1949 unconformity-bounded units were christened 'sequences' but this

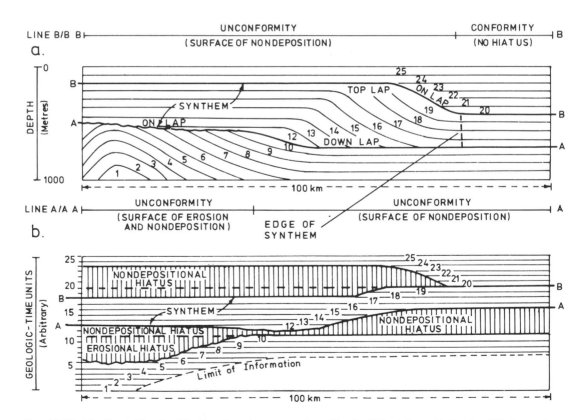

Fig. 11.12 Stratigraphic classification based on unconformities is distinct from that using lithology, time,
biological content, magnetic reversals, etc. Synthems are bounded top and bottom by unconformities. (a) A
section based on lithostratigraphy and, despite the vertical stretch, is equivalent to a normal cross-section.
(b) A chronostratigraphic version of (a) which shows intervals of time without deposition in vertical ruling
thus emphasizing gaps in the rock record

term has been used in many ways and the subcommission decided to introduce a new term free from a confusing history.

In addition to illustrating a synthem, Figure 11.12 emphasizes several other aspects of unconformities. Surface AA always represents a hiatus whereas surface BB is conformable towards the right margin where Units 19/20 are in contact. As a synthem must be bounded top and bottom by unconformities, the point where BB becomes conformable defines the limit of the synthem. Downlap, as shown by the various relationships between sedimentary units (lithostratigraphic) above surface AA, is typical of sediments building out (prograding) into a basin or on the continental margins. Features such as toplap and downlap, will only be recognized on maps with careful analysis. Toplap is typically formed by a relative stillstand of sea-level such that there is either a period of non-deposition of sediment or perhaps minor erosion. The comparison between parts a and b of Figures 11.12 also brings home the hiatus aspect of unconformities and the amount of lost information.

12 GEOLOGICAL MAP INTERPRETATION— A SYNTHESIS

12.1 GENERAL COMMENTARY

Much of what we have done to date has dealt with principles and elements on an individual basis. In moving towards the climax of the book more and more integration was, of necessity, introduced especially in dealing with igneous and metamorphic materials and unconformities. The time has now come to firmly draw the disparate threads together. With some guidelines we should be in a position at the close of this chapter to tackle a Survey-style geological map despite its daunting wealth of data. An extremely common exercise in undergraduate map interpretation courses is the somewhat feared 'Write an account of the . . . geological map'. The following is a proforma with notes that should allow this task to be successfully tackled. In this format it is unlikely that you will ever again perform such an assignment once you have left the educational system, but you will be using the skills it demands all your geological career.

Geological maps are used all the time and in many different ways from gold to petroleum explorationists, by engineering geologists, soil scientists, land-use planners, hydrogeologists, etc. At the first phase of exploration, regional maps are consulted to determine the most favourable combination of controls on economic concentration. More detailed ground acquisition is based on larger-scale maps, though in both phases other techniques (e.g. geophysical, remote sensing) may be as important depending on the commodity being sought. All the work on maps requires expertise in reading geological maps. Actual exploration generates its own maps and all members of the team must be able to interpret colleagues' maps. Mapwork is fundamental to geology. Also all that is in this book is laying the foundation for your own inevitable steps into making geological maps. Many students struggle in the field because they have not grasped the basics as presented here. Too high a proportion of undergraduate maps (and some published Geological Survey maps) can be criticized because the mapper clearly did not appreciate something as basic as the interaction of a simple geological surface with topography. A very large number of maps fall from grace because of internal inconsistencies of this type which then lead the observer to question the accuracy (truth) of the map. Clearly too many mappers leave the thinking to the evening, the weekend or at worst when back in the office. In making a map you should always be thinking on the spot about the implications of the data you are gathering. At its simplest this means that a gently dipping surface you observe on a ridge will V into the adjacent valley—a straight-line outcrop is inappropriate across varied topography. As you map, you should make sketch sections of the ground you have covered and attempt to sketch predictive sections based on extrapolating data you have on hand. Hopefully, facility at reading maps generates facility at making maps. Another contribution of this book is the reality conveyed by the remote sensing giving preparation for the real world. A written account allows you to demonstrate that you have received the message conveyed by the map. Important in this process is the integration of many different aspects (biostratigraphy, structural, igneous, sedimentary, metamorphic) all leading to an appreciation of the evolution of an area—its **geological history**. Clearly requirements will be very different depending on style, length and level of course. Reports may be based on one or two practical sessions and be limited to several hundred words. Alternatively, a much more comprehensive analysis may be asked for, requiring a modified approach. Take note of 'in-house' requirements and be prepared to be adaptable.

12.2 GRAPHICS

It has to be said that many students are reluctant to illustrate their reports. Your artistry has to be very bad for diagrams to detract from the impact of an account. **A picture is worth a thousand words**. Even if you feel art is for others there are several simple ways of producing effective diagrams. Many aspects of a map are best conveyed by reducing the published map to page size and highlighting the major features (e.g. Figure 12.1). This can be done by reducing photocopying, with a final tracing or by using the grid on the map and a scaled-down grid on an A4 page and transferring equivalent points (in a similar fashion to down-plunge profile construction—Figure 8.35). During the reduction, a lot of simplification is needed and you must choose carefully what is to be included; a wrong decision here could obscure major features of the map. Use of good-quality felt-tip pens gives clear even-weight lines, and sensible use of ornament plus colour (cf. air-photo interpretation symbols) produce good products within the reach of anyone. For complex regions each of the following topics—distribution of outcrop, structure, igneous rocks and topography—might benefit from a reduced map though combinations of two or more of these topics on one map may be appropriate. Also considering relative importance and complexity it might be sensible to show topography on a very reduced map and structure on a larger sketch-map. Enlargement maps of critical areas are also extremely useful to show key relationships between major rock units (e.g. Figures 12.2 and 12.3). Small areas can be enlarged using the map grid system or by enlarging photocopies plus tracing. These key relationships may be better handled by an enlarged vertical cross-section through a small area or a combination of section plus enlarged map.

A necessary part of the training phase, and the most common in practice, is the analysis of the map with only the legend. All other explanatory material (cross-sections, block diagrams, rock-relationship diagrams, tectonic sketch-maps, commentary of variable length, etc.) having been removed. You are then required to create these diagrams on the basis of the map information. Every account has to have a cross-section (normally vertical but a profile of the entire region might be more appropriate) and the other material may or may not be asked for. Remember that all illustrations should be adequately documented—location, scale, key, title and for a section the orientation should be clear. Hopefully, at some stage in a course, students will be given time to study complete modern maps to see the level of information now supplied.

Remember—Graphics are mandatory

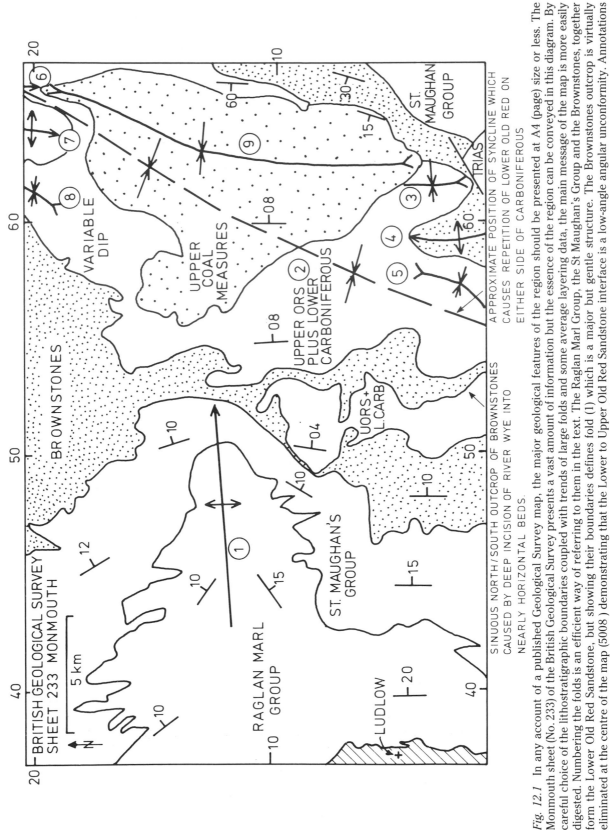

Fig. 12.1 In any account of a published Geological Survey map, the major geological features of the region should be presented at A4 (page) size or less. The Monmouth sheet (No. 233) of the British Geological Survey presents a vast amount of information but the essence of the region can be conveyed in this diagram. By careful choice of the lithostratigraphic boundaries coupled with trends of large folds and some average layering data, the main message of the map is more easily digested. Numbering the folds is an efficient way of referring to them in the text. The Raglan Marl Group, the St Maughan's Group and the Brownstones, together form the Lower Old Red Sandstone, but showing their boundaries defines fold (1) which is a major but gentle structure. The Brownstones outcrop is virtually eliminated at the centre of the map (5008) demonstrating that the Lower to Upper Old Red Sandstone interface is a low-angle angular unconformity. Annotations on the sketch-map can be very useful to amplify important aspects

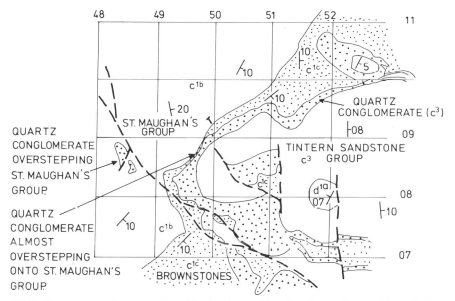

Fig. 12.2 Small areas that contain critical relationships for the understanding of the map should be given special treatment with an enlarged view. Here a very small outcrop on the Monmouth sheet of Quartz Conglomerate (Upper Old Red Sandstone) at 484087 is isolated as a partly fault bound outlier on the Brownstones (LORS). This overstep clinches the interpretation of the angular nature of the UORS/LORS interface which is suggested by the near elimination of the Brownstones outcrop seen in Figure 12.1

When writing a report the reader you should have in mind is one who does not have access to a copy of the original map. Ask yourself at the end of writing, 'Would my reader have a good picture of the region from this account?' Illustrations are crucial for a yes vote.

At the introductory level I have not given specific instructions for drawing block diagrams (Figure 12.4) or profiles of regions with significant topographic variation. These topics require slightly more advanced construction techniques and tend not to figure in beginners' courses. Block diagrams are best done by orthographic construction (see Ragan, D. M., 1985, *Structural Geology: An Introduction to Geometrical Techniques* (3rd edn), John Wiley) which requires more stereographic expertise than we cover.

12.3 ACCOUNT LAYOUT

Accounts of maps are effectively in two parts; description and then interpretation in the form of a geological history. Fact should, as much as possible, be clearly separated from inference. For example, in describing stratigraphy and rock types note features critical to an analysis of depositional environment but make the inference based on this in the history section. Generally, the account should be written in essay form unless otherwise specified. Accounts will vary depending upon the course requirements, type of map and amount of additional information to be used. However, the following core of topics will have to be covered in the majority of situations.

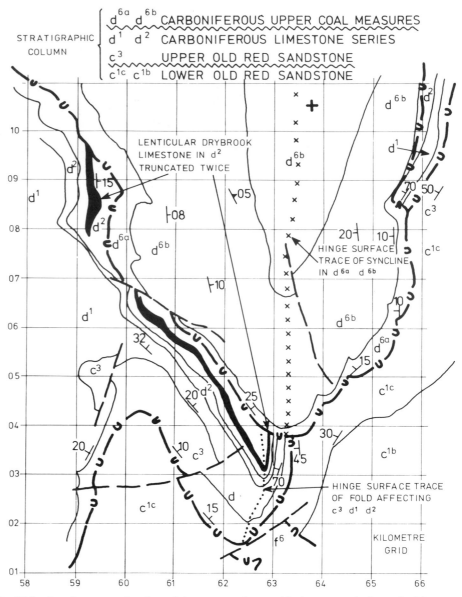

Fig. 12.3 Another example of special treatment for a critical area, again from the Monmouth sheet. The Upper Old Red Sandstone and older rocks were folded before the deposition of the Carboniferous Upper Coal Measures. This is clearly proven by the double truncation of the Drybrook Limestone. Much of the outcrop pattern of this figure could be explained by simple tilting of the pre-Coal Measures sequences and subsequent folding of two non-parallel successions. Both interpretations could account for the offset of the fold traces in the Coal Measures and the older rocks

(a) Introduction — nature of map, size of region, topography.
(b) Geological succession.
(c) Structural geology and distribution of outcrop.
(d) Igneous intrusions.
(e) Economic geology.
(f) Geological history and evolution of present-day landforms.

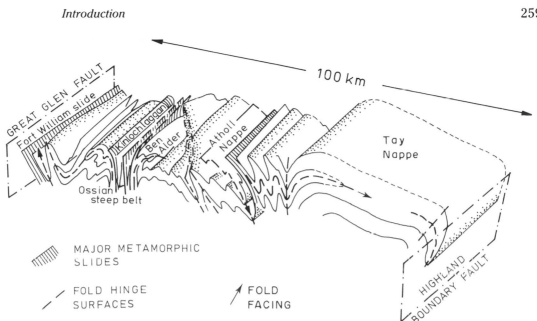

Fig. 12.4 Block diagrams are highly effective means of illustrating the geometry of a map sheet. Precise constructions are available but sketch views are still very valuable. (Reproduced by permission of the Geological Society from P. R. Thomas, New evidence for a Central Highland Root Zone, in *Special Publication No. 8*, 1979.)

If metamorphism is an important aspect of the region it could be incorporated in item (c) if regional in character, or item (d) if contact (or local) in nature.

Approaches to map analysis vary but an overview should be gained first. Perhaps even without detailed study of the key and any stratigraphic columns stand back and look at the overall map pattern. Look for any large-scale repetitions of stratigraphy and see if these are fault or fold controlled. Look for regional convergences of structural trends which delineate some kinds of hiatus — structural or unconformable. If the latter is responsible then truncations should be sought. From the key determine the continuity or otherwise of the stratigraphic record. Breaks may be schematically represented or more cryptically indicated by gaps in the geochronologic, chronostratigraphic or biostratigraphic units. A missing period or system (e.g. Cretaceous) would be obvious to all, but a missing age or stage (e.g. Campanian) might escape attention. Recognition of missing biostratigraphic zones requires detailed knowledge rarely found in the memories of non-specialists.

When the overall picture starts to fall into place, it is time to construct the cross-section if one has not been provided. This concentrates attention on a zone but also necessitates a detailed examination of the whole map to gain an appreciation of relations (time and geometric) between major rock units.

12.4 INTRODUCTION

At its most brief this section gives the nature of the map (solid, drift, metallogenic, tectonic, engineering, etc.), map scale, size and dimensions of the area covered and topography. Long accounts should start with an abstract which is a condensation and concentration of the essential information in the account. Good abstracts are hard to write because they must be intelligible without reference to the account, i.e. they must be capable of standing alone. In some cases a regional location map may be necessary or perhaps a map to show the area of interest in the tectonic context of units as large as cratons or orogenic belts (cf. Figure 10.30).

Only the major landforms (situation and magnitude) should be described. Since the account is essentially geological, this section should be as brief as possible but be specific. If discussing a ridge give its position, orientation (bearing), length + width (m or km) and height in metres above the surrounding average terrain. Position is best specified by grid references. Rather than prose, an annotated sketch-map with some contours or hachuring(?) to show slopes is often the **best** way to convey topographic information. Think of the reader who does not know the area.

12.5   GEOLOGICAL SUCCESSION

This section should be introduced by an overview statement. For example: 'A 5 kilometre Devonian and Carboniferous sedimentary sequence was thrusted and folded before deposition of Triassic and Jurassic sediments which show only slight deformation. The basal Triassic unconformity is an extremely irregular surface representing a buried topography.' Many students satisfy the requirements of this section by reproducing the stratigraphic column on the map key. This is not a very fulfilling exercise and has little merit. In keeping with the **be graphic** theme, a type of annotated stratigraphic column is very useful but this should abstract and highlight the important features not just be a clone of the key. **Oldest first** is the basic rule and all geological columns have the oldest at the bottom. Thicknesses of units or groupings of units need to be quoted and especially significant variations observed around the map. Unconformities deserve special attention. They should be classified (angular, heterolithic, non-depositional or parallel) and the surface shape described (planar, curviplanar, irregular). Detailed accounts will leave time for structure contours to be drawn on unconformities to specify the shape. In less demanding accounts it might be worth while enlarging a type area (cf. Figure 11.8) for each unconformity to show their nature (particularly by constructing structure contours for these small areas).

Because formations are often of very mixed lithology, and lithostratigraphic names do not have to say anything about lithology, rock types may only be known in the broadest of terms. What information is provided should be generally commented on in the text or on the abstracted stratigraphic column. In addition to giving a good indication of all the lithologies present, special mention must be made of environmentally significant rock types. Red beds normally mean continental conditions at least in the Phanerozoic, ooids and corals in abundance would mean shallow shelf reef situations, pillow-basalts mean subaqueous formation, stromatolites are intertidal to shallow subtidal, bentonites are altered air fall tuffs which form time horizons, etc. Some features are quite specific about the depositional environment but others give very little information. Many beginning students believe conglomerates are shoreline deposits but they can form in Andean alluvial fans down to deep-sea off-shelf fans. Mudstones are not only the product of deep quiet water, they can form in lakes, in river overbank deposits, marginal lagoons and in the oceans. Note that deep is not very tightly defined. Does it mean below wave-base, edge of the continental shelf or off-shelf on the abyssal planes? Limestones on their own are not sufficient indication of marine conditions as they can form in lakes. Use the information provided on the key carefully. You are, of course, totally dependent on the legend which at worst may tell you virtually nothing about environments of formation. If you are told to restrict the account to information only on the map do not bring in comments from your general knowledge of the region. Whenever external knowledge is introduced clearly state where the information comes from.

When dealing with formal stratigraphic terms remember the conventions relating to capital letters (see Chapter 6). Generally it is sensible to follow the lead given by the map key but beware of old terminology. Before international agreement was reached many authors used series for a sequence of formations. Current usage has groups as the next term in the hierarchy for a collection of several formations. There are also internationally agreed conventions on the way fossil names are written in texts. The International Commission on Zoological Nomenclature has an International Code which is a set of laws governing procedure. Scientific reference to an animal or plant is by a two-word **binomen**. The first name written with a capital letter denotes the genus. The species name is second and is not capitalized. In printed texts, such names are always in italics (e.g. *Homo sapiens* or *Charniodiscus concentricus*). In a hand-written or typed manuscript they must be underlined. Again take your cues from the key to the map.

12.6 STRUCTURAL GEOLOGY AND DISTRIBUTION OF OUTCROP

For fairly straightforward areas it is best to describe folds and faults separately and within these two categories each generation needs to be clearly separated. A great many regions have been affected by two or more orogenic events even if these are distal and minor effects. It is also common for structurally simple areas to have been through distinct extensional and compressional deformation events within a single orogenic cycle: the characteristics of each deformation event should be distinguished as clearly as possible.

An introductory comment giving the general structural style is useful to set the scene. This is then followed by a descriptive section. For folds and faults an account of their spatial distribution is particularly important and a sketch-map is, in most cases, the only way to proceed. Folds and faults may be dealt with individually or in related groups, but whatever approach is appropriate clearly indicate this in your report. Start each subsection with a note as to age range of rock affected.

Fold description (for each event):

1. *Orientation* — hinge lines bearing and plunge; hinge surfaces dip and strike. Note variations such as regional changes in trend or the influence of refolding.
2. *Style* — angularity; dihedral angle; symmetry or asymmetry; layer thickness variation as seen in profile.
3. *Stratigraphic relations* — fold facing, are antiforms anticlines or synclines?; effects of folding on distribution of outcrop.

Fault description (for each event):

1. *Attitude*—dip and strike; strike, oblique, or dip, faults in relation to attitude of stratigraphy; dips greater or less than layering; dips with or against dip of layering.
2. *Separation geometry*—dip and strike separations and others as appropriate. This gives effect of faulting on outcrop pattern.
3. *Slip data*—first give any calculations that can be determined, then any striation information, then any minimum estimates of slip. State if rotational or translational displacements can or cannot be identified.
4. *Patterns*—fault patterns can give an indication of likely displacement orientations, e.g. conjugate near-vertical faults are probably strike slip. Note antithetic and synthetic relations. Determine if extensional or contractional.

Needless to say this simple recipe cannot cover all situations. If a map is cut by a major strike-slip fault, such that there is no correlation of stratigraphy or structure from one side to another, then you almost have two map accounts. In a region dominated by thrust faults, all the folds may well be secondary products of the thrusting and a rigid description keeping apart folds and faults may not be sensible. A gneissic terrain may be dominated by ductile shear zones and these structures will then occupy much of the account. Regional metamorphism, if present, should be included in this section.

12.7   IGNEOUS INTRUSION

This will only be a separate section if intrusive rocks are an important feature of the region. Volcanic rocks are sensibly dealt with in the stratigraphic account though in an active volcanic region or one dominated by volcanics, the topic will require special consideration. Some map sheets may have several intrusive events perhaps separated by significant time intervals. Each of these episodes might produce more than one style or form of intrusion, and in single magmatic cycles multiple intrusions may be quite complex. Unless the account calls for concentration of attention on the intrusive activity, deal with the broad aspects of each magmatic cycle rather than individual intrusive bodies. The following is a general check list:

1. Rock type—magma type is a general control on nature of intrusive style.
2. Form—concordant/discordant; planar (give dip and strike), ring dyke, cone sheet, laccolith, pluton, stock, batholith, etc.; dimensions; contact metamorphism.
3. Distribution and patterns—describe distribution of outcrop; dyke swarms, radial dykes, linear arrangement of plutons, laccoliths related to stocks, etc.

12.8 ECONOMIC GEOLOGY

Any aspect of making money from rocks can be considered here including metallic mineralization, hydrocarbons (oil and gas), coal-mining, and mining of non-metallic non-energy minerals (road aggregate, building stone, clay, gypsum, salt, sulphur, micas, etc.). Probably the most obvious signs of economic activity will be quarries or opencast workings and vein-type mineralization. The nature of stratiform massive sulphide deposits may be apparent only on very detailed maps. Porphyry copper mineralization normally involves the quarrying of hundreds of millions of tonnes of ore at less than 1 per cent copper. Such operations have a massive impact on a locality. These and similar deposits may have large hydrothermal alteration haloes where original rock textures and chemical compositions are severely modified; this information may be shown on the map and described in the key.

12.9 GEOLOGICAL HISTORY

Such an account, from all available evidence, will provide a chronological history of the region from the time of formation of the oldest rocks up to and including the present-day situation, e.g. currently active flood plains depositing alluvium. Many items of the geological history will have been referred to in the previous descriptive sections. The aim now is to utilize all the facts in an attempt to synthesize an overall chronological account. The successive events in time should include inferences concerning deposition of sediments, environments, folding, faulting, erosion, and metamorphic, intrusive and mineralization events. The nature of the various rocks may throw light on the palaeogeography of the area, e.g. continental conditions with alternation of marine and non-marine inundations may be indicated by the presence of some coal-bearing sequences. Be cautious with environmental interpretations using only definitive evidence which should already have been noted in the stratigraphic section. Also avoid a holiday brochure style, e.g. 'corals and ooids indicate a warm balmy clear tropical sea'! If a large region is being considered and sufficient information is available, sketch palaeogeographic maps for several time intervals may be useful. In published papers a more advanced equivalent of this is to construct block diagrams to show several points in the evolution of the region.

The final section will deal with the establishment of the present topography. Erosional or destructional landforms can be related to the control exerted by the underlying geology. The morphology can be used to discuss glacial, humid or arid influences in landscape creation. Constructional landforms will be strongly controlled by climate. In a glacially dominated region, moraines, eskers, outwash gravels, loess, drumlins, etc., contribute to the story. However, much of the glacial activity may not be shown on solid maps. Drift maps show glacial striations, glacial drainage, ice-moulded topography, drumlins, roche moutonnée, fabric in boulder clays, plus more. Many of these features are directional and may allow a partial reconstruction of the glacial history. The most recent features of an area may be raised beaches and alluvial deposits.

Basic relationships

1. **Folds** are younger than the folded rocks.
2. **Faults** are younger than the rocks they cut.
3. **Metamorphism** is younger than the rocks it affects.
4. An **unconformity** is younger than the underlying rocks and truncated structures, and older than overlying deposits.
5. **Intrusive** rocks are younger than the host. This is true for igneous intrusions, but for salt domes and sandstone dykes it is the act of intrusion that is younger than the country rock. The salt in a dome will be from a lower (older) part of the stratigraphy.

Figure 12.5 of the Wicked Hills shows some basic overprinting relationships on an outline map. Note that the mapped lithostratigraphy is not given in order, a not uncommon exercise situation in map interpretation classes. The following gives a geological history and briefly comments on how the sequence of events was recognized:

1. Deposition of impure limestone (calc-silicate), quartz sandstone (quartzite) and mudrock (kyanite schist). These are protoliths to the schist, quartzite and calc-silicate, but the stratigraphic order is not known.
2. Isoclinal folding of (1) and probable high-grade metamorphism. Layers and isoclines are truncated by Cambrian conglomerate, plus contrast in metamorphic grade from kyanite schist to slate above unconformity.
3. Fault (strike slip) in north-central part of metamorphic complex. Truncated by Cambrian.
4. Uplift and erosion to expose and erode the structure in the schist complex.
5. Deposition of Cambrian boulder conglomerate, basin enlargement and Cambrian conglomerate overlap of lowest unit. Ordovician slate then sandstone.
6. Units in (5) folded into a north-east – south-west-trending upright synformal synclines which plunges at either end away from the core in a saddle-like structure. Synformal geometry is demonstrated by Vs in the river valleys and fold facing is given by Cambrian on the fold flanks and Ordovician in the core syncline. Refolding of isoclines probably occurred at this time as second-generation folds in the metamorphic rocks have the same trends as those in the Cambrian to Ordovician succession. Folds are truncated by the granite and gently dipping younger succession.
7. Intrusion of compositionally zoned granite to diorite pluton and pegmatite dyke. Xenoliths of schist in the pluton. Pegmatite cuts all earlier fold structures and is shown as an offshoot of the granite.
8. Uplift and erosion to expose the granite, and low-grade metamorphosed Cambrian/Ordovician. High-grade metamorphic complex re-exposed. The low-grade metamorphism could have taken place during (6), (7) or (8) but was probably synchronous with the folding (6).
9. Deposition of a pebbly grit, basalt, limestone sequence of unspecified age. Basalt oversteps all previous lithological groupings.
10. Gentle tilting of 2 degrees to the north-west.
11. North-west – south-east-trending vertical dip-slip fault in north-east corner of the map. This fracture event could have been before (10).
12. Uplift and erosion to generate the present topography.

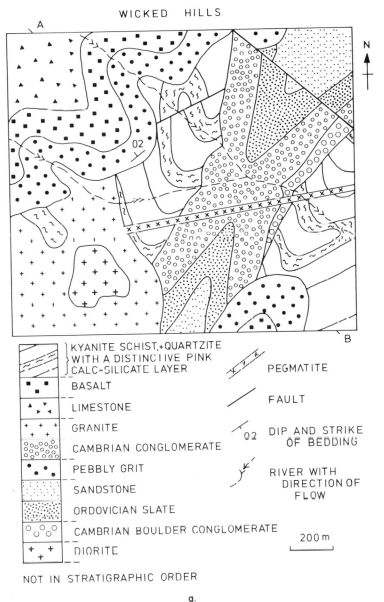

WICKED HILLS

KYANITE SCHIST,+QUARTZITE
WITH A DISTINCTIVE PINK
CALC-SILICATE LAYER

BASALT

LIMESTONE

GRANITE

CAMBRIAN CONGLOMERATE

PEBBLY GRIT

SANDSTONE

ORDOVICIAN SLATE

CAMBRIAN BOULDER CONGLOMERATE

DIORITE

NOT IN STRATIGRAPHIC ORDER

PEGMATITE

FAULT

DIP AND STRIKE
OF BEDDING

RIVER WITH
DIRECTION OF
FLOW

200 m

a.

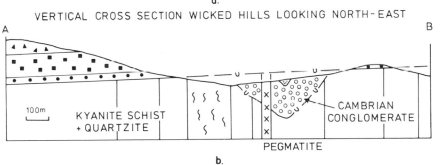

VERTICAL CROSS SECTION WICKED HILLS LOOKING NORTH-EAST

A   B

100m

KYANITE SCHIST
+ QUARTZITE

CAMBRIAN
CONGLOMERATE

PEGMATITE

b.

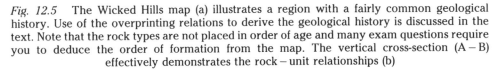

Fig. 12.5 The Wicked Hills map (a) illustrates a region with a fairly common geological history. Use of the overprinting relations to derive the geological history is discussed in the text. Note that the rock types are not placed in order of age and many exam questions require you to deduce the order of formation from the map. The vertical cross-section (A – B) effectively demonstrates the rock – unit relationships (b)

Reminder

A. During the analysis of a problem, attempt to **visualize** the 3-D situation you are handling.
B. Avoid using exaggerations in constructing cross-sections, they give a false impression of an area.
C. In presenting an account of an area **clearly separate fact from interpretation**.
D. Having been given the method for tackling a particular situation you may be asked to work from a different starting-point — be **adaptable**.

12.10   WRITING STYLE

'I would have written a shorter letter if I had had more time.' B. Franklin.

Regrettably, few tertiary education establishments pay enough attention to the teaching of report writing. Considering the number of reports, projects, theses, dissertations to be written in a course and the number that a professional will write in a career, such training would seem necessary. I do not plan to fill this vacuum but will make a plea for some extra reading. Everyone should read Strunk and White (1959). It is a short book on style and is effectively a 'good practice' manual (Strunk, W. and White, E. B. 1959, *The Elements of Style*. Macmillan, New York, 71p). Many books on writing in science are aimed at the researcher publishing in journals (e.g. Booth, V. (1985) *Communicating in Science: Writing and Speaking*. Cambridge University Press, 68p or Bates, J. D. 1980, *Writing with Precision: How to Write so that you Cannot Possibly be Misunderstood* (3rd edn). Acropolis Books, Washington, D.C., 226p). A growing number of books are now available catering to the general student market (e.g. Ellis, R. and Hopkins, K. 1985, *How to Succeed in Written Work and Study*. Collins, London, 210p). Books aimed at the geological market tend to be for researchers or those in geological surveys. Both *Methods in Field Geology*, by F. Moseley (Freeman, San Francisco, 1981, 211p), and *Geology in the Field*, by R. R. Compton (John Wiley, New York 1985, 398p) have good sections on writing map reports. Moseley in particular gives many practical tips on how to present material.

Once you have taken the trouble to learn something of style and technique, a bit of work is required — practice, practice, practice — revise, revise, revise!!

APPENDIX 1: STEREOGRAPHIC AND RELATED PROJECTION

Quantitative treatment of 3-D geometry is commonly awkward in several respects. Trigonometric formulae have a habit of becoming very long very rapidly. As a result, their manipulation is prone to human error and an understanding of what the equations are doing is often somewhat remote. Fortunately, to help us we have fairly simple yet powerful graphical techniques which come under the heading of stereographic and related projections. As the name suggests the approach is similar to that adopted in making map projections especially in the way that 3-D is represented in 2-D. A crucial requirement for operating the method is that you have a good appreciation of the projection process. If you relax your hold on visualizing the method and become an automaton, horrendous errors become possible. To create an image of the map projection process in your head may appear difficult at first but a little practice generates considerable expertise. The method is ideally suited to analysing angular relations and displaying orientation data. Typical applications include determining angles between pairs of lines, pairs of planes and lines plus planes. It is also very useful for the analysis of the form and attitude of curved surfaces (folds) and can easily handle large amounts of data.

BASIC PRINCIPLES OF STEREOGRAPHIC PROJECTION

The projection process is thought of as occurring inside a projection sphere (Figure A1.1). Planes and lines within this sphere are projected on to the horizontal plane which passes through the centre. This is the equivalent of the earth's equatorial plane and is known as the projection plane. Where the horizontal plane is in contact with the sphere it defines the primitive circle. In practice we work on a circular graph which represents the projection and so we are very much in the business of portraying 3-D in 2-D. To project variably oriented planes or lines, they have to be placed within the sphere such that they pass through its centre. A free choice of position in the sphere would lead to innumerable projections for any one attitude, hence the need for a standardized approach.

In this process we are dealing only with angular relations and geographic relationships are not represented. Two planes of the same dip and strike, one measured at Timbuktu and the other at Widgiemooltha, will have the same projection. Likewise, we could plot a differently oriented plane from the Urals together with the first two planes and calculate their angular relations even though they will never intersect.

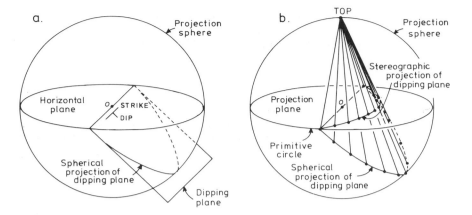

Fig. A1.1 Basic principles of the projection process. (a) Structural elements (planes or lines), passing through the centre of a sphere, intersect the sphere as a spherical projection. (b) Points on the spherical projection are joined to the top of the sphere. Where these joins pierce the projection plane defines the stereographic projection

The dipping plane of Figure A1.1 (north—south strike, dipping steeply east) intersects the lower half of the projection sphere in a line as shown. Take a large number of points on this line and project each one to the top of the sphere. Where each one of these joins pierces the projection plane, a point is defined on the stereographic projection of the dipping plane; eventually the number of points generates a continuous curved line which is the stereographic projection of the plane. Because of the geometry of the projection, this line is an arc of a circle and is thus referred to as a cyclographic trace. The same line is more commonly referred to as a great circle, though strictly speaking this term should be reserved for the trace of the plane on the surface of the sphere not its projection. Great circle routes for aircraft follow the imaginary traces on the earth's surface of planes passing through the centre of the earth which also contain the departure and destination airports. Their commercial advantage is that they give the shortest distance between two points on a sphere.

It is important to note that we have restricted ourselves to the lower hemisphere. The dipping plane could have been extended to intersect the upper hemisphere and then projected to the South Pole. The two projections are mirror images of each other about the line of strike. Fortunately in structural geology we only use the lower hemisphere though some old publications may present data in the upper. It is best to state which you are using when writing reports. Most people can more readily visualize the projection process by imagining themselves looking down on the projection sphere. To help the 3-D thinking, the hands may be cupped to create a crude lower hemisphere and in the mind differently oriented planes and lines may be inserted in the sphere and their approximate projections appreciated. At this stage you should be aware that a gently dipping plane with the same strike as Figure A1.1 would have a cyclographic trace close to the edge of the projection plane. A steeper dip would project nearer the diameter. Such visualization is critical if blunders are to be avoided, and in the early days you should carefully develop a feel for what you are doing in 3-D. Equally critical for your appreciation of the projection process is the loss of a dimension for whatever structural element is being considered: planes plot as lines and lines go to points.

PLOTTING A PLANE

In practice we do not draw great circles by defining a series of dots, we in fact make use of a plotting device, the stereographic net (Figures A1.2 and A1.3), which is something like a graph. Because it is mainly concerned with angular relations, the stereonet could be thought of as a protractor. It has been named the Wulff Net after an influential crystallographer. On the net there is a family of north/south great circles plotted at dip intervals of 2° with heavy lines every 10° to allow easier counting. In projection, the great circles can be thought of as being the equivalent of lines of longitude on the earth (Figure A1.2). The lines of greater curvature running approximately east/west are known as small circles. These may be generated in two ways; they are projections of cones of varying apical angles with north – south axes or of planes that do not pass through the centre of the sphere but which are perpendicular to the north south horizontal diameter. In cartographic terms, small circles are the projections of planes of equal latitude. Their 3-D significance becomes important when rotational displacements are studied on the net. For our purposes the small circles divide up, into 2° intervals, the 180° arc of each great circle which represents the straight angle defined by the strike. From each end of the strike to the dip direction, measured on the plane, is 90° and the total angle from strike to strike is 180°. Measuring angles on great circles is very important when dealing with line/ line, plane/plane, and line/plane angular relations.

A sheet of tracing paper is used as an overlay to the stereonet and fixed at the centre by a drawing-pin to allow for rotation. A piece of masking tape at the back of the net's centre is useful to reduce wear and tear on the hole. North – south-striking planes are easily drawn on the overlay using the printed great circles. A plane 50/180 dips 50° towards the east. To plot its stereographic projection, count 50° from the eastern end of the east—west diameter towards the centre and then draw in the great circle at this point. Dips are counted from the edge of the net as should be suggested to you by 3-D visualization. For strikes other than north/south the general procedure, using the right-hand rule, is (Figure A1.3):

1. Mark on the overlay, at the edge of the stereonet, the strike quoted in the reading and north as a reference. All overlays **Must** have north labelled.
2. Rotate the overlay anticlockwise to take the marked strike to the north of south point of the stereonet whichever comes first.
3. The great circle to be plotted will fall in the half of the net which is anticlockwise from the strike mark in its position at step 2.
4. From the outside edge of the east—west diameter, in the anticlockwise half of the net, count off the dip of the plane and draw in the great circle from this point. Remember that every 10° of dip is marked by a heavier line weight great circle.
5. Reorientate the overlay such that its north point is aligned with the north point of the stereonet and you have plotted the stereographic projection of the plane.

Before steps 1 – 5 are implemented it is as well to place a flattened hand over the net in the approximate attitude of the plane to be plotted. This gives the general form and location of the great circle you are after. Without understanding the projection method, the great circle may well be put in the wrong half of the stereonet.

If you record measurements following dip and dip direction convention, an equally systematic plotting procedure may be established. More *ad hoc* arrangements (e.g.

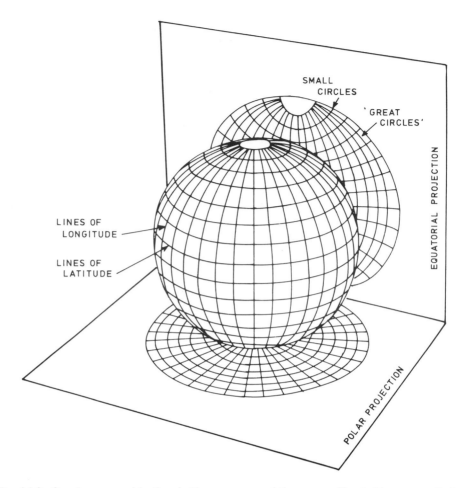

Fig. A1.2 Graphs are used in the plotting process and they are calibrated by curves that are the equivalent of projecting lines of latitude and longitude on the earth. In this text we use only the equatorial projection. Of several styles of equatorial projection we use the Wulff Net which preserves angular relations. On all equatorial projections, great circles are the projection of planes that pass through the centre of the sphere: small circles are projections of planes perpendicular to the north—south axis of the sphere and only the equatorial plane passes through the sphere centre. This diagram is schematic and does not show the projection geometry from the sphere to final graph

N40° E, 50° SE) lead to haphazard non-systematic plotting methods which slow down the process. One selling-point for the right-hand rule and dip plus dip direction methods is that they follow rules that can be implemented by a machine whereas other 'conventions' cannot. On a typical published map there may be hundreds or thousands of readings and their orientation analysis is greatly speeded up by computer manipulation. The computer can study patterns and calculate angular relations quickly, but often the slowest step is typing in the data. All other modes of recording planes have to be converted to the two systematic methods before entering into the computer so it makes sense to adopt one of thse early on in your career. Hopefully, in the professional world you will have ready access to computers to aid your anaysis, but everyone has to go through material of this appendix (plus some more) before they can appreciate what the computer is doing.

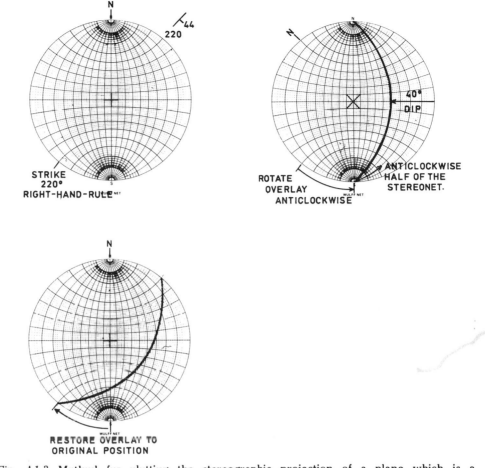

Fig. A1.3 Method for plotting the stereographic projection of a plane which is a cyclographic trace (great circle)

PLOTTING A LINE

The basics are the same as for plotting a plane. The line must be envisaged as passing through the centre of the sphere extending downwards until it touches the lower hemisphere at a point (Figure A1.4). This is then projected back to the top of the full sphere and the stereographic projection of the line is located where this join crosses the projection plane, again as a point. Because lines are recorded as plunge and plunge direction, there can be no confusion over what direction has been measured as can be the case with the strike of a plane. To help create a mental image of the plotting process before it is carried out, a pencil should be held over the net in the given orientation to provide an estimate of how the plot will finally appear. Steeply plunging lines plot close to the centre of the net and gently plunging lines plot near the edge. The procedure is as follows (Figure A1.5):

1. On an overlay, at the edge of the stereonet, mark the bearing of the line and north as a reference.

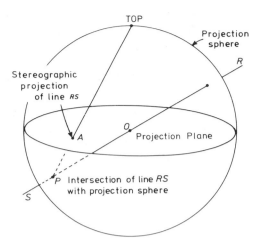

Fig. A1.4 The 3-D view of a linear structure within a projection sphere

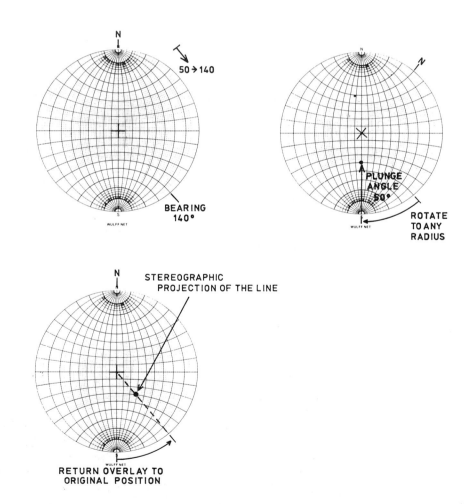

Fig. A1.5 Method for plotting the stereographic projection of a linear structure which plots as a single point

2. Revolve the overlay such that the bearing mark is on a north − south or east − west radius (whichever is the most convenient). Count off the plunge along this radius from the edge of the net and put a point. This is the stereographic projection of the line.
3. Return the north point of the overlay to the north point of the stereonet.

A single point does not look much but it is the stereographic projection of a line and could represent an enriched linear ore shoot in a gold-mine! Hopefully the stereographic method will reinforce the meaning of a line's bearing which is the strike of the vertical plane that contains the line. The dashed line on the final part of Figure A1.5 is half the stereographic projection of this vertical plane.

Structure contours on a surface are linear features whose bearing is the strike of the plane. By definition they are non-plunging lines and, for example, the full specification of a north − south strike in terms of orientation is 00 → 360 or 00 → 180. These two lines plot as points in stereographic projection at the north and south points of the stereonet and they are in fact representing the same line. When plotting a plane it is worth remembering that the points where the great circle intersects the primitive (the edge of the net) represent the projection of the plane's strike.

PLOTTING PITCHES

If a line is contained within a plane (e.g. ripple crests on bedding, hinge line on a hinge surface, net slip on a fault surface) then its orientation may be specified either by plunge and bearing or by pitch which involves quoting the attitude of the plane. A north—south-striking plane will be used to illustrate the principle of pitch representation in stereographic projection (Figure A1.6). Given a plane 30/360 with a line pitching 60° from 360° (remember the pitch could have been quoted 120° from 180°). Using the 10° subdivisions created by the small circles, count 60° from 360° along the great circle 30/360 and place a point which is the projection of the line. Pitch is the angular relation between two linear features on the plane, one of which is the strike (structure contours) and the other is the line in question. A simple extension of the procedure shows the stereonet to be a very quick method for determining angular relations between any pair of lines. In the example above the line pitching at 60° from 360° plunges 26° towards 303°, which shows pitch and plunge for the same line may be very different quantities.

The general plotting procedure for pitches is (Figure A1.7):

1. Plot the great circle for the plane, but when marking the strike at the edge of the stereonet label its bearing. Also whilst drawing the great circle label the other end of the strike, this is, strike + 180°.
2. Whilst the overlay is in position to plot the great circle, from whichever end of the strike the pitch is quoted, count off the pitch along and plane's great circle and place a point.
3. Return the overlay north to match north on the stereonet.
4. If the plunge and bearing of the line are required, simply move the point to any

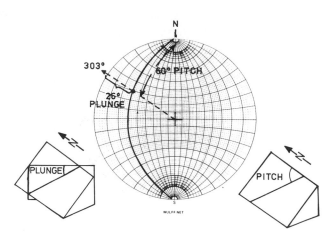

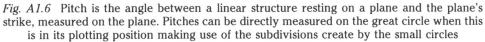

Fig. A1.6 Pitch is the angle between a linear structure resting on a plane and the plane's strike, measured on the plane. Pitches can be directly measured on the great circle when this is in its plotting position making use of the subdivisions create by the small circles

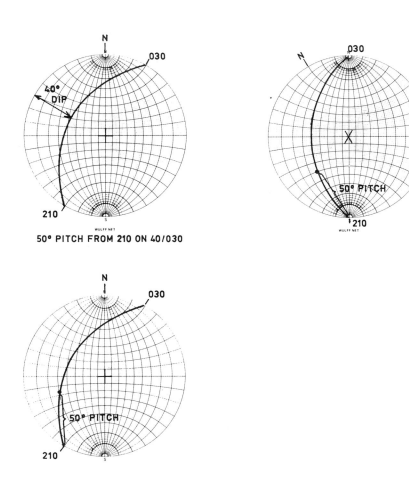

Fig. A1.7 Method for plotting the pitch of a linear structure on a non-north – south-striking plane

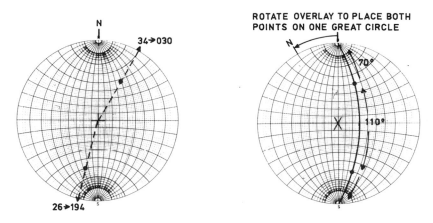

Fig. A1.8 Method for measuring the angle between two linear structures

radius, mark the end of the radius and count the plunge from the edge inwards to the point. Returning the overlay north to the stereonet north will allow the bearing to be read off.

ANGULAR RELATIONSHIP BETWEEN TWO LINES

This exercise in principle is very like the process involved in representing pitch. The angle between two lines is measured on an imaginary plane that contains the lines and the stereonet is ideally suited to this work. Given two lines in terms of plunge and bearing the procedure is as follows (Figure A1.8):

1. Plot both lines as points.
2. Rotate the overlay until both points fall on the same great circle; this is the stereographic projection of the imaginary plane that contains the lines.
3. Keep the overlay in position. On the great circle, using the small circle subdivisions, count off the angular distance between the lines. In the example shown the angle measured along the great circle is 110° but counting the opposite way, through the horizontal, is equally valid to give 70°. The two angles must add to 180° because the straight angle defined by the structure contours is 180° and usually the smaller angle would be quoted.

TRUE AND APPARENT DIPS

Several situations provide two or more apparent dips on a surface but do not allow direct measurement of the surface itself. Cliffs on a coastline or gorge may be too steep to climb to measure bedding, but the cliff faces and bedding trace attitudes may

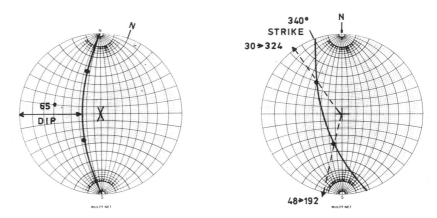

Fig. A1.9 From two apparent dips (given as plunge and bearings), the true dip and strike can be calculated

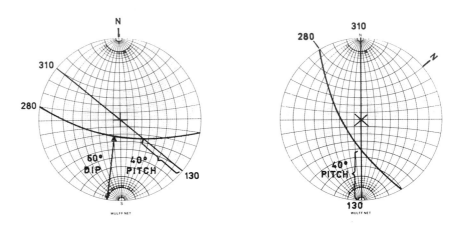

Fig. A1.10 Calculation of the apparent dip of a bedding surface (60/280) on a vertical cross-section striking 310/130

be recorded. At least two differently oriented sections are required. Map analysis of folds may provide two apparent dips of the hinge surface in the form of its strike (as the hinge surface trace on the map and the plunge and bearing of the fold hinge line). Again the stereonet is a very rapid means of calculating solutions to these problems. Apparent dips are linear features and their format will either be pitches or plunges and bearings. Given two plunges and bearings the plotting is similar to calculating the angle between two lines (Figure A1.9):

1. Plot the plunge and bearings as points.
2. Rotate the overlay until the points (two or more) lie on one great circle. Note that in real life with more than two points they will rarely all fall on one great circle and a best fit must be made.
3. Draw in the great circle and count off its dip from the edge of the stereonet.
4. Return the north mark of the overlay to stereonet norrth and read off the strike (right-hand rule to give the true dip and strike (65/340 in this example).

If the cliff- or gorge-face type of data are presented, then the first plotting step involves dealing with pitches. The great circle of each face is plotted together with its own pitch and then the procedure is as before.

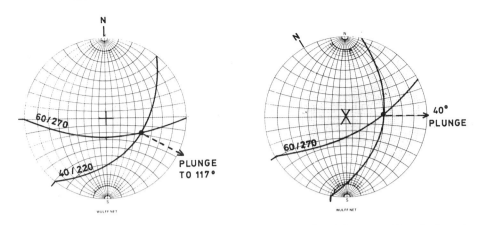

Fig. A1.11 Two planes intersect in a line. The attitude of this line (plunge and bearing) can be read off on the stereonet

To show something of the versatility of the stereographic net we can rearrange the problem somewhat. When drawing a cross-section oblique to the dip direction, the true dips and strikes from the map have to be converted to apparent dips as would be seen in the section. At first the method may seem slow but a little practice transforms a plod into a gallop!

The general method is (Figure A1.10):

1. Plot the great circle representing the dip and strike of the section (it is a planar feature usually vertical, 90/310 or 90/130 in this case).
2. Plot the great circle representing the dip and strike of the bedding surface to be plotted on the section (here 60/280).
3. The point defined by the intersection of these two great circles represents a line which is the apparent dip of the bedding plane on the section (40° in the example, quite different to its true dip of 60°). The apparent dip is read as a pitch on the section great circle from its strike. The line may also be thought of as the trace of the plane on the section. In another way of looking at it, the line represents the line of intersection of the two planes.

TWO INTERSECTING PLANES

The geometry of two intersecting planes is encountered in many situations. It is fairly common for two mineralized veins to show marked enrichment along their line of intersection and hence it is useful to be able to predict the orientation of such a bonanza if an exploratory drilling programme is to have any chance of success. Other important pairs of planes are dykes intersecting bedded sequences, angular unconformities, fold dihedral angles and conjugate faults. Besides the orientation of lines of intersection, we often need to know the angle between the planes. This is measured in their profile plane which is at right angles to their line of intersection. The latter is plotted as follows (Figure A1.11):

1. Plot the two planes as great circles.

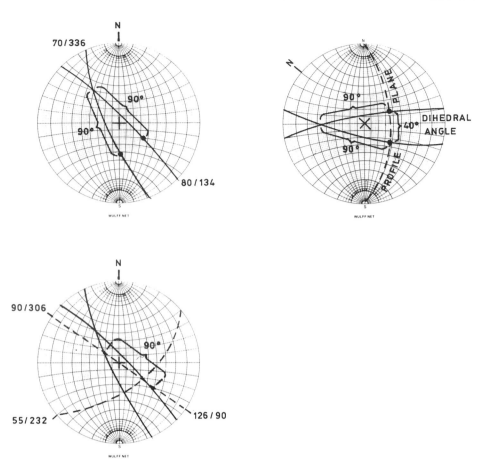

Fig. A1.12 The dihedral angle between two planes must be measured in the profile plane. From the line of intersection of the planes, count 90° on each and place a point. Find the great circle that contains these two points and this is the stereographic projection of the profile plane. Count the angular distance between the two planes along the profile plane (**Not** on the edge of the stereonet or anywhere else)

2. The point where the two great circles cross is the projection of their line of intersection (remember the loss of a dimension). Its plunge and bearing may be read by placing the point on a convenient radius, marking the radius end, counting off the plunge, and then returning overlay north to stereonet north.

As a variation on the above, it may be useful to quote the attitude of the line of intersection as a pitch on one or both of the planes. If asked for both the answer becomes somewhat lengthy but remember the need for unequivocal specification of pitch.

The angle between any two specified planes can be almost anything you want and hence a standard direction, the profile plane, has been agreed. The profile is perpendicular to the line of intersection. To calculate the angle between two planes (Figure A1.12):

1. Plot the two planes as great circles and mark the point where they cross, this is the projection of their line of intersection.
2. On each great circle, count 90° from the intersection point and place another point. This requires each plane's great circle to be returned to the position where

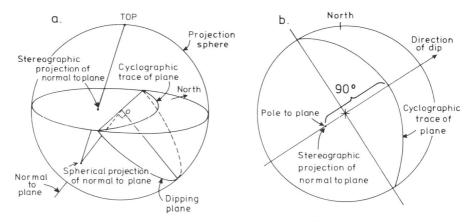

Fig. A1.13 (a) A 3-D view of a plane and its normal within a projection sphere. The stereographic projection of the plane normal (its pole), is sufficient to represent the plane. (b) A view of the stereonet showing the relation between the cyclographic trace (great circle) representation of a plane and its pole

it was plotted and the 90° is counted as in the exercise involving the angle between two lines.

3. Find the great circle which contains these two new points by rotating the overlay as in the apparent dip, and angle between two lines, exercises. Draw in this great circle which is the projection of the profile plane.

4. Along the profile plane, count off the angular distance between the great circles representing the two planes. This is the dihedral angle.

For the same two planes used in the illustration, apparent dihedral angles range from 0 to 90° but the true dihedral angle is 40°. If the planes shown were cut by a vertical section striking 126°/306°, then the angle between the traces of the two planes on the section would be 90°. Any section which passes through the line of intersection would show the two planes as parallel traces, a dihedral angle of 0°. The stereonet gives the opportunity of simply determining other possible apparent dihedral angles and it is worth investigating some of the other possibilities yourself.

NORMALS TO PLANES

A great circle representation of a plane is obviously a convenient method in many situations. However, if the orientation of many planes is being studied, a plot crowded with great circles becomes cluttered and perhaps indecipherable. The solution to this problem is to use the stereographic projection of the normal to each plane rather than its cyclographic trace (Figure A1.13). Because the normal is a line, its projection is a point and a large number of planes may be portrayed with ease. The projection of the normal is referred to as a pole and each pole is a unique representation of one dip and strike orientation. To represent a plane by only a point is at first an unusual concept, but in more advanced work the benefits are enormous. Visualization is clearly important in the initial stages of handling poles and this may

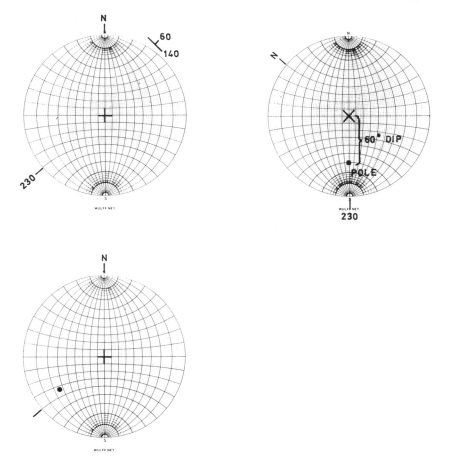

Fig. A1.14 Method for plotting the pole to a plane on the stereonet

be aided by holding a pencil, pointing downwards between the fingers at right angles to your flattened hand. If you put your hand over the stereonet in the approximate orientation of the plane, you should obtain a general impression of where the pole is going to plot.

The right-angle relationship between normal and plane should always be remembered. This geometry means that if the plane dips at $X°$ then the normal plunges at 90-$X°$. Hence a vertical plane has horizontal normal which plots at the edge of the stereonet. A flat-lying surface has a normal plunging at $90°$ and, in projection, the pole plots at the centre. The plane/normal orthogonal relationship means that the normal's bearing is $90°$ from the strike in a direction opposite to the dip of the plane. With the right-hand rule plotting poles is simple (Figure A1.14):

1. Add $90°$ to the strike of the plane and mark this on the overlay at the edge of the stereonet.
2. Rotate this mark to any radius and count the dip from the centre of the stereonet out along the radius towards the mark on the overlay. Return the overlay north to stereonet north.
3. The point so located is the pole to the plane.

With the first few examples you try, it is useful to plot the associated great circles to the planes to show the relations between the two elements. This plotting is easily done by placing the pole on the east—west diameter and counting 90° along the diameter; this gives the location of the appropriate great circle. The reverse case is equally simple if you have a great circle and wish to locate the pole. With the great circle in the position where it was plotted count 90° along the east—west diameter and mark the pole's position. An overlay of poles is often referred to as a PI diagram and it is in effect a scattergram.

FOLD GEOMETRY IN STEREOGRAPHIC PROJECTION

Because this text is aimed at an introductory level, only the elements of this study will be given. The simplest use of projection in fold analysis is to determine whether the fold is conical or cylindrical and, for the latter, the fold axis orientation. Two separate techniques may be employed both using the same raw data. Dip and strike measurements of the form surface are made around the fold structure (see Figure 8.5). In the simplest case these readings will be of bedding, and in real life the more the merrier to hopefully swamp small-scale irregularities in both bedding and the shape of the fold.

One method plots each dip and strike reading as a great circle and is known as the Beta plot. For a perfectly cylindrical fold, each of the cyclographic traces intersects at a common point which is the stereographic projection of the fold axis. Hence the plunge and bearing may easily be read off. The method works because, in cylindrical fold, the fold axis when moved parallel to itself around the fold always stays in contact with the fold surface. Therefore, any small planar portion of the form surface contains the fold axis as an apparent dip. Any number of differently orientated planes sharing a common apparent dip will intersect in that one common orientation.

The second method plots each dip and strike reading as a pole and is referred to as a PI plot. For a perfect cylinder all of the poles should lie on one great circle, that is, all be parallel to one plane which is the fold's profile plane. The great circle is known as the PI girdle. Remember the stereonet deals with orientation not position and we are interested in the profile's attitude not its location. Figure 8.5 shows the relationship between poles to the form surface and the profile plane. The PI method is less direct than the Beta plot because the fold axis is the pole to the PI girdle. However, by plotting poles it is much easier to handle large amounts of data, and there are other disadvantages of the Beta method.

On the PI diagram, non-clindricity is recognized by poles showing a non-plane distribution. If, however, a reasonable (but non-great circle) curve is defined you may be dealing with a cone but analysis of this nature is left to more advanced work. The Beta-style diagram is less easy to analyse because small irregularities create many great circle intersections. This technique is rarely used in practice though it has its place in illustrating fundamental principles.

RELATED PROJECTIONS

We have not investigated the properties of the Wulff Net in detail but basically it is an equal-angle net in that it preserves angular relations. All projections involve some sort of trade-off and in the case of the Wulff Net it distorts areas. Equal areas on the surface of the projection sphere are unequal when projected on to the stereonet. This is best seen in a 10° latitudinal belt either side of the east – west diameter. A $10° \times 10°$ area at the centre of the net is much smaller than its equivalent at the margin. As a result, a randomly oriented set of lines, when plotted on a Wulff Net would show an apparent concentration towards the centre. In some studies (folds, faults, joints) it becomes important to study variations in angular distribution and to assess their significance. This can only be done if the plotting device does not introduce false concentrations. Such work uses the equal-area projection named the Schmidt Net after the person who introduced it into structural geology. To create an equal-area projection an extra step is needed, and strictly speaking it is no longer a stereographic projection though not everyone applies the terminology rigorously. However, the difference between the two projection processes is relatively small. All the procedures given in this chapter may be followed on either net to give the same results without any need to change any aspects of the methods. The differences in projection also have no effect on visualizing the process to get a general idea of how the plot should turn out. The one thing you cannot do is change from one type of net to the other half-way through a calculation.

APPENDIX 2: GEOLOGICAL SYMBOLS

This set of symbols should be regarded as an example of the general style and approach to symbology. Each national Geological Survey, large exploration company, research organization, etc., has its own reference set of symbols. There is no such thing as a standard for symbols, though within each category (bedding, foliation, lineation, etc.) there is reasonable conformity in choice of basic symbol style. There are, however, many variations on the theme and, unfortunately, the same symbols may have different meanings from scheme to scheme. The tremendous diversity of geological maps cannot be covered here, for example maps of recent volcanics, basement terrains or engineering geology, will have very different sets of symbols.

GEOLOGICAL SYMBOLS

BEDDING

Younging information (or that there is a lack of it) is often not clearly portrayed on maps and three different styles of conveying such data are given (Table A2.1).

Table A2.1

	Dip and strike Younging unknown	Bedding overturned	Beds known to be right-way-up
I	60	60	60
II	60	60 used with symbol for younging direction O⟩⟶ Y or OLD ⟩⟶ Y	60
III	60	60	60

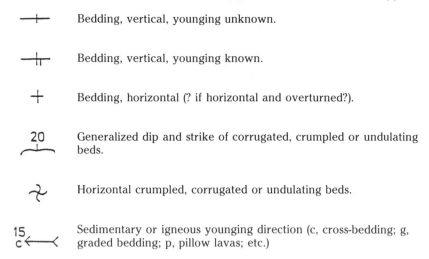

—+— Bedding, vertical, younging unknown.

—+r— Bedding, vertical, younging known.

+ Bedding, horizontal (? if horizontal and overturned?).

20 Generalized dip and strike of corrugated, crumpled or undulating
beds.

Horizontal crumpled, corrugated or undulating beds.

15 Sedimentary or igneous younging direction (c, cross-bedding; g,
c←——< graded bedding; p, pillow lavas; etc.)

FOLIATION AND BANDING

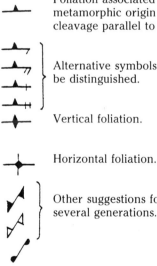

Foliation associated with compositional layering of possible
metamorphic origin (e.g. in a granitic gneiss). Could be used for
cleavage parallel to bedding in meta-sediment.

Alternative symbols if more than one generation of foliation has to
be distinguished.

Vertical foliation.

Horizontal foliation.

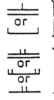

Other suggestions for symbols that could be used for foliations of
several generations.

CLEAVAGE AND SCHISTOSITY

Slaty cleavage or first schistosity (S_1).

Second cleavage or schistosity (S_2), etc.

Types of planar fabric may be indicated by lettering e.g. CR,
crenulation; SC, spaced cleavage; GN, gneissosity.

FOLDS AND LINEATIONS

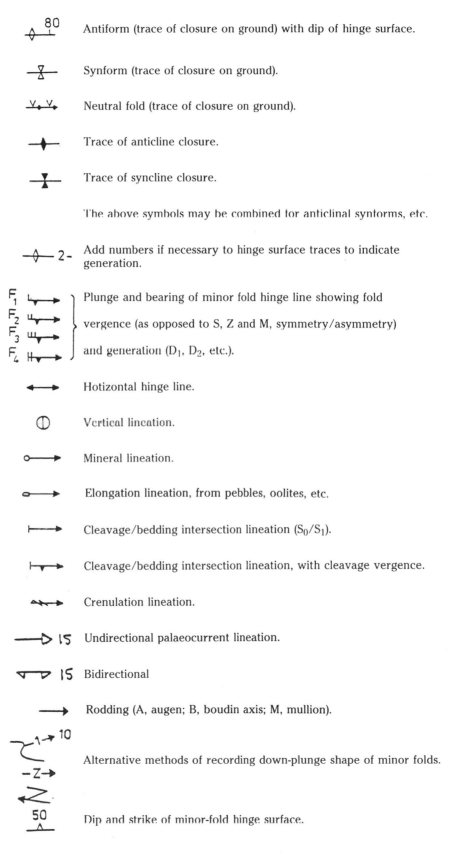

Antiform (trace of closure on ground) with dip of hinge surface.

Synform (trace of closure on ground).

Neutral fold (trace of closure on ground).

Trace of anticline closure.

Trace of syncline closure.

The above symbols may be combined for anticlinal synforms, etc.

Add numbers if necessary to hinge surface traces to indicate generation.

Plunge and bearing of minor fold hinge line showing fold vergence (as opposed to S, Z and M, symmetry/asymmetry) and generation (D_1, D_2, etc.).

Hotizontal hinge line.

Vertical lineation.

Mineral lineation.

Elongation lineation, from pebbles, oolites, etc.

Cleavage/bedding intersection lineation (S_0/S_1).

Cleavage/bedding intersection lineation, with cleavage vergence.

Crenulation lineation.

Undirectional palaeocurrent lineation.

Bidirectional

Rodding (A, augen; B, boudin axis; M, mullion).

Alternative methods of recording down-plunge shape of minor folds.

Dip and strike of minor-fold hinge surface.

JOINTS

<u>50</u> ─■─   Joint, dip and strike.

─■─   Joint, vertical.

╋   Joint, horizontal.

Shear and tension joints could be distinguished by letters S and T if necessary.

GEOLOGICAL BOUNDARIES

50 ↑ ──────   Geological boundary, definite. Showing dip direction at a specific locality.

↑↓ ── ── 50   Geological boundary, inferred. With observed sense of overturning.

─ ─ ─ ─ ─   Geological boundary, position conjectural.

- - - - - -   Drift boundary.

············
─ · ─ · ─ · }   Other categories of geological boundary: metamorphic isograds, structural subareas, etc.
× × × × × × ×

⌄⌄⌄⌄⌄⌄   Unconformity, top of 'U' is towards the younger rocks.

MISCELLANEOUS

⊗ S$_n$ Mineralization, symbol for element or letter for mineral species.

△   Fault breccia.

∼∿   Crush or shatter zone.

m   Mylonite.

⇉   Ductile shear zone, showing sense of movement and orientation.

─▭─   Flow banding in lava flow, showing dip and strike.

✳ ⑤   Fossil locality.

Symbols may be needed for: mine adits, mine shafts, water wells, dip and strike of foreset, volcanic centres, glacial striations, position of diamond drill holes, stream sediment sample sites, etc.

FAULTS

Fault, non-specific, well located but not exposed.

Fault, non-specific, located approximately.

Fault, non-specific, assumed (existence uncertain) — poor exposure, fault needed to explain outcrop pattern.

Fault, concealed beneath mapped units.

75° 50-65° ~60° Fault, showing dip, range of dips and approximate dip.

70° 40°   70° 40° Fault, exposed, locally exposed, with striations (note these may relate to only part of the displacement history).

Contractional fault, teeth on hanging wall.

or   Extensional fault, blocks or bar + ball in hanging wall.

SEPARATION CLASSIFICATION

Dip and strike of fault must be known and stratigraphy must contain distinctive layers to allow offset recognition. Symbols above relate to where these conditions do not hold.

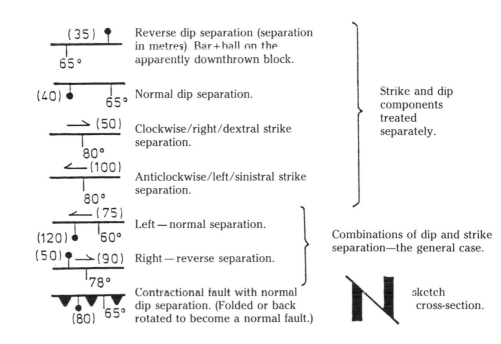

(35) 65° Reverse dip separation (separation in metres). Bar + ball on the apparently downthrown block.

(40) 65° Normal dip separation.

(50) 80° Clockwise/right/dextral strike separation.

(100) 80° Anticlockwise/left/sinistral strike separation.

Strike and dip components treated separately.

(75) (120) 60° Left — normal separation.

(50) (90) 78° Right — reverse separation.

Combinations of dip and strike separation—the general case.

(80) 65° Contractional fault with normal dip separation. (Folded or back rotated to become a normal fault.)

sketch cross-section.

SLIP CLASSIFICATION

These symbols will not appear on many maps because of the large amount of data required to calculate net slip. Open triangles and blocks show that the slip has been quantified.

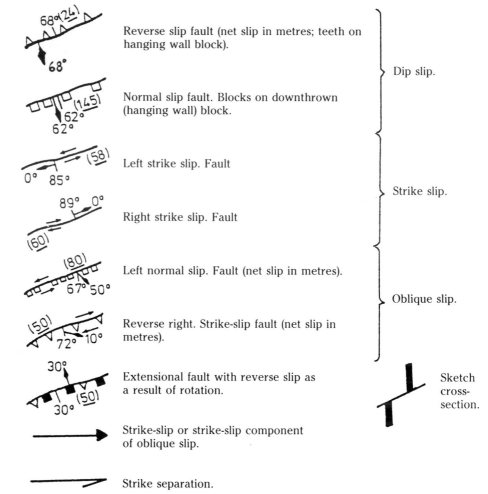

Reverse slip fault (net slip in metres; teeth on hanging wall block).

Normal slip fault. Blocks on downthrown (hanging wall) block.

Left strike slip. Fault

Right strike slip. Fault

Left normal slip. Fault (net slip in metres).

Reverse right. Strike-slip fault (net slip in metres).

Extensional fault with reverse slip as a result of rotation.

Strike-slip or strike-slip component of oblique slip.

Strike separation.

Dip slip.

Strike slip.

Oblique slip.

Sketch cross-section.

INDEX